KINETIC ANATOMY

THIRD EDITION

Robert S. Behnke, HSD

Professor Emeritus, Indiana State University

Human Kinetics

Library of Congress Cataloging-in-Publication Data

Behnke, Robert S., 1938-
Kinetic anatomy / Robert S. Behnke. -- 3rd ed.
    p. ; cm.
Includes bibliographical references and index.
ISBN 978-1-4504-1055-7 (soft cover) -- ISBN 1-4504-1055-3 (soft cover)
I. Title.
[DNLM: 1. Movement. 2. Musculoskeletal System--anatomy & histology. 3. Musculoskeletal Physiological Phenomena. WE 103]
611'.7--dc23
                    2011052403

ISBN-10: 1-4504-1055-3 (print)
ISBN-13: 978-1-4504-1055-7 (print)

Acquisitions Editors: Loarn D. Robertson, PhD, and Melinda Flegel; **Developmental Editor:** Amanda S. Ewing; **Assistant Editors:** Kali Cox, Tyler Wolpert, and Jan Feeney; **Copyeditor:** Patricia MacDonald; **Indexer:** Nancy Ball; **Permissions Manager:** Dalene Reeder; **Graphic Designer:** Robert Reuther; **Graphic Artist:** Dawn Sills; **Cover Designer:** Robert Reuther; **Photograph (cover):** © Human Kinetics; **Photographs (interior):** © Human Kinetics; **Photo Production Manager:** Jason Allen; **Art Manager:** Kelly Hendren; **Associate Art Manager:** Alan L. Wilborn; **Illustrations:** © Human Kinetics, unless otherwise noted; **Printer:** Versa Press

Printed in the United States of America   10   9   8   7   6   5   4   3   2   1

The paper in this book is certified under a sustainable forestry program.

Human Kinetics
Website: www.HumanKinetics.com

United States: Human Kinetics
P.O. Box 5076
Champaign, IL 61825-5076
800-747-4457
e-mail: humank@hkusa.com

Canada: Human Kinetics
475 Devonshire Road Unit 100
Windsor, ON N8Y 2L5
800-465-7301 (in Canada only)
e-mail: info@hkcanada.com

Europe: Human Kinetics
107 Bradford Road
Stanningley
Leeds LS28 6AT, United Kingdom
+44 (0) 113 255 5665
e-mail: hk@hkeurope.com

Australia: Human Kinetics
57A Price Avenue
Lower Mitcham, South Australia 5062
08 8372 0999
e-mail: info@hkaustralia.com

New Zealand: Human Kinetics
P.O. Box 80
Torrens Park, South Australia 5062
0800 222 062
e-mail: info@hknewzealand.com

E5449

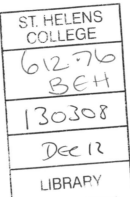

How to access the supplemental web resource

We are pleased to provide a one-year subscription to a web resource that supplements your textbook, *Kinetic Anatomy, Third Edition.* This resource includes hundreds of 3-D images of the human body, a regional review of human anatomy, and pretest and posttest evaluations to test retention.

Accessing the web resource is easy!
Follow these steps if you purchased a new book:

1. Visit www.HumanKinetics.com/MusculoskeletalAnatomyReview.

2. Click

3. Click   ve an
   accou

4. If the   y Items
   box   box.
   Enter   . Click
   the S

5. After   roduct
   will   do is
   sign   the left
   men

6. Your   r after
   you

→ Click the   ance
   along th

How to
used bo

You may pur   e,
www.Huma   he
following:

800-747-445   istomers
800-465-730   istomers
+44 (0) 113   istomers
08 8372 099   istomers
0800 222 06   istomers
217-351-507   istomers

For technica
support@hkusa.com . . . . U.S. and international customers
info@hkcanada.com . . . . Canadian customers
academic@hkeurope.com . . . . European customers
keycodesupport@hkaustralia.com . . . . . Australian and New Zealand customers

HUMAN KINETICS
The Information Leader in Physical Activity & Health

06-2012

Product: Kinetic Anatomy, Third Edition, web resource
Key code: MAR-JZRFXI-OSG

HUMAN KINETICS WEB RESOURCE

This unique code allows you access to the web resource.

Access is provided if you have purchased a new book. Once submitted, the code may not be entered for any other user.

At one time or another in our lives, we come in contact with someone whom we consider our teacher, supervisor, mentor, or role model. In many instances, such a person may also become our friend. I was fortunate during my professional preparation to have all of these and more in one individual. During my professional preparation, this person emphasized the importance of knowing human anatomy because he believed it was a keystone to understanding athletic performance and to preventing, recognizing, treating, and rehabilitating athletic trauma. This emphasis has inspired me throughout the preparation of this book. I'd like to thank Robert Nicolette, former head athletic trainer (1957–1969) at the University of Illinois, on behalf of all of us who were fortunate enough to know him. He has touched everyone we work with as a result of our association with him. I dedicate this book to him to express how much I appreciated him.

CONTENTS

Some may say the human body is the most fascinating machine ever designed. Science has long studied it and attempted to improve it through various methods, even going so far as trying to make parts interchangeable or to create new synthetic parts. Learning about oneself through the study of the human body could lead a person to a longer and healthier life. This can all begin with a basic understanding of the various elements making up the human body.

If this exposure to the study of human anatomy is a one-time experience, *Kinetic Anatomy* provides a good overview of the human body's various structures. For the student who seeks further study of human anatomy, *Kinetic Anatomy* provides the basics that can facilitate more in-depth study, in particular of the human body's physiological functions involving the anatomical structures presented in this text.

Goals of the Text

The goals of *Kinetic Anatomy* are (1) to familiarize students with the vocabulary of human anatomy, (2) to describe the essentials of human anatomy for movement, and (3) to provide students with the knowledge needed to pursue healthy living.

Having a firm understanding of the vocabulary of human anatomy allows you to communicate effectively with colleagues, physicians, therapists, educators, coaches, allied health personnel, and others using a universal language of human anatomy.

This text also gives readers a firm concept of how the human body is constructed and how it moves by discussing bones, tying the bones together to make articulations (joints), placing muscles on the bones (crossing joints), and then observing how the joints move when the muscles contract. The book also discusses the nerves (including the central nervous system's brain and spinal nerves and the peripheral nervous system) and blood vessels (including the heart) as well as the lungs, all of which provide elements essential for skeletal movement, but the main emphasis is on putting together the human body for the purpose of studying movement. Knowing what structures are involved and how they should function allows you to identify problems and correct them to enhance physical activity.

Finally, this book imparts knowledge that allows the pursuit of healthy living. Knowing about your body can alert you to potential problems and, with other acquired information, help you prevent or resolve those problems and lead a healthful lifestyle.

Organization of the Text

The text and illustrations are devoted to the structures that play a primary role in moving the human body: bones, ligaments, joints, muscles, and the nerves and blood vessels supplying innervations and circulation to those structures. This edition also addresses anatomical structures not often considered when studying the anatomy of movement: the brain, the heart, and the lungs. The purpose of these additions is to provide entry-level students with further understanding of anatomical structures involved in movement. Although the bones, ligaments, muscles, nerves, and blood vessels are the primary structures that create motion in the human body, other structures of the nervous system (brain, peripheral nervous system), the heart, and the respiratory system are introduced to show how these structures contribute to human movement.

To that end, this text is organized into four parts. Part I discusses the basic concepts of anatomy. The remainder of the text, like many textbooks in the areas of kinesiology and biomechanics, divides the body into the upper extremity (part II of this text); the head (brain), spinal column, pelvis, and thorax (heart and lungs) (part III); and lower extremity (part IV). Each anatomical chapter in parts II, III, and IV follows the same format: bones, joints and ligaments, muscles, and, where appropriate, the inclusion of three major organs also essential for movement (the brain, the heart, and the lungs). Parts II, III, and IV also include summary tables for muscles, bones, joints, ligaments, movements, nerves, and blood vessels, and these tables have been supplemented to include structures not found in previous editions of *Kinetic Anatomy*.

Updates to the Third Edition

The third edition of *Kinetic Anatomy* includes the following anatomical structures: the head, the brain, the heart, and the lungs. These structures, while not as obvious as bones, joints, and muscles, play major roles in human movement. The central nervous system (brain and spinal nerves), the peripheral nervous system, the heart, and the lungs all function to allow muscles to move bones and create motion in joints.

With more and more people participating in organized sports and personal fitness activities, there has been an increased interest in a possible unfortunate aspect of this participation: head trauma. *Kinetic Anatomy* looks at the anatomy of the head and brain, including the central and peripheral nervous systems as well as the blood vessels of the circulatory system. The vast network of blood vessels (numerous arteries and veins with multiple branches) is discussed, with identification of names and anatomical areas. In-depth investigation of both the nervous system and circulatory system is encouraged, requiring advanced anatomical study far beyond the entry-level information provided in *Kinetic Anatomy*.

In addition to the new material just mentioned, further discussion is presented regarding joint strength and movement, the function of muscles (agonists, antagonists, fixers or stabilizers, synergists), levers, and exercise. These additions are presented to enhance your understanding of muscle function if future

study in kinesiology and human biomechanics is desired.

The third edition of *Kinetic Anatomy* also grants students access to a new web component, *Musculoskeletal Anatomy Review*. See page xiv for more information on this resource.

These updates to *Kinetic Anatomy* make it a more inclusive entry-level text for undergraduate and secondary students and others seeking basic information about the anatomical structures of the human body in relation to movement.

Key Features of the Text

When one studies human anatomy, many devices are available to supplement learning. Human cadavers; audiovisual aids including photos, illustrations, models, and software programs; and numerous other means are provided to assist learning. *Kinetic Anatomy* additionally facilitates learning by providing a cost-free and readily available aid for comprehending how the body utilizes various aspects of human anatomy to allow movement: the hands-on experience. Throughout the book, readers will find "Hands On" boxes that provide instructions for feeling specific anatomical structures either on themselves or on a partner. In a very basic way, this makes the study of human anatomy a personal and practical experience available to everyone.

This text also provides an extensive listing of terms. Key terms are set in bold throughout the text and listed at the end of each chapter. This is important because it gives readers the opportunity to review what they were exposed to in the text. An understanding of the key terms helps ensure that readers have obtained the information about the anatomical structures presented in the chapter.

Detailed anatomical illustrations show readers the key structures that contribute to human movement in the anatomical areas discussed in any particular chapter. The artist has made every effort to accurately present these structures as they appear in the human body. Extensive use of cadaver photography would obviously produce a more exact illustration of the structures, but the expense of such reproduction would take the cost far beyond what might be considered an entry-level textbook.

To enhance understanding, the text also features photographs that illustrate movements resulting from the activity of the anatomical structures discussed, utilizing the old adage that a picture is worth a thousand words. These illustrations include appropriate labels to help readers find the structures presented in the text. The photographs help readers further understand what the structures being discussed actually do when they create movement.

"Focus on . . ." sidebars are presented throughout the book to illustrate circumstances in everyday activity that relate to the specific anatomical structures in the text. Health conditions commonly mentioned in everyday life are discussed to hopefully advance readers' understanding of these conditions. References to these various conditions in the print and electronic media should be more meaningful to readers as a result of these sidebars.

Each chapter ends with a set of learning aids, including a review of the key terms used in the chapter, suggested learning activities for students to complete, a set of multiple-choice questions, and a set of fill-in-the-blank questions. (Answers to the questions are provided at the end of the book.) Students can use these learning aids to ensure they have a firm grasp of the key points of the chapter content as well as to prepare for tests and quizzes. Additionally, functional movement exercises at the end of several chapters challenge readers' knowledge of the various functions of muscles. Although examples of possible answers are presented at the end of the book, there are many, many alternative answers, and readers are encouraged to use the text, the *Musculoskeletal Anatomy Review* web resource, the instructor, and fellow students if enrolled in an entry-level human anatomy course to seek additional answers to these functional movement exercises.

Finally, each part ends with summary tables. These summary tables provide a quick resource when seeking the components of a particular joint, its type, bones, ligaments, and movements as well as the components of a muscle including its origin, insertion, action, nerve supply, and blood supply. Whether students are answering questions posed in the text or preparing a paper or presentation on a particular anatomical structure or human movement, the summary tables can assist as a quick reference.

In addition to these text features, the book is also accompanied by the *Musculoskeletal Anatomy Review* web resource. More information on this resource can be found on page xiv. Students can access the *Musculoskeletal Anatomy Review* by visiting www.HumanKinetics.com/MusculoskeletalAnatomyReview.

Instructor Resources

Instructors have access to a full array of ancillary materials that support the text.

- **Image bank.** The image bank includes all the figures, tables, and photos from the text. Instructors can use these images to supplement lecture slides, create handouts, or develop other teaching materials for their classes.

- **Instructor guide.** The instructor guide includes many valuable tools to help instructors build a lecture. For each chapter, instructors will find an overview of the chapter, the chapter objectives, a lecture outline, lecture aids (additional items that would be useful to have on hand when covering a chapter's content), and additional activities that students can complete during class to enhance their learning experiences through doing and seeing.

- **Test package.** The test package includes more than 600 multiple-choice, true-or-false, and fill-in-the-blank questions. Instructors can use these questions to create or to supplement tests or quizzes.

Instructors can access these ancillary resources by visiting www.HumanKinetics.com/KineticAnatomy.

ACKNOWLEDGMENTS

A group of people at Human Kinetics (HK) has been responsible for the guidance needed to bring this edition of *Kinetic Anatomy* to completion. Dr. Loarn Robertson, former senior acquisitions editor, was responsible for deciding a new edition of *Kinetic Anatomy* would be a worthy addition to the entry-level study of human anatomy. Upon his retirement, this project was assumed by Melinda Flegel, Human Kinetics' new senior acquisitions editor. Inheriting this project in midstream, with my ideas and the senior acquisition editor's ideas already being enacted, was a task she not only accepted but also graciously guided to a successful completion. Amanda Ewing, developmental editor, took over the task of making sure I put together a textbook and ancillary materials that accomplished my goals for the text in an accurate and attractive format that would appeal to anyone interested in seeking an entry-level experience for learning human anatomy. Her comments, suggestions, and questions along with the ability to keep me on task played a major role in the completion of this edition.

The new illustrations in this edition are the result of the efforts of Joanne Brummett, the design, art, and photo coordinator at Human Kinetics. Her contributions and those of her outstanding staff in finding new artwork and additional photographs have made the illustrations supporting the written word an excellent adjunct to this edition's new subject matter.

I must thank Dr. Rainer Martens, HK founder, for approving the project that has resulted in the creation of *Kinetic Anatomy*. His contributions in the areas of sport, physical education, health education, and recreation have received worldwide recognition and appreciation by authors, teachers, and students everywhere. His thoughts and actions in the publishing business opened avenues in these areas at a time when it was sorely needed and now is so widely accepted.

These people have made working with Human Kinetics a pleasure and, hopefully, have produced a publication that will make the study of human anatomy enjoyable for anyone interested in learning about the human body and how it moves.

CREDITS

Figure 1.14 Adapted, by permission, from W.C. Whiting and S. Rugg, 2006, *Dynatomy: Dynamic human anatomy* (Champaign, IL: Human Kinetics), 15.

Figure 1.22b Adapted, by permission, from National Strength and Conditioning Association, 2008, Biomechanics of resistance exercise, by E. Harman. In *Essentials of strength training and conditioning*, 3rd ed., edited by T.R. Baechle and R.W. Earle (Champaign, IL: Human Kinetics), 70.

Figure 1.23b Adapted, by permission, from National Strength and Conditioning Association, 2008, Biomechanics of resistance exercise, by E. Harman. In *Essentials of strength training and conditioning*, 3rd ed., edited by T.R. Baechle and R.W. Earle (Champaign, IL: Human Kinetics), 70.

Figure 7.1 Adapted, by permission, from J. Watkins, 2010, *Structure and function of the musculoskeletal system,* 2nd ed. (Champaign, IL: Human Kinetics), 30.

Figure 7.2 Adapted, by permission, from L.A. Cartwright and W.A. Pitney, 2011, *Fundamentals of athletic training,* 3rd ed. (Champaign, IL: Human Kinetics), 55.

Figure 7.3 Adapted, by permission, from J. Watkins, 2010, *Structure and function of the musculoskeletal system,* 2nd ed. (Champaign, IL: Human Kinetics), 29.

Figure 7.4 Adapted, by permission, from L.A. Cartwright and W.A. Pitney, 2011, *Fundamentals of athletic training,* 3rd ed. (Champaign, IL: Human Kinetics), 67.

Figure 7.6 Adapted, by permission, from J. Loudon, M. Swift, and S. Bell, 2008, *The clinical orthopedic assessment guide,* 2nd ed. (Champaign, IL: Human Kinetics), 28.

Figure 7.7 Reprinted, by permission, from J. Watkins, 2010, *Structure and function of the musculoskeletal system,* 2nd ed. (Champaign, IL: Human Kinetics), 30.

Figure 7.8 Reprinted, by permission, from J. Loudon, M. Swift, and S. Bell, 2008, *The clinical orthopedic assessment guide,* 2nd ed. (Champaign, IL: Human Kinetics), 20, 24.

Figure 7.9 Adapted, by permission, from J. Watkins, 2010, *Structure and function of the musculoskeletal system,* 2nd ed. (Champaign, IL: Human Kinetics), 30.

Figure 7.12 Reprinted, by permission, from L.A. Cartwright and W.A. Pitney, 2011, *Fundamentals of athletic training,* 3rd ed. (Champaign, IL: Human Kinetics), 68.

Figure 9.15 Adapted, by permission, from W.L. Kenney, J.H. Wilmore, and D.L. Costill, 2012, *Physiology of sport and exercise,* 5th ed. (Champaign, IL: Human Kinetics), 144.

Figure 9.17 Adapted, by permission, from W.L. Kenney, J.H. Wilmore, and D.L. Costill, 2012, *Physiology of sport and exercise,* 5th ed. (Champaign, IL: Human Kinetics), 145.

Figure 9.19 Adapted, by permission, from W.L. Kenney, J.H. Wilmore, and D.L. Costill, 2012, *Physiology of sport and exercise,* 5th ed. (Champaign, IL: Human Kinetics), 165.

Figure 9.20 Adapted, by permission, from W.L. Kenney, J.H. Wilmore, and D.L. Costill, 2012, *Physiology of sport and exercise,* 5th ed. (Champaign, IL: Human Kinetics), 165.

Figure 10.2a,b Reprinted, by permission, from R.S. Gotlin, 2008, *Sports injuries guidebook* (Champaign, IL: Human Kinetics), 62.

Figure 10.3 Reprinted, by permission, from W.L. Kenney, J.H. Wilmore, and D.L. Costill, 2012, *Physiology of sport and exercise,* 5th ed. (Champaign, IL: Human Kinetics), 165.

Figure 10.4 Adapted, by permission, from W.L. Kenney, J.H. Wilmore, and D.L. Costill, 2012, *Physiology of sport and exercise,* 5th ed. (Champaign, IL: Human Kinetics), 165.

Musculoskeletal Anatomy Review includes hundreds of 3-D images of the human body to aid students in their study of anatomy. This engaging supplement to the text offers a regional review of structural anatomy with exceptionally detailed, high-quality graphic images—the majority provided by Primal Pictures. Students can mouse over muscles and click for muscle identification. This online feature offers students a self-paced and self-directed review of the musculoskeletal anatomy, providing an intensely visual interface through which students may gain a clear understanding.

Each chapter of *Musculoskeletal Anatomy Review* features a pretest and posttest evaluation to help students pinpoint knowledge gaps and test their retention. The pretest can be taken multiple times and is generated randomly so it will never be the same, but the posttest may be taken only once. Test results can be printed and turned in so instructors have the option to use the tests as a grading tool.

As students proceed through this review of musculoskeletal anatomy, they will encounter interactive learning exercises that will quiz them on key concepts and help them apply what they've learned about manual muscle testing or range of motion assessment in helping a virtual client.

There may be concepts presented in *Musculoskeletal Anatomy Review* that students have not learned in the past. Whenever possible, a learning aid will be provided to assist students in retention of the material. The learning and review aids may be mnemonics, simple organization of a group of muscles, or just a way to understand the terminology and locations of structures. Please take time to learn using the aids provided; if you do, your retention of the material is apt to surprise you.

Students can access *Musculoskeletal Anatomy Review* by going to www.HumanKinetics.com/MusculoskeletalAnatomyReview.

PART I

General Concepts of Anatomy

Structures

Human anatomy has been defined simply as the structure of organisms pertaining to humankind. A structure is, by one definition, something composed of interrelated parts to form an organism, and an organism is simply defined as a living thing. The body is made up of four different types of tissues (a collection of a similar type of cells). **Connective tissue** makes up bone, cartilage, and soft tissue such as skin, fascia, tendons, and ligaments. **Muscle tissue** is divided into three types: skeletal, which moves the parts of the skeleton; cardiac, which causes the pumping action of the heart; and smooth, which lines arterial walls and other organs of the body. **Nerve tissue** is divided into neurons, which conduct impulses involving the brain, the spinal cord, spinal nerves, and cranial nerves, and **neuroglia**, which are specifically involved in the cellular processes that support the neurons both metabolically and physically. The fourth type of tissue is known as **epithelial tissue**. There are four varieties, and all are involved with the struc-

tures of the respiratory, gastrointestinal, urinary, and reproductive systems.

The study of human anatomy as it pertains to movement concentrates on the bones, joints (and associated ligaments), and muscles responsible for the human body's movement. Additionally, the role of the nervous system in stimulating muscle tissue; the role of the vascular system in providing the muscle tissue with energy and removing by-products; the bone, joint, and muscle components of the body's lever systems; and the effects of exercise need to be studied. *Kinetic Anatomy* presents brief overviews of the respiratory system, the circulatory system, and the autonomic nervous system. Although human anatomy also includes other structures such as the endocrine system, digestive system, reproductive system, urinary system, and sensory organs, this text concentrates specifically on those anatomical structures chiefly responsible for producing movement of the human organism.

Proper vocabulary is extremely important when discussing anatomy. Common terms make communication with others (physicians, coaches, therapists, athletic trainers) much easier, and it is essential that a student of human anatomy become familiar with standard terminology presented in this chapter. Knowledge of the structures and common terms used to describe movement anatomically also facilitates the use of specific coaching principles; the use of therapeutic techniques involving human movement for prevention, treatment, and rehabilitation of various physical conditions; and the application of scientific principles to human movement.

Although all systems of the human organism can be said to contribute in some unique way to movement, this text emphasizes those systems (skeletal, articular, muscular, nervous, and circulatory) that directly accomplish movement. Primary concentration is on the following structures: bones, ligaments, joints, and muscles producing movement, with additional comments about the nerves and blood vessels in each specific anatomical area.

Bones

The body contains 206 bones. Bones have several functions, such as support, protection, movement, mineral storage, and blood cell formation. Arrangements of bones that form joints and the muscular attachments to those bones determine movement. Bones are classified by their shapes into four groups: long bones, short bones, flat bones, and irregular bones. Some authors also distinguish a fifth type of bone, known as sesamoid bones, which are small, nodular bones embedded in a tendon (figure 1.1). The bones that provide the framework for the body and that make movement possible are classified as **long bones** (figure 1.2). A long bone has a shaft, known as the **diaphysis**, and two large prominences at either end of the diaphysis, known as the **epiphyses**. Early in life

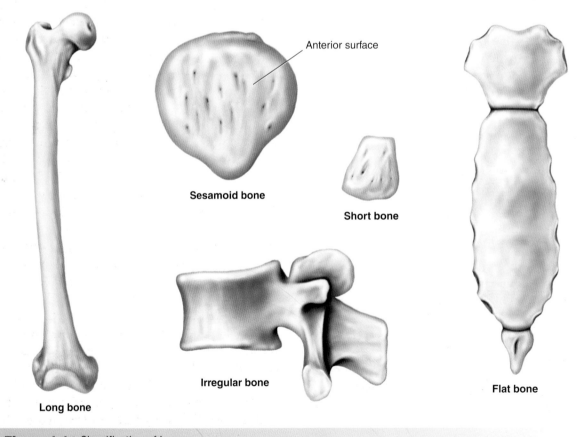

Anterior surface

Sesamoid bone

Short bone

Irregular bone

Flat bone

Long bone

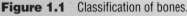

Figure 1.1   Classification of bones.

the epiphysis is separated from the diaphysis by a cartilaginous structure known as the **epiphyseal plate**. It is from these epiphyseal plates at both ends of the diaphysis that the bone grows; thus, this area is often referred to as the growth plate. Once a bone has reached its maximum length (maturity), the epiphyseal plate "closes" (bone tissue has totally replaced the cartilaginous tissue), and the epiphysis and diaphysis become one continuous structure. Around the entire bone is a layer of tissue known as the **periosteum**, where bone cells are produced. Additionally, the very ends of each bone's epiphyses are covered with a material known as **articular cartilage**. This covering provides for smooth movement between the bones that make up a joint and protects the ends of the bones from wear and tear.

Short bones differ from long bones in that they possess no diaphysis and are fairly symmetrical. Bones in the wrist and ankle are examples of short bones. Flat bones, such as the bones of the head, chest, and shoulder, get their name

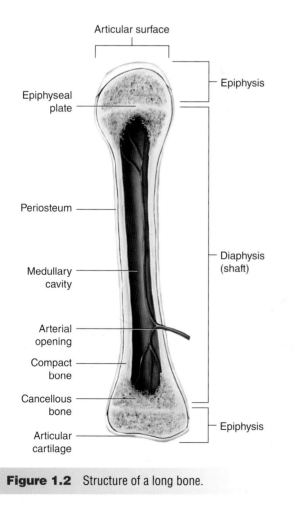

Figure 1.2 Structure of a long bone.

FOCUS ON

Osteoporosis

The loss of calcium and other minerals resulting from the natural aging process can cause bones to become porous and brittle. This condition is known as **osteoporosis** and can lead to broken bones and postural disfigurement. Approximately 50% of people older than 60 years of age have this condition.

from their flat shape. Irregular bones are simply bones that cannot be classified as long, short, or flat. The best example of an irregular bone is a vertebra of the spinal column. An additional classification that some anatomists recognize is sesamoid (sesame seed–shaped) bones. These oval bones are free-floating bones usually found within tendons of muscles. The kneecap (patella) is the largest sesamoid bone in the body; others are found in the hand and the foot.

Several terms are commonly used to describe the features of bones. These features are usually referred to as **anatomical landmarks** and are basic to one's anatomical vocabulary. A **tuberosity** on a bone is a large bump (figure 1.3). A **process** is a projection from a bone (figure 1.3). A **tubercle** is a smaller bump (figure 1.4). All three of these bony prominences usually serve as the attachment for other structures. A **spine**, or **spinous process**, is typically a longer and thinner projection of bone, unlike any of the previously mentioned prominences (figure 1.5). The large bony knobs at either end of a long bone are known as the **condyles** (figure 1.6). The part of the condyle that articulates (joins) with another bone is known as the **articular surface** (figure 1.2). Smaller bony knobs that sometimes appear just above the condyles of a bone are known as **epicondyles** (figure 1.4). A **fossa** is a smooth, hollow surface on a bone and usually functions as a source of attachment for other structures (figure 1.3). A smaller and flatter smooth surface is a **facet** (figure 1.7). Facets also serve as attachments for other structures. A **notch** is an area on a bone that appears to be cut out and allows for the passage of other structures such as blood vessels or nerves (see figure 1.8). Similar in function to a notch but appearing as a hole in a bone is a **foramen** (figure 1.5).

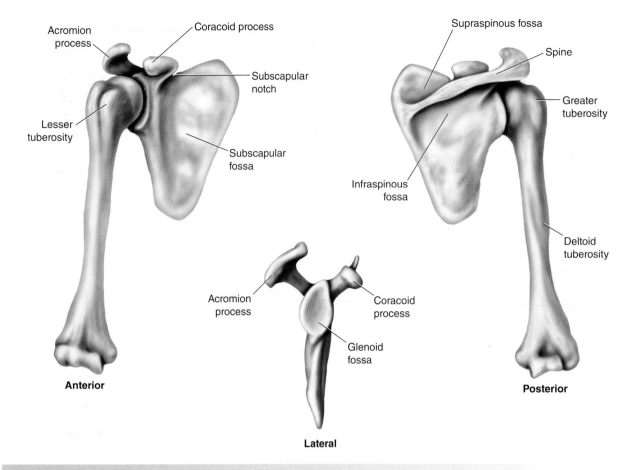

Figure 1.3 Landmarks of the shoulder bones: anterior (front), posterior (back), and lateral (side) views.

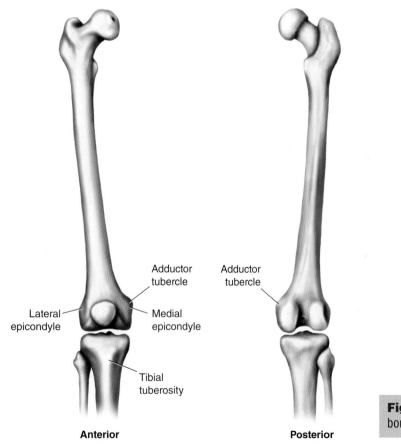

Figure 1.4 Landmarks of the thigh and leg bones, anterior and posterior views.

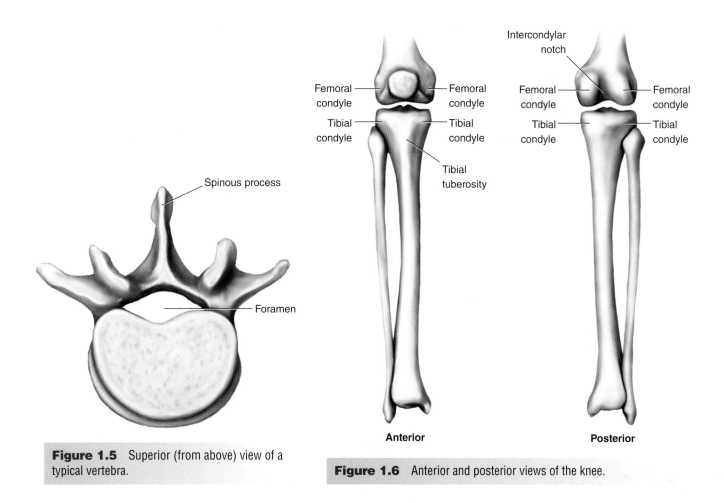

Figure 1.5 Superior (from above) view of a typical vertebra.

- Spinous process
- Foramen

Figure 1.6 Anterior and posterior views of the knee.

Femoral condyle
Femoral condyle
Tibial condyle
Tibial condyle
Tibial tuberosity
Anterior

Intercondylar notch
Femoral condyle
Femoral condyle
Tibial condyle
Tibial condyle
Posterior

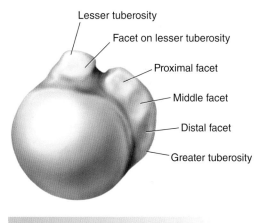

Lesser tuberosity
Facet on lesser tuberosity
Proximal facet
Middle facet
Distal facet
Greater tuberosity

Figure 1.7 Superior view of the humerus.

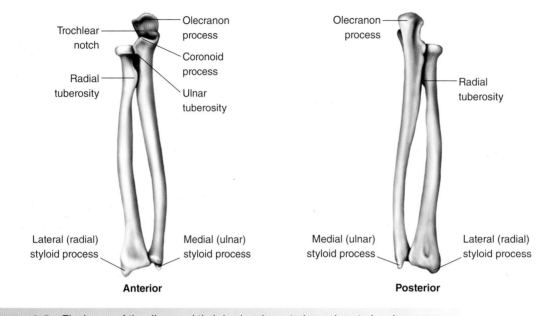

Figure 1.8 The bones of the elbow and their landmarks, anterior and posterior views.

Joints

The place where two or more bones join together anatomically is referred to as an **articulation**. The terms **joint** and *articulation* are interchangeable, and the study of joints is known as **arthrology**. Tying bones together at articulations are structures of dense, fibrous connective tissue known as **ligaments** (figure 1.9). A ligament is a cord, band, or sheet of strong, fibrous connective tissue that unites the articular ends of bones, ties them together, and facilitates or limits movements between the bones. Ligaments are not the sole support for the stability of joints. The muscles that cross the joint and the actual formations of the articulating bones also contribute to joint stability.

There are two major forms of joints: diarthrodial and synarthrodial. **Diarthrodial joints** are distinguished by having a separation of the bones and the presence of a joint cavity. These joints are divided into six subdivisions by their shape (figure 1.10). The **hinge joint** has one concave surface, with the other surface looking like a spool of thread. The elbow joint is an example of a hinge type of diarthrodial joint. The **ball-and-socket** type of diarthrodial joint consists of the rounded head of one bone fitting

into the cuplike cavity of another bone. Both the hip joint and the shoulder joint are examples of the ball-and-socket type of diarthrodial joint. The **irregular** type of diarthrodial joint consists of irregularly shaped surfaces that are typically either flat or slightly rounded. The joints between the bones of the wrist (carpals) are an example of this type of joint. Gliding movement occurs between the carpal bones. The **condyloid joint** consists of one convex surface fitting into a concave surface. Although the description of the condyloid joint is similar to that of the ball-and-socket joint, the difference is that the condyloid joint is capable of movement in only two planes about two axes, whereas the ball-and-socket joint is capable of movement in three planes about three axes. (Note: Planes and axes are discussed in chapter 2.) An example of a condyloid joint is where the metacarpal bones of the hand meet the phalanges of the fingers. The **saddle joint** is often considered a modification of the condyloid joint. Both bones have a surface that is convex in one direction and concave in the opposite direction, like a saddle. These joints are rare, and the best example is the joint between the wrist and the thumb (carpometacarpal joint). In the **pivot joint**, one bone rotates about the other bone. The radius bone (of the forearm) rotating

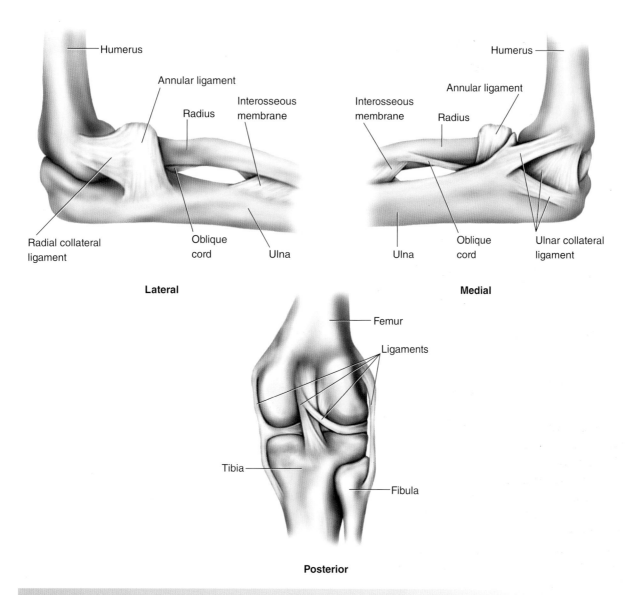

Figure 1.9 Lateral and medial views of the major ligaments of the elbow; posterior view of the ligaments of the knee.

on the humerus (upper-arm bone) is an example of a pivot joint.

All of the diarthrodial joints are considered **synovial joints**. The synovial joints are where the greatest amount of movement occurs. They are characterized by a space between the articulating surfaces (figure 1.11); a synovial membrane lining the joint secretes synovial fluid for lubrication and provides nutrients to joint structures. Synovial joints are surrounded by a joint (articular) **capsule**. These joints are classified into four categories by the type of movement they permit in planes and about axes (figure 1.12).

Joints between bones that allow only a gliding type of movement over each other are known as **nonaxial joints**, such as are found in the wrist and the foot. **Uniaxial joints**, such as the elbow joint, permit movement in only one plane about one axis. A **biaxial joint**, such as the wrist, permits movement in two planes, about two axes. A **triaxial joint** allows movement in three planes, about three axes, illustrated by the movements of the shoulder joint and the hip joint, which are both ball-and-socket joints.

Synarthrodial joints have no separation or joint cavity, unlike the diarthrodial joints.

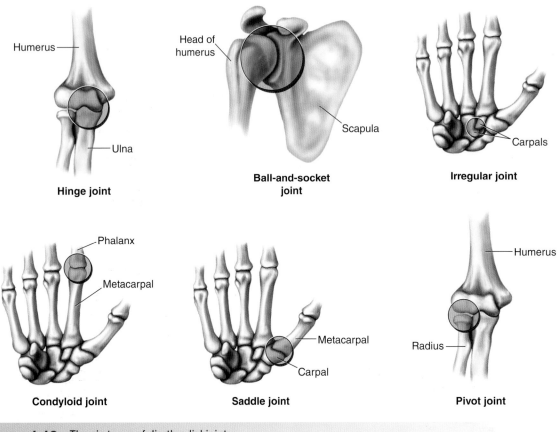

Hinge joint

Ball-and-socket joint

Irregular joint

Condyloid joint

Saddle joint

Pivot joint

Figure 1.10 The six types of diarthrodial joints.

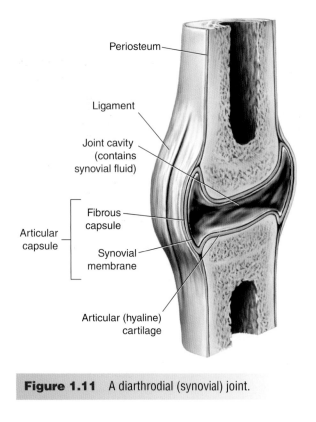

Figure 1.11 A diarthrodial (synovial) joint.

There are three subdivisions of synarthrodial joints (figure 1.13): suture, cartilaginous, and ligamentous. The **suture joint** has no detectable movement and appears to be sewn (sutured) together like a seam in clothing. The bones of the skull are the classic examples of suture joints. There is no movement in these joints. **Cartilaginous joints** allow some movement, but other than those found in the spinal column, they do not play a major role in movement. A cartilaginous joint contains **fibrocartilage** that deforms to allow movement between the bones and also acts as a shock absorber between them. Examples include the intervertebral joints, the pubic symphysis, and the sacroiliac joints. The **ligamentous joints** tie together bones where there is very limited or no movement. The joints between two structures of the same bone (e.g., the coracoid process and acromion process of the scapula) and between the shafts of the forearm and lower-leg bones are examples of the ligamentous form of a synarthrodial joint.

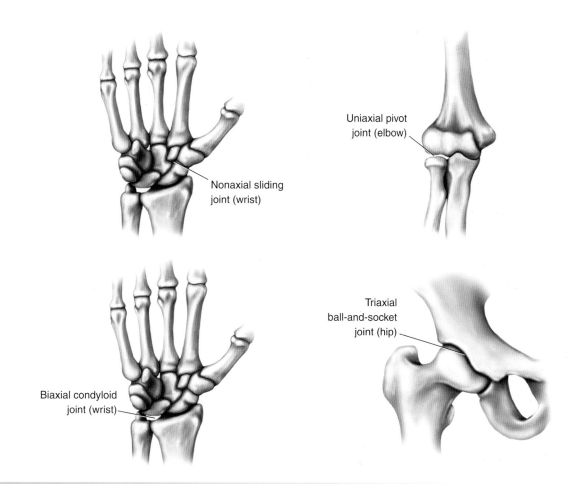

Nonaxial sliding
joint (wrist)

Uniaxial pivot
joint (elbow)

Biaxial condyloid
joint (wrist)

Triaxial
ball-and-socket
joint (hip)

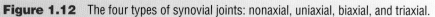

Figure 1.12 The four types of synovial joints: nonaxial, uniaxial, biaxial, and triaxial.

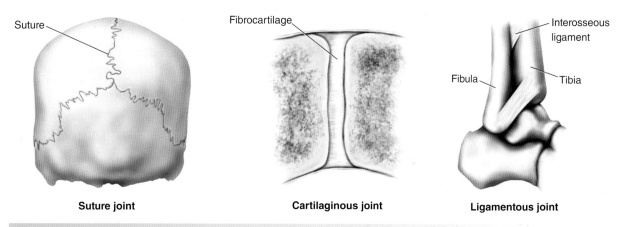

Suture

Fibrocartilage

Interosseous
ligament

Fibula

Tibia

Suture joint

Cartilaginous joint

Ligamentous joint

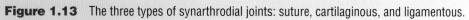

Figure 1.13 The three types of synarthrodial joints: suture, cartilaginous, and ligamentous.

Joint strength is determined by a number of factors:

- The physical structure of the bones contributes to joint strength (e.g., both the shoulder joint and the hip joint are classified as ball-and-socket joints, but the deeper structure of the hip joint makes it stronger than the shoulder joint).

- The strength, number, and anatomical position of ligaments also determine joint strength, as do the number and strength of muscles and tendons crossing a joint. The strength of the muscles and tendons is extremely important for injury prevention and rehabilitation.

- To a lesser extent, other structures (blood vessels, nerves, skin, and fascia) crossing a joint also contribute to joint strength.

The degree of movement in joints varies, not only from person to person but also within a particular person. The following factors influence degree of movement in joints:

- The bones involved (joint structure can limit range of motion; consider elbow movement versus hip movement)

- The thickness and laxity of the ligamentous structures

- The amount of fat and muscular tissue around a joint

- The strength and flexibility of the muscle tissue crossing the joint

- The resistance of other structures (mentioned previously as structures contributing to joint strength: blood vessels, nerves, skin, and fascia crossing the joint)

Muscles

The study of muscle is known as **myology**. Chemically, muscles consist of water and solids; those solids are proteins, carbohydrates, inorganic salts (including calcium, chloride, iron, magnesium, phosphorus, potassium, and sodium), enzymes, fat globules, nitrogenous extractives (e.g., creatine, uric acid), and nonnitrogenous extractives (e.g., lactic acid, glycogen). Further analysis of the chemical composition of muscle is found in the study of human physiology.

The number of muscles in the human body depends on a number of factors. Not all people have exactly the same number of muscles. Some muscles may appear on one side of the body and not the other (e.g., psoas minor). Some muscles are totally absent in some people (e.g., palmaris longus). Various texts may list certain muscles separately, while others might consider certain muscles as part of a larger muscle (e.g., flexor hallucis brevis and flexor digitorum brevis). Most authorities agree on the number 680 for the total muscles in the human body, with approximately 240 that have separate names. Muscles are named using the following criteria:

- Action, such as flexor or extensor
- Attachment to bones, such as sternocleidomastoid
- Direction of pull, such as oblique or rectus
- Location, such as tibialis or ulnar
- Size, such as maximus or minor
- Shape, such as teres or trapezius
- Structure, such as triceps or quadriceps
- Some combination of these criteria, such as flexor digitorum brevis

Muscle tissue is often categorized into three types: **smooth**, which occurs in various internal organs and vessels; **cardiac**, which is unique to the heart; and **skeletal**, which causes movement of the bones and their joints. Unlike skeletal muscles, smooth muscles (figure 1.14a) are not organized into motor units, and they receive nervous control via the autonomic nervous system. Smooth muscles are typically found in blood vessels, where they increase and decrease the lumen (opening) of the vessels to assist in the flow of blood through the circulatory system; in hollow organs such as the stomach and bladder; and in the alimentary (digestive) tract, where the muscles create a rhythmic type of contraction. In the intestines this rhythmic movement is known as peristalsis. Cardiac muscle fibers (figure 1.14b) are associated with the heart. This type of muscle creates a rhythmic contraction consisting of two phases, systole and diastole. **Systole** refers to contraction of cardiac muscle, and **diastole** is the period when the cardiac muscle relaxes. Both smooth muscle and cardiac muscle are structurally different from skeletal muscle.

For the purpose of looking at anatomy and movement, this text concentrates on the skeletal muscle. Skeletal muscle has the ability to stretch (extensibility), to return to its original length when stretching ceases (elasticity), and to shorten (contractility) when stimulated. The various forms of skeletal muscle are **fusiform**, **quadrate**, **triangular**, **unipennate**, **bipennate**, **multipennate**, and **longitudinal** (figure 1.15). Most skeletal muscles are either fusiform or pennate. Fusiform muscles are formed by long, parallel fibers and typically are involved

in movements over a large range of motion. Pennate muscles consist of short, diagonal fibers and are involved in movements that require great strength over a limited range of motion.

The strength a particular muscle can generate is dependent on several factors:

- Cross-sectional area: All else being equal, a larger cross section means more muscle fibers and therefore greater strength.
- Length: Longer fibers typically generate a force creating more motion since they shorten over a greater distance.
- Texture: A muscle with less noncontractile tissue will have more muscle fibers per area and, therefore, is capable of generating a greater force.

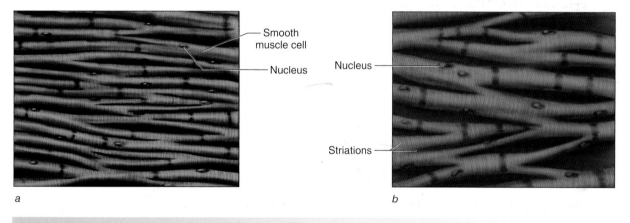

a   *b*

Figure 1.14 *(a)* Smooth and *(b)* cardiac muscle cells.

Adapted by permission from Whiting and Rugg 2006.

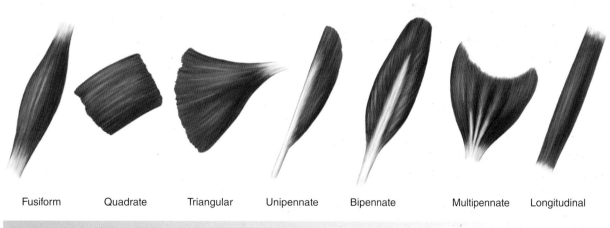

Fusiform   Quadrate   Triangular   Unipennate   Bipennate   Multipennate   Longitudinal

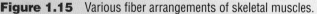

Figure 1.15 Various fiber arrangements of skeletal muscles.

- Specificity: The chemical structure of muscle (e.g., viscosity of fluids present, amount of sarcoplasm, number of amino acids) can affect muscle actions.
- Tension: A muscle generates more force when it is placed on stretch, or under tension.
- Coordination: Poor coordination between muscles can create friction between muscle fibers, reducing the force that can be generated.

The fibers of a muscle form the muscle belly. At either end of the belly, a unique form of connective tissue, a **tendon**, attaches the muscle to the bones. Tendons are extensible and elastic, like skeletal muscle, but they are not contractile. They are similar to ligaments in that both are dense, fibrous connective tissues. The main difference is that tendinous tissue does not have as much elasticity as ligamentous tissue. Tendons of skeletal muscles are usually defined as either tendons of origin or tendons of insertion (figure 1.16). **Tendons of origin** are usually longer and are attached to the proximal (closest to the center of the body) bone of a joint, which is typically the less mobile (fixed) of the two bones of a joint. The origin of a muscle is usually the most stable attachment. The **tendons of insertion** are shorter and are attached

to the more distal (farther from the center of the body) bone of a joint, which is typically the more movable (unstable) of the two bones of a joint. While most tendons of muscles are attached to bone, they may also be found attached to other tendons, fascia, ligaments, or even skin. Additionally, because tendons cross bony areas or need to be confined to certain areas, the tendons are covered by connective tissue known as **tendon sheaths** (figure 1.17) to protect them from wear and tear from the bony structures they cross.

Contraction of a skeletal muscle usually results in movement of bones in some direction. The movement (action) is typically described as one or more of the following:

- Flexion
- Extension
- Abduction
- Adduction
- Lateral rotation
- Medial rotation
- Pronation
- Supination
- Circumduction

These actions are described in more detail in chapter 2.

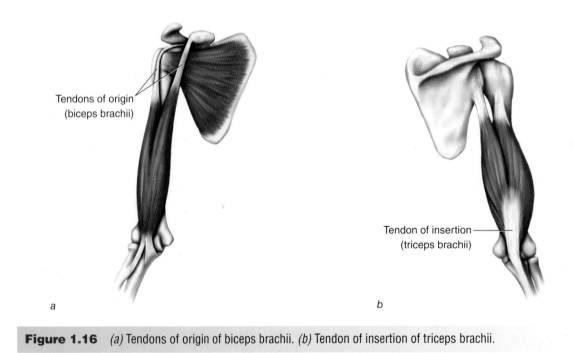

Tendons of origin (biceps brachii)

Tendon of insertion (triceps brachii)

a

b

Figure 1.16   *(a)* Tendons of origin of biceps brachii. *(b)* Tendon of insertion of triceps brachii.

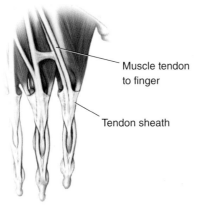

Figure 1.17 Tendon sheaths.

In almost every movement, more than one muscle is involved in producing the motion. The relationship of the skeletal muscles involved in a movement can be described by the actual function a particular muscle provides. The muscle identified as the main muscle producing a particular movement is known as the **prime mover**, or **agonist**. Any muscle helping a prime mover accomplish its action is known as a **synergist**. The prime mover is typically opposed by another muscle. The term of the opposing muscle is **antagonist**. At the elbow joint the triceps brachii is the antagonist to the biceps brachii

(agonist) during elbow flexion. A muscle may also simply hold a bone in place while another muscle performs its function. This action is known as *fixation*, and the muscle is called a **fixator**.

Skeletal muscle has several functions. (1) The obvious function of skeletal muscle is to provide movement. (2) Skeletal muscles provide protection from external trauma and act as shock absorbers for underlying bones and internal organs. (3) Support to joints through the tension of muscles and tendons is another function of skeletal muscle. A prime example of joint support by muscles and tendons is body posture. During sitting, standing, walking, running, and almost every activity of the body, muscles and tendons crossing joints provide support. (4) An often overlooked function of skeletal muscle is heat production. The process of muscle contraction includes the liberation of heat. A good example is the simple act of shivering.

The general structure of skeletal muscle is shown in figure 1.18. Skeletal muscle is encased by a form of connective tissue known as the **epimysium**. Within the epimysium are numerous **bundles** of **muscle fibers** that are individually wrapped in a fibrous sheath known as the **perimysium**. Within the perimysium, the muscle fibers are in turn enclosed in a connective sheath known as the **endomysium**. A muscle

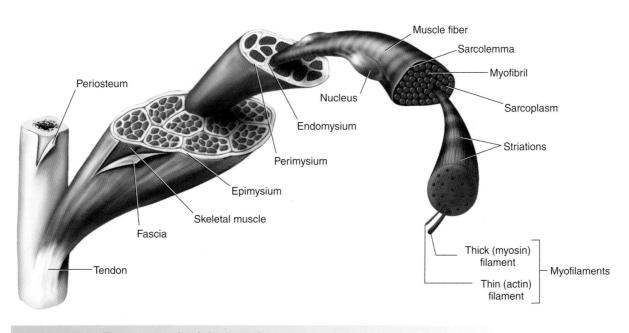

Figure 1.18 The structure of a skeletal muscle.

fiber consists of a number of **myofibrils**, which are the contractile elements of muscle. Some muscle fibers are large enough to be seen, while others are visible only through a microscope. Individual myofibrils are enclosed by a viscous material known as **sarcoplasm** and wrapped in a membrane known as the **sarcolemma**. Lengthwise, myofibrils consist of bands of alternating dark and light filaments of contractile proteins known as **actin** and **myosin** (figures 1.18 and 1.19). This alternating pattern produces a striped (striated) appearance when viewed under a microscope.

A myofibril is divided into a series of **sarcomeres**, which are considered the functional units of skeletal muscle (figure 1.19). Sarcomeres contain an **I-band** (**isotropic band**), the light-colored portion made up mostly of the protein filament actin, and an **A-band (anisotropic band)**, the dark-colored area made up mostly of the protein filament myosin (figure 1.19). A sarcomere is that portion of a myofibril that appears between two **Z-lines** (Z-lines bisect I-bands). Actin is also found in A-bands. As the actin filaments extend into the A-band, they overlap with the myosin filaments, contributing to the darker appearance at the edges of the A-band. The lighter-colored central portion of

the A-band is known as the **H-zone**. This region is lighter in color because actin does not extend into this area and because the myosin filament is thinner in the middle than at its outer edges. The two protein filaments, actin and myosin, are the site of muscular movement (contraction). The myosin filament has **cross-bridges** (small extensions) that reach out at an angle toward the actin filaments (figure 1.19).

There are two primary types of skeletal muscle fibers, commonly known as **fast-twitch** and **slow-twitch fibers** (figure 1.20). Most muscles contain both types of fibers, but depending on heredity, function, and, to a lesser degree, training, some muscles contain more of one type of fiber than the other. Fast-twitch fibers are large and white and appear in muscles used to perform strength activities. The slow-twitch fibers are small and darker (red) than the fast-twitch fibers (primarily because they have a greater supply of myoglobin). Slow-twitch fibers are slow to fatigue and are prevalent in muscles involved in performing endurance activities. A runner with a higher percentage of slow-twitch fibers in lower-extremity muscles is more likely to develop into a distance runner, whereas a runner with a higher percentage of fast-twitch fibers in lower-extremity muscles is more likely to become a sprinter.

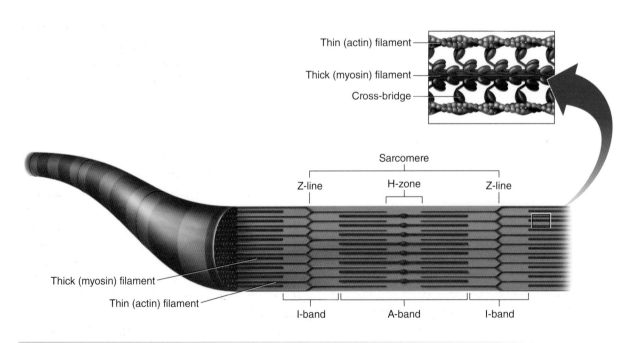

Figure 1.19   A muscle fiber and its myofibrils.

Muscle Viscosity

Viscosity is more easily understood if you consider motor oil used in an automobile. The viscosity (thickness) of the oil depends on the temperature: The oil either thins (as the temperature increases) or thickens (as the temperature decreases). Viscosity resists molecular rearrangement caused when a muscle contracts, and therefore, more energy is needed to move quickly (and overcome resistance) than slowly because of muscle viscosity. Viscosity tends to slow the speed of muscle contraction, so depending on the activity, lowering muscle viscosity can affect performance. Some authorities believe that one of the benefits of a warm-up before physical activity is that muscle viscosity changes with the increased temperature within the muscle tissue, which makes the muscle more able to endure the stress of the physical activity.

The study of muscle physiology and exercise physiology reveals other values of warming up. Anyone involved in prescribing physical activity (physicians, coaches, therapists, athletic trainers, and personal trainers) needs to understand the physiological factors involved in the warm-up, including the effect on muscle viscosity.

Figure 1.20 Fast-twitch (light) and slow-twitch (dark) muscle fibers.

Some of the major factors distinguishing *red* muscle fibers from *white* muscle fibers are that (1) red muscles are typically smaller, (2) have more endurance, (3) contract more slowly, (4) contain more myoglobin, (5) have greater density, and (6) have a greater amount of sarcoplasm.

Various forms of exercise have numerous effects on muscles. Exercise increases muscle size by increasing the size of fibers and thickening the sarcolemma and connective tissue. It increases muscle strength through this increase in fiber size and by training the body to activate more muscle fibers. Over a period of time, endurance training activities will increase blood flow to the muscle tissue being exercised, resulting in an increased ability to continue working (endurance). While this training does not alter the composition of the muscle, it does enhance what is already there.

Exercise improves neuromuscular control by improving the transmission of nerve impulses to muscles, resulting in increases in both endurance and strength. Through the process of training, exercise improves muscular coordination. It also increases blood flow to muscles, which increases the amount of myoglobin present. The combination of increased blood flow and a greater presence of myoglobin in muscle increases the amounts of nutrients and oxygen present for the muscle to produce work.

Before discussing the innervation of muscles, which allows them to function, a simple explanation is needed of how the structures presented so far—bones, joints, and muscles—combine to produce movement through a system of levers.

Levers

Bones, ligaments, and muscles are the structures that form levers in the body to create human movement. In simple terms, a joint (where two or more bones join together) forms the axis (or fulcrum), and the muscles crossing the joint apply the force to move a weight or resistance. Levers are typically labeled as first class, second class, or third class. All three types are found in the body, but most levers in the human body are third class.

A **first-class lever** has the axis (fulcrum) located between the weight (resistance) and the force (figure 1.21*a*). An example of a first-class lever is a pair of pliers or scissors. First-class levers in the human body are rare. One example is the joint between the head and the first vertebra (the atlantooccipital joint) (figure 1.21*b*). The weight (resistance) is the head, the axis is the joint, and the muscular action (force) come from any of the posterior muscles attaching to the skull, such as the trapezius.

In a **second-class lever**, the weight (resistance) is located between the axis (fulcrum) and the force (figure 1.22*a*). The most obvious example is a wheelbarrow, where a weight is placed in the bed of the wheelbarrow between the wheel (axis) and the hands of the person using the wheelbarrow (force). In the human body, an example of a second-class lever is found in the lower leg when someone stands on tiptoes (figure

1.22*b*). The axis is formed by the metatarsophalangeal joints, the resistance is the weight of the body, and the force is applied to the calcaneus bone (heel) by the gastrocnemius and soleus muscles through the Achilles tendon.

In a **third-class lever**, the most common in the human body, force is applied between the resistance (weight) and the axis (fulcrum) (figure 1.23*a*). Picture someone using a shovel to pick up an object. The axis is the end of the handle where the person grips with one hand. The other hand, placed somewhere along the shaft of the handle, applies force. At the other end of the shovel (the bed), a resistance (weight) is present. There are numerous third-class levers in the human body; one example can be illustrated in the elbow joint (figure 1.23*b*). The joint is the axis (fulcrum).

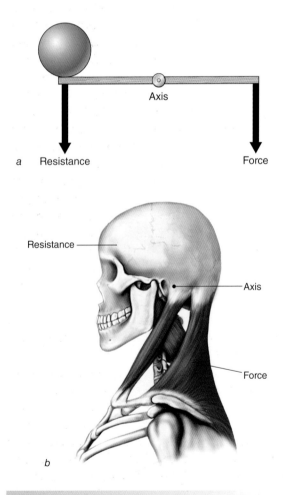

Figure 1.21 *(a)* A first-class lever; *(b)* a first-class lever in the human body.

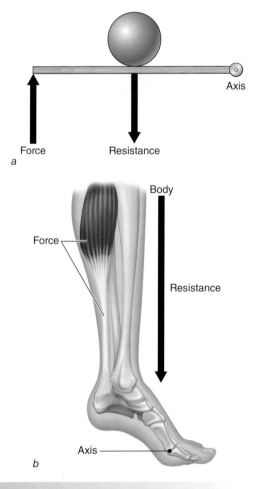

Figure 1.22 *(a)* A second-class lever; *(b)* a second-class lever in the human body.

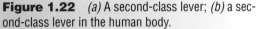

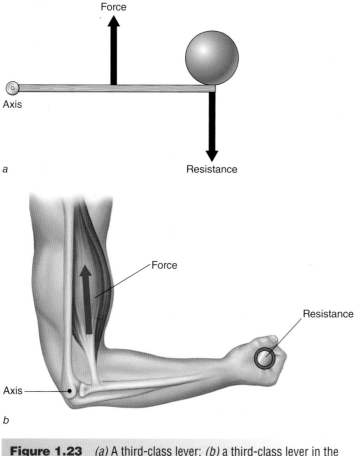

Figure 1.23 *(a)* A third-class lever; *(b)* a third-class lever in the human body.

Part *b* adapted by permission from NSCA 2008.

The resistance (weight) is the forearm, wrist, and hand. The force is the biceps muscle when the elbow is flexed.

Nerves

The body has three main nervous systems: an **autonomic nervous system**, a **central nervous system**, and a **peripheral nervous system**. The autonomic nervous system controls the glands and smooth muscle of the body. This system is often divided into the **parasympathetic system** (cranial and sacral portions) and the **sympathetic system** (thoracic and lumbar portions) (see chapter 10). The central nervous system consists of the brain and spinal cord. The brain is divided into the cerebrum (frontal, parietal, occipital, and temporal lobes), the brain stem, and the cerebellum. The outer layer of the cerebrum, which contains many cell bodies and dendrites, is often referred to as the gray matter; the inner portion of the cerebrum (the white matter) contains primarily axons (see figure 1.25). The peripheral nervous system consists of 12 pairs of cranial nerves and 31 pairs of spinal nerves. The cranial nerves are both sensory and motor in nature and typically receive a particular sensory stimulus (externally such as smell, sight, temperature, pain, or pressure and internally such as hunger, thirst, fatigue, or balance) and convert it into a nerve impulse, which can result in an appropriate effect (response). The spinal nerves, divided in plexuses (networks of peripheral nerves), innervate (stimulate) the muscles to create movement. The major plexuses are the cervical, brachial, lumbar, sacral, and, to a limited extent, pudendal (coccygeal) (figure 1.24). Spinal column levels are typically referred to by specific vertebra. For example, C5 is the fifth cervical vertebra, T8 is the eighth thoracic vertebra, and L2 is the second lumbar vertebra.

The **nerve** (figure 1.25), or **neuron**, consists of a nerve cell body and projections from it, which are known respectively as the **axon** and the **dendrite**. In a motor nerve, the dendrite receives information from surrounding tissue and conducts the nerve **impulse** *to* the nerve's **cell body** (responsible for neuron nutrition), and the axon conducts the nerve impulse *from* the cell body to the muscle fibers. A nerve innervating a muscle is referred to as a **motor neuron**, and a motor neuron plus all the muscle fibers it innervates is known as a **motor unit**. Another structural component of a motor nerve is the **myelin sheath** that insulates the axon. The gaps in the myelin sheath are known as the **nodes of Ranvier**; impulses "leap" along the myelin sheath (from node to node), allowing the impulses to travel at higher speeds than they would across an unmyelinated axon. At the end of the axon is a structure known as the **motor end plate**, which consists of **end brushes** (terminal branches) that are in very close proximity to the

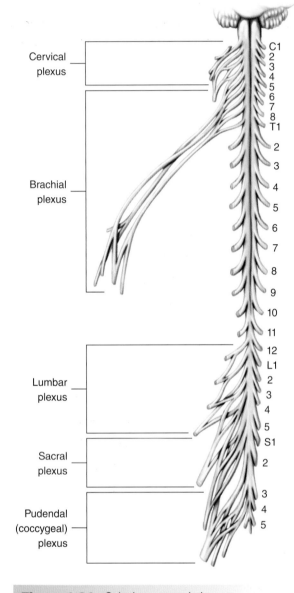

Figure 1.24   Spinal nerves and plexuses.

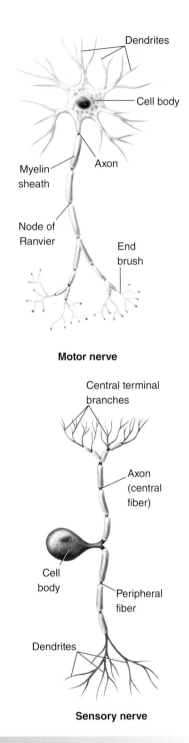

Figure 1.25   A motor (efferent) nerve and a sensory (afferent) nerve.

muscle fibers. This connection between the nerve fibers and the muscle fibers is referred to as the **myoneural junction**. **Motor nerves** carry impulses *away* from the central nervous system to the muscle tissues, whereas **sensory nerves** (not discussed in this chapter) carry impulses from muscles, ligaments, tendons, and other tissues *to* the central nervous system. Motor nerves are also referred to as **efferent nerves**; sensory nerves are also referred to as **afferent nerves**. All skeletal motor nerves innervate muscles and are connected indirectly or directly with the motor area of the brain known as the cerebral cortex (see chapter 10).

Blood Vessels

The **blood vessels** bring nutrients to the muscle tissue and carry away the waste products produced as the muscle tissues expend energy. When the heart pumps, blood moves out of the heart into a huge vascular tree consisting of **arteries**, arterioles (smaller arteries), **capillaries**, **veins**, and **venules** (small veins). There are three tissue layers (tunics) of the walls of arteries, veins, and capillaries (tunica intima, tunica media, and tunica adventitia). The middle layer (tunica media) contains various quantities of smooth muscle fibers depending on the type of vessel. The arteries and arterioles (figure 1.26) distribute

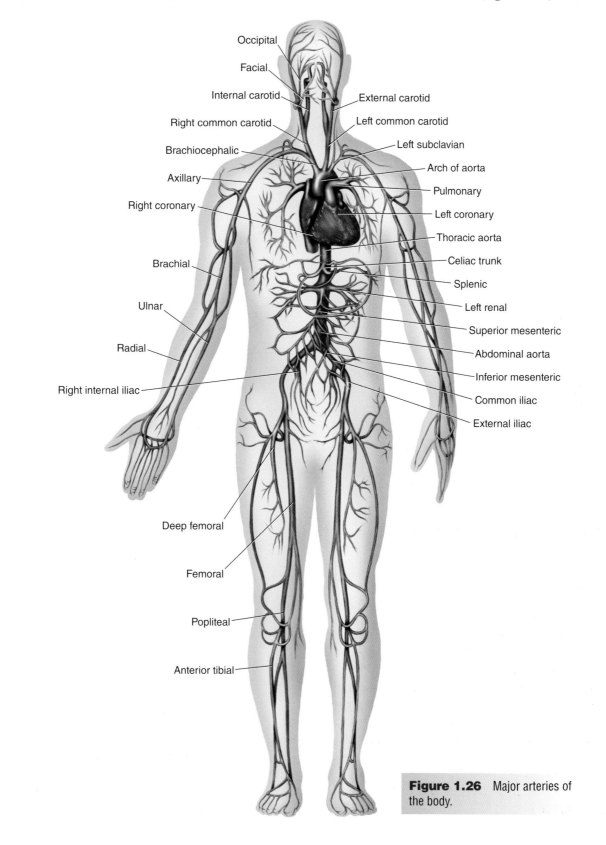

Occipital
Facial
Internal carotid
Right common carotid
Brachiocephalic
Axillary
Right coronary
Brachial
Ulnar
Radial
Right internal iliac
Deep femoral
Femoral
Popliteal
Anterior tibial

External carotid
Left common carotid
Left subclavian
Arch of aorta
Pulmonary
Left coronary
Thoracic aorta
Celiac trunk
Splenic
Left renal
Superior mesenteric
Abdominal aorta
Inferior mesenteric
Common iliac
External iliac

Figure 1.26 Major arteries of the body.

blood to the tissues, where capillaries provide the blood directly to the cells. The veins and venules (figure 1.27) collect the blood from the capillaries and return it to the heart. The middle wall of the arteries contains a vast amount of smooth muscle that contracts with the heart to pump the blood throughout the body. The veins contain small valves that permit blood to flow in only one

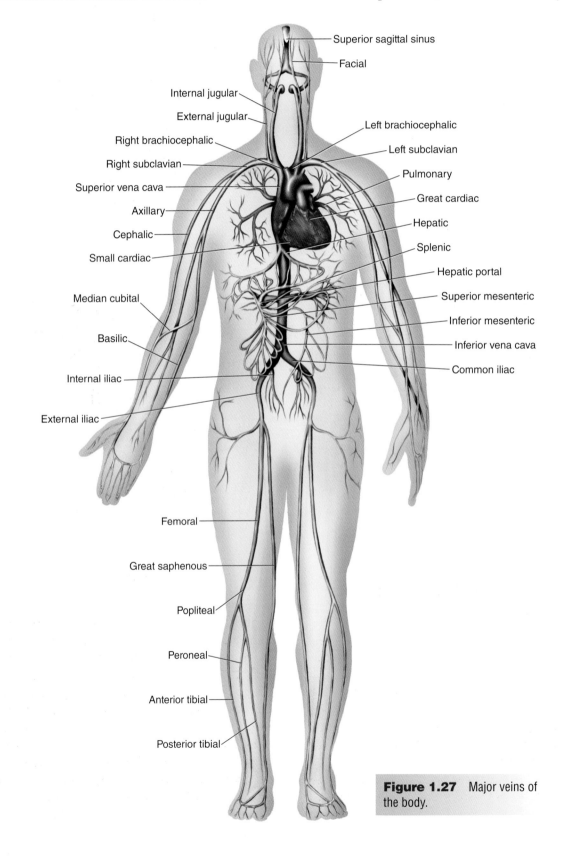

Figure 1.27 Major veins of the body.

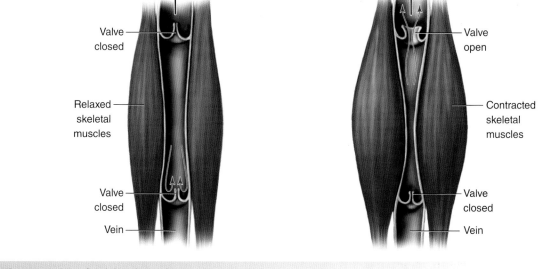

To heart

Valve
closed

Relaxed
skeletal
muscles

Valve
closed

Vein

To heart

Valve
open

Contracted
skeletal
muscles

Valve
closed

Vein

Figure 1.28 A vein's valve action.

direction (toward the heart). All three layers of tissue in the veins are much thinner as compared to the arteries. As a result, smooth muscle fibers are either absent or minimal in veins, with only a few thin fibers found in the tunica media. For this reason, skeletal muscles assist in returning blood to the heart as they contract and squeeze the veins between muscles or between muscles and bones (figure 1.28). The skeletal muscles act as muscular venous pumps that squeeze blood upward past each valve. Gravity also assists venous return in veins that are found above the heart. There are more valves in the veins of the extremities, where upward blood flow is opposed by gravity.

Other Tissues

Other types of tissues associated with bones, joints, and muscles are fascia and bursa. **Fascia** is another form of fibrous connective tissue of the body that covers, connects, or supports other tissues. One form of fascia, the sarcolemma of muscle, has already been discussed. A **bursa** (figure 1.29) is a saclike structure that contains bursa fluid and protects muscles, tendons, ligaments, and other tissues as they cross the bony prominences described earlier. Bursae provide lubricated surfaces to allow movements

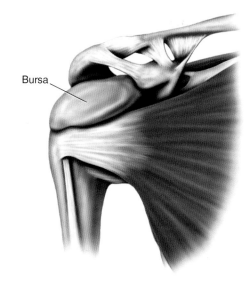

Bursa

Figure 1.29 A typical bursa.

of muscles and tendons directly over structures such as bone without being worn away over time from friction. Trauma to a bursa can inflame it and create a condition known as **bursitis**. This trauma can be from an infection, pressure, or a direct blow to the area. Bursae are identified by their position in the body. There are subfascial bursae located beneath the fascia, subcutaneous bursae beneath the skin, submuscular bursae beneath and between muscles, and subtendinous bursae beneath and between tendons.

depends on two major factors: (1) the number of motor units recruited and (2) the frequency at which they are stimulated. As the force required increases, more motor units are called into action. Additionally, they are stimulated more frequently. If impulses are sent rapidly enough to muscle fibers that they contract before totally relaxing from the previous contraction, a greater force of contraction can occur (up to a point). Once the muscle is receiving impulses at such a rate that it cannot relax, it reaches a state of continuous contraction known as tetanus (figure 1.35). The application of the term *tetanus* to this state of continuous contraction—the result of physical effort—should not be confused with another usage of this term, an infectious disease that can cause involuntary muscular contractions.

Courses in biology, human physiology, exercise physiology, kinesiology, biomechanics, and

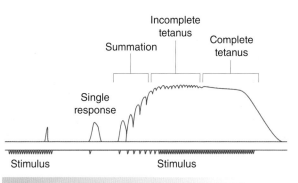
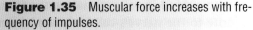

Figure 1.35 Muscular force increases with frequency of impulses.

other fields examine the generation of a nerve impulse and the resulting contraction of muscle in detail. The preceding material should be considered an introductory overview and by no means a detailed analysis of the neuromuscular system and its motor units.

LEARNING AIDS

REVIEW OF TERMINOLOGY

The following terms are discussed in this chapter. Define or describe each term, and where appropriate, identify the location of the named structure either on your body or in an appropriate illustration.

A-band	capsule	fibrocartilage
acetylcholine	cardiac muscle	first-class lever
actin	cartilaginous joint	fixator
afferent nerve	cell body	foramen
agonist	central nervous system	fossa
anatomical landmark	condyle	fusiform
antagonist	condyloid joint	hinge joint
arteries	connective tissue	H-zone
arterioles	cross-bridge	I-band
arthrology	dendrite	impulse
articular cartilage	diaphysis	irregular joint
articular surface	diarthrodial joint	joint
articulation	diastole	ligament
autonomic nervous system	efferent nerve	ligamentous joint
axon	end brush	long bone
ball-and-socket joint	endomysium	longitudinal
biaxial joint	epicondyle	motor end plate
bipennate	epimysium	motor nerve
blood vessel	epiphyseal plate	motor neuron
bundle	epiphysis	motor unit
bursa	epithelial tissue	multipennate
bursitis	facet	muscle fiber
calcium	fascia	muscle tissue
capillaries	fast-twitch fiber	myelin sheath

myofibril	process	synergist
myology	quadrate	synovial joint
myoneural junction	saddle joint	systole
myosin	sarcolemma	tendon
nerve	sarcomere	tendon of insertion
nerve tissue	sarcoplasm	tendon of origin
neuroglia	sarcoplasmic reticulum	tendon sheath
neuromuscular system	second-class lever	third-class lever
neuron	sensory nerve	threshold
node of Ranvier	skeletal muscle	transverse tubule
nonaxial joint	slow-twitch fiber	triangular
notch	smooth muscle	triaxial joint
osteoporosis	spine	tubercle
parasympathetic system	spinous process	tuberosity
perimysium	suture joint	uniaxial joint
periosteum	sympathetic system	unipennate
peripheral nervous system	synapse	veins
pivot joint	synaptic cleft	venules
prime mover	synarthrodial joint	Z-line

SUGGESTED LEARNING ACTIVITIES

1. Make a fist. List all the anatomical structures (starting with the brain) that were used for that action to occur.

2. Either at the dinner table or in a grocery store, look at the poultry and explain why a particular fowl (turkey, chicken) has meat of different colors in its various parts (legs, thighs, breasts, wings).

   a. Did the fowl's normal activities dictate more or less effort of certain body parts?

   b. What type of muscle fibers likely dominate the muscles of these various parts? Why?

3. From a standing position, rise up on your toes and stand that way for a few minutes (or as long as you can).

   a. What type of leg muscle (fast-twitch or slow-twitch) was primarily responsible for initially getting you into the toe-standing position?

   b. What type of leg muscle (fast-twitch or slow-twitch) was primarily responsible for sustaining you in the toe-standing position?

MULTIPLE-CHOICE QUESTIONS

1. A junction of two or more bones forming a joint is also known as

   a. an epiphysis

   b. a fossa

   c. an articulation

   d. a diaphysis

2. Which of the following terms does *not* appropriately fit with the other three?

   a. notch

   b. process

   c. tubercle

   d. tuberosity

3. The functional unit of skeletal muscle is known as

   a. a myofibril

   b. a sarcomere

   c. the A-band

   d. the I-band

4. A series of sarcomeres linked together is known as

   a. a myofibril

   b. a muscle

   c. actin

   d. myosin

(continued)

MULTIPLE-CHOICE QUESTIONS *(continued)*

5. The release of which of the following substances causes the cross-bridges to move, which in turn causes a sarcomere to shorten?
   a. actin
   b. myosin
   c. calcium
   d. sarcoplasm

6. A bundle of fibers within a muscle is wrapped in a fibrous sheath known as the
   a. endomysium
   b. epimysium
   c. perimysium
   d. sarcolemma

7. The origin of a muscle is most likely defined as being located at the
   a. distal end of a joint
   b. distal end of a tendon
   c. most stable bony attachment
   d. most movable bony attachment

8. A third-class lever places the resistance
   a. between the axis and the force
   b. outside of the force and the axis
   c. on the opposite side of the axis as the force
   d. on the axis

FILL-IN-THE-BLANK QUESTIONS

1. Joints with no observable movement are known as _____ joints.

2. Saclike structures that protect soft tissues as they pass over bony projections are known as _____.

3. A motor unit is a motor neuron and all the _____ it supplies.

4. Dendrites conduct nerve impulses _____ a cell body.

5. Axons conduct nerve impulses _____ a cell body.

6. The term _____ refers to the contraction of cardiac muscle.

7. _____ resists molecular rearrangement when a muscle contracts.

8. The most common type of lever found in the human body is the _____ lever.

Movement

Now that you understand the structures involved in movement (bones, ligaments, muscles) and the terms used to describe them, let us look at the universal language that describes the movements performed by these structures.

When we describe a human movement, there is an anatomical point that is universally accepted as being the position all movements start from: the **anatomical position**. In this position, all joints are considered to be in a neutral position, or at 0°, with no movement having yet occurred. Occasionally, you might also hear the term **fundamental position**. Note carefully the only difference between the two positions (figure 2.1). The anatomical position is preferred to the fundamental position for any discussion of human movement because the hand position in the fundamental position makes certain upper-extremity movements impossible. In the following sections, the description of any movement starts from the anatomical position.

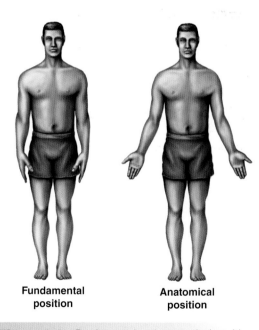

Fundamental
position

Anatomical
position

Figure 2.1 Fundamental and anatomical positions.

Anatomical Locations

Several terms are considered universal for discussing the spatial relationship between one anatomical structure and another. The term **superior** refers to something that is above or higher than another structure (e.g., your head is superior to your chest). The opposite term, **inferior**, means something is below or lower than another structure (e.g., your chest is inferior to your head). **Lateral** refers to something farther away from the midline of the body than another structure (e.g., your arms are lateral to your spinal column). **Medial** means a structure is closer to the midline of your body than another structure (e.g., your nose is medial to your ears). **Anterior** refers to a structure that is in front of another structure (e.g., your abdomen is anterior to your spinal column). **Posterior** refers to a structure that is behind another structure (e.g., your spinal column is posterior to your abdomen).

The terms **proximal** (close) and **distal** (far) are usually used in reference to structures of the extremities (arms and legs). Proximal means closer to the trunk, and distal means farther from the trunk (e.g., your knee is proximal to your ankle, and your hand is distal to your wrist). The term **dorsal** indicates the top side of an animal, such as the dorsal fin on the top of a fish, or the posterior of the human body. (The dorsal aspect of your hand is commonly called the back of your hand.) The term **volar** refers to the down side, or bottom aspect, of a structure. (The volar aspect of your wrist or hand is also referred to as the **palmar** aspect, whereas the volar aspect, or sole, of the foot is referred to as the **plantar** aspect.) Two terms refer to actions of the forearm and foot. The term **pronation** refers to the turning of the forearm toward the body, resulting in the volar, or palmar, surface of the hand facing the body, or if the elbow is flexed, palm down. Turning your foot downward (**plantar flexion**) and inward (**inversion**) toward the other foot is referred to as pronation of the foot. **Supination**, the reverse of pronation, refers to turning the forearm outward and palm upward from the pronated position and to the upward (**dorsiflexion**) and outward (**eversion**) of the foot away from the other foot.

Planes and Axes

Human movement that takes place from the starting (anatomical) position is described as taking place in a **plane** (a flat surface) about an **axis** (a straight line around which an object rotates). Muscles create movements of body segments in one or more of three planes that divide the body into different parts. These three specific planes are perpendicular (at right angles) to each other (figure 2.2). The **sagittal plane** (also known as the anteroposterior plane) passes from the front

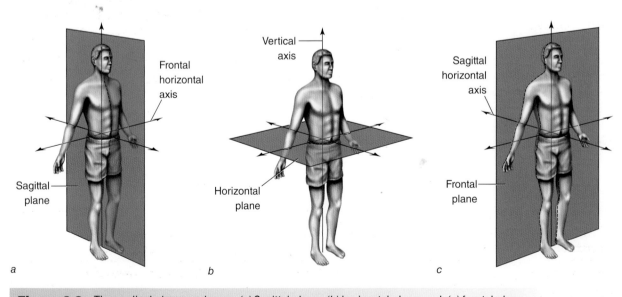

a   *b*   *c*

Figure 2.2   The cardinal planes and axes. *(a)* Sagittal plane, *(b)* horizontal plane, and *(c)* frontal plane.

through the back of the body, creating a left side and a right side of the body. There could be any number of sagittal planes; however, there is only one cardinal sagittal plane. The term *cardinal* refers to the one plane that divides the body into equal segments, with exactly one half of the body on either side of the cardinal plane. Therefore, the **cardinal sagittal plane** divides the body into two equal halves on the left and right. The term *cardinal plane* appears in some texts as the *principal plane*. The terms are interchangeable.

The **horizontal plane** (also known as the transverse plane) passes through the body horizontally to create top and bottom segments of the body. There could be any number of horizontal planes, but there is only one **cardinal horizontal plane**, which divides the body into equal top and bottom halves.

The **frontal plane** (also known as the lateral plane) passes from one side of the body to the other, creating a front side and a back side of the body. Again, there could be any number of frontal planes, but there is only one **cardinal frontal plane**, which divides the body into equal front and back halves.

The point at the intersection of all three cardinal planes is the body's **center of gravity**. When all segments of the body are combined and the body is considered one solid structure in the anatomical position, the center of gravity lies approximately in the low back area of the spinal column. As body parts move from the anatomical position or as weight shifts through weight gain, weight loss, or by carrying loads, the center of gravity also shifts. No matter what the body's position or weight distribution, however, half of the weight of the body (and its load) will always be to the left and right, in front and behind, and above and below the center of gravity. The center of gravity of the body constantly changes with each movement, each change in weight distribution, or both.

Earlier we defined an axis as a straight line that an object rotates around. In the human body, we picture joints as axes and bones as the objects that rotate about them in a plane perpendicular to the axis. There are three main axes, and rotation is described as occurring in a plane about the axis that is perpendicular to the plane (figure 2.2). The sagittal plane rotates about a **frontal horizontal axis** (figure 2.2*a*).

Hands On

The knee joint is a frontal horizontal axis, and the lower leg is the object that moves in the sagittal plane when you bend your knee.

The horizontal plane rotates about a **vertical (longitudinal) axis** (figure 2.2*b*).

Hands On

As you turn your head to the left and right as if to silently say no, your head rotates in a horizontal plane about the vertical axis created by your spinal column.

The frontal plane rotates about the **sagittal horizontal axis** (figure 2.2*c*).

Hands On

When you raise your arm to the side, your shoulder joint is the sagittal horizontal axis, and your arm is the object moving in the frontal plane.

For a summary of the relationship between anatomical planes and associated axes, see table 2.1.

TABLE 2.1

Planes, Axes, and Fundamental Movements

Plane	Axis	Movements
Sagittal (anteroposterior)	Frontal horizontal	Flexion and extension
Frontal (lateral)	Sagittal horizontal	Abduction and adduction
Horizontal (transverse)	Vertical	Rotation

Fundamental Movements

Again, remember that movement takes place in a plane about an axis. There are three planes and three axes with two **fundamental movements** possible in each plane. In the sagittal plane, the fundamental movements known as flexion and extension are possible. **Flexion** is defined as a reduction of the angle formed by the bones of the joint (figure 2.3). In flexion of the elbow joint, the angle between the forearm and upper arm decreases. **Extension** is defined as an increase of the joint angle (figure 2.4). Returning a joint in flexion to the anatomical position is considered extension. Further extension beyond the anatomical position is referred to as **hyperextension**. Fundamental movements in the frontal plane are known as abduction and adduction. **Abduction** is defined as movement away from the midline of the body (figure 2.5). As you move your arm away from the side of your body in the frontal plane, you are abducting the shoulder joint. Movement toward the midline of the body is defined as **adduction** (figure 2.6). Returning your arm from an abducted shoulder

position to the anatomical position is adduction. The fundamental movement in the horizontal plane is simply defined as **rotation** (figure 2.7). The earlier example of shaking your head no is rotation of the head. For describing movement in the upper (arm) and lower (leg) extremities, the terms *external rotation* and *internal rotation* are often used (figure 2.8). When the anterior (front) surface of the arm or leg rotates laterally (away from the midline of the body), this is defined as **external rotation** (or lateral rotation). When the anterior surface of the arm or leg rotates medially (toward the midline of the body), this is defined as **internal rotation** (or medial rotation).

Joints capable of creating movement in two (biaxial) or three (triaxial) planes are also capable of another movement, **circumduction**, which, because it combines two or more fundamental movements, is *not* considered a fundamental movement of any joint. When movement occurs in two or three planes in a sequential order, the joint is said to be circumducting. Moving your arm at the shoulder joint in a "windmill" motion is an example of circumduction (figure 2.9).

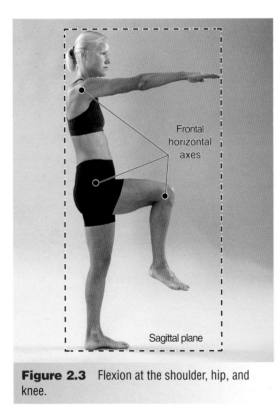

Figure 2.3 Flexion at the shoulder, hip, and knee.

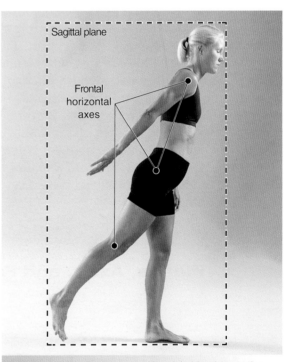

Figure 2.4 Extension at the shoulder, hip, and knee.

Figure 2.5 Abduction at the shoulder and hip.

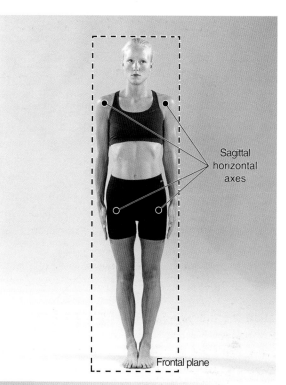

Figure 2.6 Adduction at the shoulder and hip as a return from the abducted position in figure 2.5.

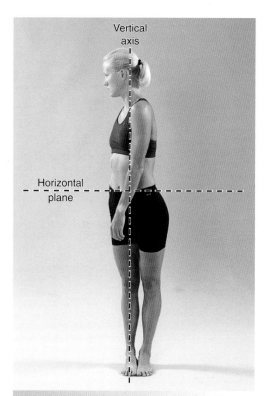

Figure 2.7 Rotation as a twisting movement of the spine.

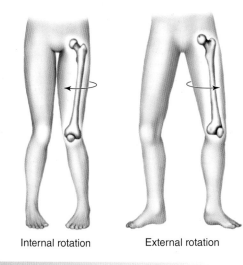

Internal rotation   External rotation

Figure 2.8 Internal rotation and external rotation of the lower extremity.

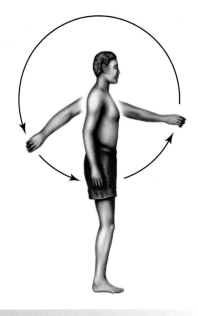

Figure 2.9   Circumduction at the shoulder.

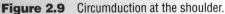

LEARNING AIDS

REVIEW OF TERMINOLOGY

The following terms are discussed in this chapter. Define or describe each term, and where appropriate, identify the location of the named structure either on your body or in an appropriate illustration.

abduction	extension	palmar
adduction	external rotation	plane
anatomical position	flexion	plantar
anterior	frontal horizontal axis	plantar flexion
axis	frontal plane	posterior
cardinal frontal plane	fundamental movement	pronation
cardinal horizontal plane	fundamental position	proximal
cardinal sagittal plane	horizontal plane	rotation
center of gravity	hyperextension	sagittal horizontal axis
circumduction	inferior	sagittal plane
distal	internal rotation	superior
dorsal	inversion	supination
dorsiflexion	lateral	vertical (longitudinal) axis
eversion	medial	volar

SUGGESTED LEARNING ACTIVITIES

1. Stand in the anatomical position.

a. Flex your knee joint. In what plane did movement occur? About what axis did movement occur?

b. Abduct your hip joint. In what plane did movement occur? About what axis did movement occur?

c. Rotate your head so your chin touches your left shoulder. In what plane did movement occur? About what axis did movement occur?

2. The body's center of gravity is considered the point where the three cardinal planes of the body intersect.

a. In what direction would your center of gravity shift if you were to put a backpack full of books on your shoulders?

b. In what direction would your center of gravity shift if you were to carry a large book in both hands in front of you below your waist?

c. In what direction would your center of gravity shift if you were to carry a heavy briefcase in your left hand with your arm fully extended at your side?

MULTIPLE-CHOICE QUESTIONS

1. Which of the following movements is defined as movement in the frontal plane toward the midline of the body?
   a. abduction
   b. adduction
   c. flexion
   d. extension

2. For movement of the shoulder joint to occur in the frontal plane, which of the following joint actions must take place?
   a. internal rotation
   b. circumduction
   c. flexion
   d. abduction

3. A motion occurring in the horizontal (transverse) plane about a vertical axis is known as
   a. abduction
   b. rotation
   c. adduction
   d. extension

4. Movement taking place in the frontal plane occurs about the
   a. frontal horizontal axis
   b. sagittal horizontal axis
   c. vertical axis
   d. horizontal axis

FILL-IN-THE-BLANK QUESTIONS

1. When the angle formed at a joint diminishes and the movement takes place in the sagittal plane, the movement is known as _____.

2. Joint motion is typically described as taking place about an axis and within a _____.

3. An axis of the body that passes through the body horizontally from front to back is known as a _____ axis.

4. When referring to a structure of an extremity being closer to the trunk than another structure, we say that it is _____ to the other structure.

Upper Extremity

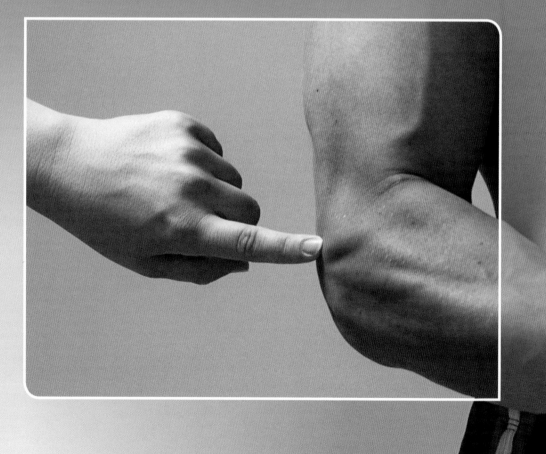

The Shoulder

Any discussion of the shoulder must start with the fact that the shoulder is actually two distinct anatomical structures: the shoulder girdle and the shoulder joint. The **shoulder girdle** consists of the clavicle and scapula bones, whereas the **shoulder joint** is formed by the scapula and the humerus bones. The primary function of the shoulder girdle is to position itself to accommodate movements of the shoulder joint.

Bones of the Shoulder Girdle

Two bones make up the structure known as the shoulder girdle: the clavicle and the scapula (figure 3.1). The **clavicle** is a long, slender, S-shaped bone that attaches to the **sternum** (breastbone) at the medial end and to the scapula at the lateral end. The clavicle is often referred to as the collarbone. It is the only bony attachment that the upper extremity has to the trunk.

Because of its shape; the fact that it is held in place at either end by strong, unyielding ligaments; and the fact that it has little protection from external forces (with the exception of skin), the clavicle is an often-fractured bone.

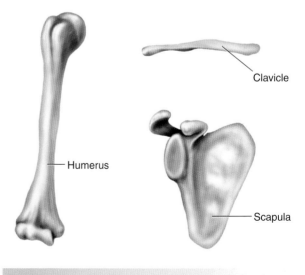

Figure 3.1 The bones of the shoulder, anterior view.

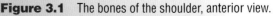

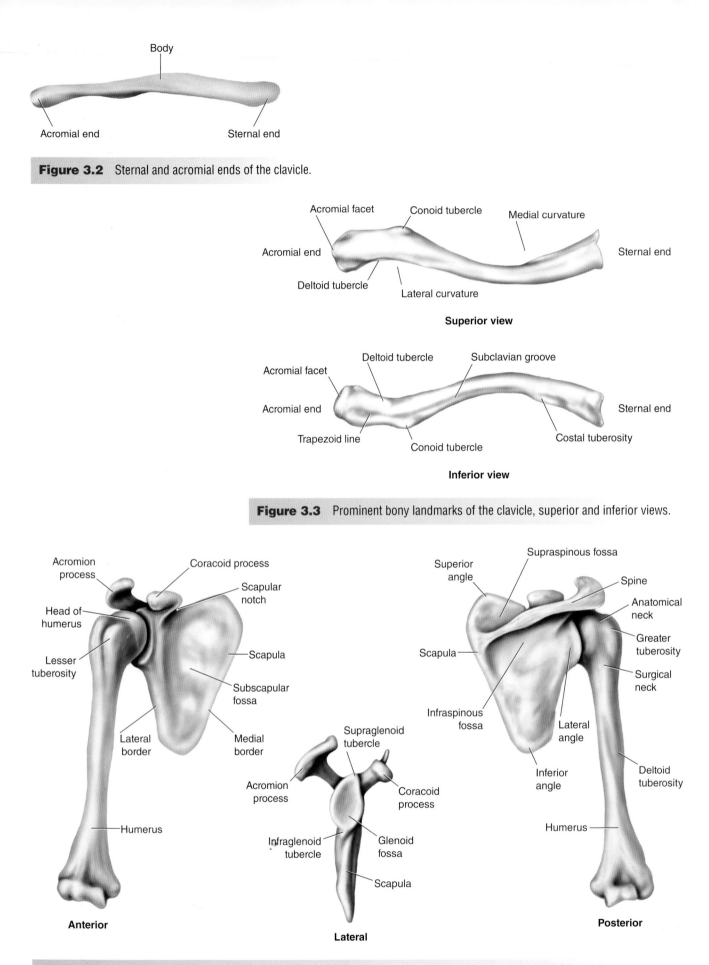

Figure 3.2 Sternal and acromial ends of the clavicle.

Figure 3.3 Prominent bony landmarks of the clavicle, superior and inferior views.

Figure 3.4 Landmarks of the shoulder bones; anterior, posterior, and lateral views.

The lateral end of the clavicle is referred to as the **acromial end**, and the medial end is referred to as the **sternal end** (figure 3.2). Prominent bony landmarks are observed on the superior and inferior views (figure 3.3) and include the **deltoid tubercle**, the **conoid tubercle**, the **trapezoid line**, the **costal tuberosity**, and the **subclavian groove**. These structures are important as places of attachment for soft tissue.

The **scapula** is the large, triangular, wing-like bone in the upper posterior portion of the trunk. This bone, sometimes referred to as the shoulder blade, is considered a bone of both the shoulder girdle and the shoulder joint. Figure 3.4 illustrates the many bony prominences of the scapula, including the **lateral** and **medial borders** and the **inferior angle** at the junction of the two borders.

👋 Hands On

You can locate these structures on a partner (figure 3.5).

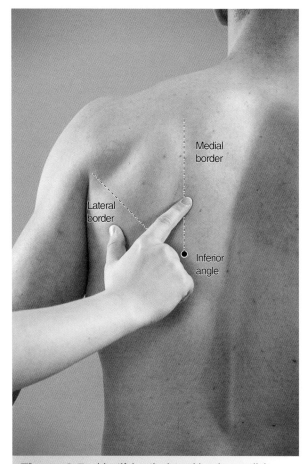

Figure 3.5 Identifying the lateral border, medial border, and inferior angle of the scapula.

Additionally, the **superior border** at the medial end of the scapula becomes the **superior angle** and at its lateral end has a notch known as the **scapular notch**. Two large bony prominences at the superior lateral portion of the scapula are known as the **coracoid process** (anterior) and the **acromion process** (posterior).

👋 Hands On

Either on yourself or a partner, palpate (touch) both the coracoid and acromion processes (figures 3.6 and 3.7).

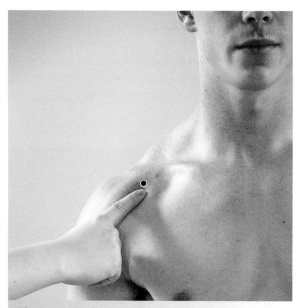

Figure 3.6 Locating the coracoid process.

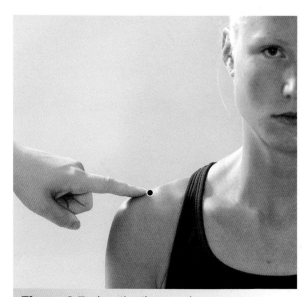

Figure 3.7 Locating the acromion process.

The acromion process is the lateral expansion of a ridge of bone approximately one-third of the way down the posterior aspect of the scapula. This ridge of bone is known as the **spine** of the scapula.

👋 Hands On

Apply pressure with your index and middle fingers along the upper third of your partner's scapula and you'll feel this spine (figure 3.8).

Most laterally, the scapula forms a smooth, round, slightly depressed surface known as the **glenoid fossa**. This cavity forms the socket for the shoulder joint. Above and below the glenoid fossa are two bony prominences known, respectively, as the **supraglenoid** and **infraglenoid tubercles**. The smooth area of bone between the lateral and medial borders of the scapula on the anterior surface is known as the **subscapular fossa**. On the posterior surface of the scapula, the smooth bony surfaces above and below the spine are known, respectively, as the **supraspinous fossa** and **infraspinous fossa** (see figure 3.4).

Bones of the Shoulder Joint

The shoulder joint is the articulation between the scapula and the **humerus** (bone of the upper arm). The joint is known as the **glenohumeral (GH) joint** because of the two articulating bony surfaces. The prominent structure of the scapula in terms of the shoulder joint is the anatomical area labeled the glenoid fossa. The shoulder joint is classified as a ball-and-socket joint, and the glenoid fossa, although somewhat shallow, is considered the socket of the joint.

The "ball" of the shoulder joint is the structure known as the **head** of the humerus. This chapter discusses the humerus only because it is part of the shoulder joint (the proximal end; see figure 3.9). The humerus at its distal end is discussed further in the chapter on the elbow joint.

The head of the humerus is separated from the shaft of the bone by two necks. The **anatomical neck** is located between the head of the humerus and two bony prominences known as the **greater** and **lesser tuberosities**. The **surgical neck** (see figure 3.4) is actually the upper portion

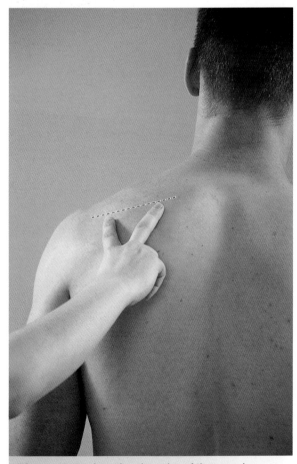

Figure 3.8 Locating the spine of the scapula.

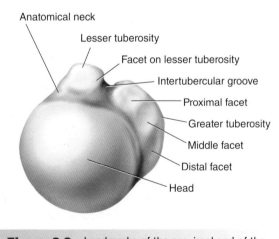

Anatomical neck
Lesser tuberosity
Facet on lesser tuberosity
Intertubercular groove
Proximal facet
Greater tuberosity
Middle facet
Distal facet
Head

Figure 3.9 Landmarks of the proximal end of the humerus.

of the shaft of the humerus. A groove known as the intertubercular (bicipital) groove is created by the greater and lesser tuberosities. Atop both the lesser and greater tuberosities appear four flat surfaces known as facets. The lesser tuberosity has one facet, whereas the greater tuberosity has three: a **proximal**, **middle**, and **distal facet**. Approximately halfway down the shaft of the humerus, on the lateral surface, is a bony prominence known as the **deltoid tuberosity** (see figure 3.4).

🖐 Hands On

Apply pressure to either your own or your partner's humerus, approximately halfway between the head and the distal end at the elbow, on the lateral aspect; you'll feel the bump of the deltoid tuberosity (figure 3.10).

Joints and Ligaments of the Shoulder Girdle

The shoulder girdle has two joints, one at either end of the clavicle, known as the **acromio-clavicular (AC)** and **sternoclavicular (SC) joints**. Movement in the SC joint is slight in all directions and of a gliding, rotational type. The joint receives its stability both from its bony arrangement, because the sternal end of the clavicle lies in the **clavicular notch** of the manubrium of the sternum (see chapter 9), and from the ligamentous arrangement that ties the clavicle and sternum together.

Three primary ligaments are responsible for the SC articulation (figure 3.11). The **sternoclavicular ligament**, with **anterior**, **superior**, and **posterior fibers**, and two other ligaments help to stabilize the SC articulation: the **costoclavicular ligament**, which secures the sternal end of the clavicle to the first rib, and the **interclavicular ligament**, which secures the sternal ends of both clavicles into the clavicular notch of the manubrium of the sternum. Also, an **articular disc** is present between the sternal end of the clavicle and the clavicular notch of the manubrium of the sternum.

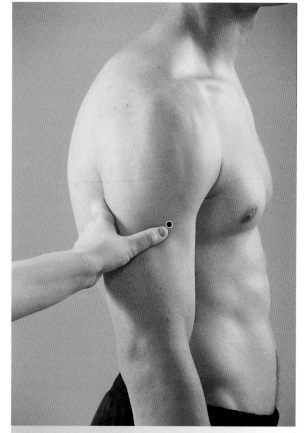

Figure 3.10   Locating the deltoid tuberosity.

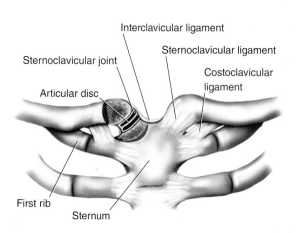

Figure 3.11   Sternoclavicular, costoclavicular, and interclavicular ligaments and the articular disc.

✋ **Hands On** ▬▬▬▬

Place your index finger in the space between each of your clavicles; you will feel the clavicular notch formed between both SC joints (figure 3.12).

Figure 3.12 Locating the sternoclavicular joint.

The joint between the lateral (acromial) end of the clavicle and the scapula is divided into two separate areas: the AC joint and the **coracoclavicular joint**. Simply described, the lateral end of the clavicle articulates with both the acromion process and the coracoid process of the scapula (figure 3.13).

✋ **Hands On** ▬▬▬▬

Palpate the area at the lateral end of the clavicle (figure 3.14). You should both feel and see the bump that is the AC joint.

The AC joint is the articulation between the acromion process of the scapula and the acromial end of the clavicle. There is a slight gliding type of movement between the two bones of this joint when elevation and depression of the acromial end of the clavicle and the acromion process of the scapula take place. The **acromioclavicular ligament** functions as the joint capsule, tying together and totally surrounding the lateral end of the clavicle and the acromion process of the scapula.

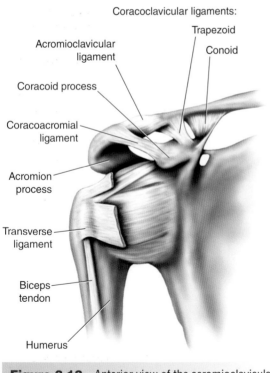

Coracoclavicular ligaments:
Trapezoid
Conoid
Acromioclavicular ligament
Coracoid process
Coracoacromial ligament
Acromion process
Transverse ligament
Biceps tendon
Humerus

Figure 3.13 Anterior view of the acromioclavicular and coracoclavicular ligaments.

Figure 3.14 Locating the acromioclavicular joint.

The other shoulder girdle joint, the coracoclavicular joint, is sometimes considered a component of the acromioclavicular joint and sometimes treated as a separate joint. The joint is the articulation between the lateral (acromial) end of the clavicle and the coracoid process of the scapula. Two ligaments run between the coracoid process of the scapula and the inferior surface of the clavicle. The two ligaments, the conoid and the trapezoid, are often referred to as a single ligament, the **coracoclavicular ligament**. Although some people do not consider the coracoclavicular joint a true joint, slight movement occurs in all directions in the articulation. The **trapezoid ligament** is the more lateral component of the coracoclavicular ligament and runs from the superior aspect of the coracoid process of the scapula to the anterior inferior aspect of the clavicle. It opposes forward, upward, and lateral movement of the lateral aspect of the clavicle. The **conoid ligament** is the medial component of the coracoclavicular ligament and runs from the superior aspect of the coracoid process of the scapula to the posterior inferior aspect of the clavicle. It opposes backward, upward, and medial movement of the lateral aspect of the clavicle. The coracoclavicular ligament is a strong supporter of the acromioclavicular ligament. Loss of these ligaments results in separation of the upper extremity from the trunk of the body.

Ligaments of the Shoulder Joint

The shoulder joint is the articulation between the head of the humerus and the glenoid fossa of the scapula. The ligaments of the shoulder joint (figure 3.16) include the **capsular ligament**, the **glenohumeral ligament** (superior, inferior, and middle sections), and the **coracohumeral ligament**. The capsular ligament

FOCUS ON

Shoulder Separation

A sprain (partial or complete tearing) of the acromioclavicular and coracoclavicular ligaments results in a visible gap between the clavicle and the scapula, a classic illustration of what is known as a shoulder separation (figure 3.15). The separation is actually a widening of the space between the lateral end of the clavicle and the acromion process of the scapula. The weight of the upper extremity often reveals this gap when the affected shoulder girdle is compared with the unaffected one. When you hear the term *shoulder separation*, know that it more correctly refers to the acromioclavicular joint of the shoulder girdle.

Falling out of a tree, being tackled in a football game, being taken down to the mat in a wrestling match, or slipping on ice and falling on the shoulder can all lead to spraining the acromioclavicular joint. To further study the topic of sprains, consult athletic training or sports medicine texts, and talk with athletic trainers and physicians who specialize in sports medicine. Knowledge of the basic anatomy of the shoulder will enhance your ability to understand the literature and communicate with these specialists.

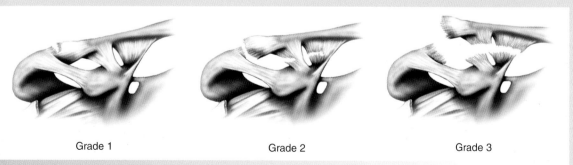

Grade 1   Grade 2   Grade 3

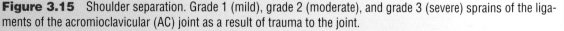

Figure 3.15 Shoulder separation. Grade 1 (mild), grade 2 (moderate), and grade 3 (severe) sprains of the ligaments of the acromioclavicular (AC) joint as a result of trauma to the joint.

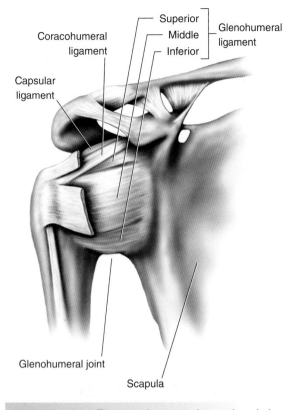

Figure 3.16 The capsular, coracohumeral, and glenohumeral ligaments.

Other Ligaments of the Shoulder

In addition to ligaments of the shoulder joint and girdle, other ligaments of the shoulder include those specific to the scapula and the humerus.

Ligaments of the Scapula

Although we typically think of ligaments as tying bones together to form articulations, some ligaments run from one aspect of a bone to another aspect of the same bone, serving some function other than forming joints. Four such ligaments are prominent on the scapula (figure 3.17). The **superior transverse scapular ligament** crosses the scapular notch, converting the notch into a foramen through which the suprascapular nerve passes. The **inferior transverse scapular ligament** crosses from one edge to the other of the **great scapular notch**. This ligament forms a tunnel for the passage of the suprascapular nerve that innervates (stimulates), and the transverse scapular blood vessels that

attaches the anatomical neck of the humerus and the circumference of the glenoid of the scapula. The glenohumeral ligaments are located beneath the anterior surface of the joint capsule and reinforce the capsule. The **superior glenohumeral ligament** runs between the upper surface of the lesser tuberosity of the humerus and the superior edge of the glenoid of the scapula. The **middle glenohumeral ligament** runs between the anterior surface of the lesser tuberosity of the humerus and the anterior edge of the glenoid of the scapula. The **inferior glenohumeral ligament** runs between the lower anterior surface of the lesser tuberosity of the humerus and the lower anterior edge of the glenoid of the scapula. The coracohumeral ligament runs between the anatomical neck of the humerus, near the greater tuberosity, and the lateral aspect of the coracoid process of the scapula.

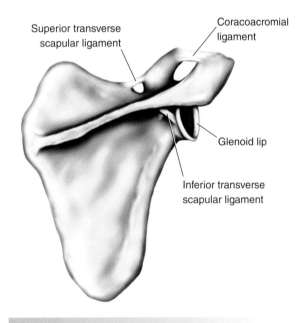

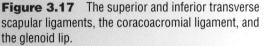

Figure 3.17 The superior and inferior transverse scapular ligaments, the coracoacromial ligament, and the glenoid lip.

supply blood to, the infraspinatus muscle. The **coracoacromial ligament** crosses between the coracoid process and the acromion process of the scapula. Although a ligament of the scapula, this ligament limits superior movement of the humeral head. The **glenoid lip** (also known as the **glenoid labrum**) (figure 3.18) is a ligament that forms an edge around the entire circumference of the glenoid of the scapula. The glenoid lip helps deepen the glenoid fossa for the head of the humerus to add to the stability of the shoulder joint.

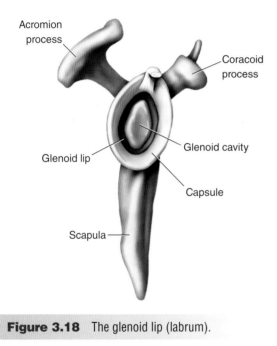

Figure 3.18 The glenoid lip (labrum).

Ligament of the Humerus

On the anterior surface of the proximal end of the humerus, two structures were discussed: the greater and lesser tuberosities. Also mentioned was the **intertubercular groove** formed between these structures. Crossing the intertubercular groove is a ligament known as the **transverse humeral ligament** (figure 3.19). This ligament has one function: to hold the tendon of origin of the long head of the biceps brachii muscle in the groove.

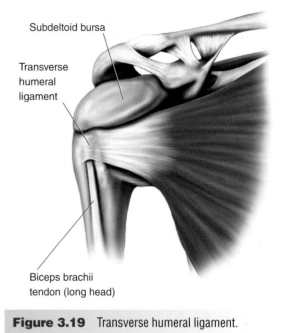

Figure 3.19 Transverse humeral ligament.

Fundamental Movements and Muscles of the Shoulder Girdle

In this chapter and all succeeding chapters, the fundamental movements involving the anatomical area under discussion are introduced first. A discussion of the muscles involved in those movements is then presented.

Movements of the Shoulder Girdle

Movements of the shoulder girdle are primarily for the purpose of accommodating shoulder joint movement through changing positions of the glenoid of the scapula. Although it was stated earlier that fundamental movements are those confined to a single plane about a single axis, the shoulder girdle is an exception. Remember that the starting position for description of all fundamental movements of any joint is the anatomical position. There are four fundamental movements of the shoulder girdle: **elevation**, **depression**, **abduction**, and **adduction**. Because of the relationships between the clavicle and scapula and between the scapula and the thorax, on which it is positioned, movement exclusively in one plane about one axis is not always possible. Fundamental movements of the shoulder girdle are described relative to the direction in which the scapula moves.

Elevation of the shoulder girdle is defined as a superior (upward) movement of the scapula in the frontal plane (figure 3.20).

Depression of the shoulder girdle may be described as inferior (downward) movement of the scapula in the frontal plane (figure 3.21) but should more correctly be described as return from elevation. Because of the anatomical starting position, depression of the shoulder girdle is not possible. Eccentric contraction (lengthening) of muscles that concentrically contracted (shortened) to cause elevation of the shoulder girdle results in depression of the shoulder girdle (return to the anatomical position).

Abduction of the shoulder girdle cannot be simply defined as movement away from the midline of the body by the scapula. Because the scapula is tied to the clavicle (at the AC joint) by ligaments and to the chest (**thorax**) by muscle tissue, the scapula cannot move purely laterally away from the midline of the body. The scapula must rotate about the distal end of the clavicle (at the AC joint) and tilt as it glides along the chest (thorax). Thus, abduction of the shoulder girdle is more correctly defined as **upward rotation** and **lateral tilt** of the scapula (figure 3.22). The upward rotation is defined as the upward movement of the glenoid of the scapula that accommodates shoulder joint movement. Additionally, the scapula tilts laterally as it glides along the curvature of the chest (thorax). This movement, when performed by both scapulae, is also referred to as **protraction**. Hugging another person by placing your arms around that person requires you to protract both of your shoulder girdles.

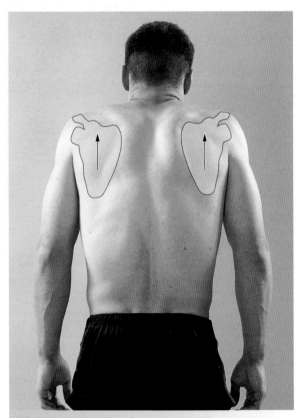

Figure 3.20   Elevation of the scapulae.

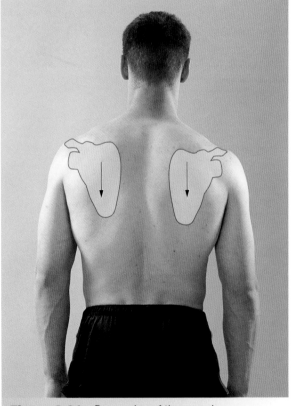

Figure 3.21   Depression of the scapulae.

Adduction of the shoulder girdle cannot be simply defined as movement toward the midline of the body by the scapula. Again, because the scapula is tied to the clavicle (at the AC joint) by ligaments and to the chest (thorax) by muscle tissue, the scapula cannot move purely medially toward the midline of the body. The scapula rotates about the distal end of the clavicle (at the AC joint) and tilts as it glides along the chest (thorax). Thus, adduction of the shoulder girdle is more correctly defined as **downward rotation** and **medial tilt** of the scapula (figure 3.23). The downward rotation is defined as the downward movement of the glenoid of the scapula that accommodates shoulder joint movement. Additionally, the scapula tilts medially as it glides along the curvature of the chest (thorax). This movement, when performed by both scapulae, is also referred to as **retraction**. Bringing your shoulders back, as when standing at attention, requires you to retract both of your shoulder girdles.

Anterior Muscles of the Shoulder Girdle

Six muscles are primarily involved in producing the fundamental movements of the shoulder girdle. Three muscles are anatomically anterior to the shoulder girdle bones, and three are posterior to it. The anterior muscles of the shoulder girdle include the pectoralis minor, serratus anterior, and subclavius.

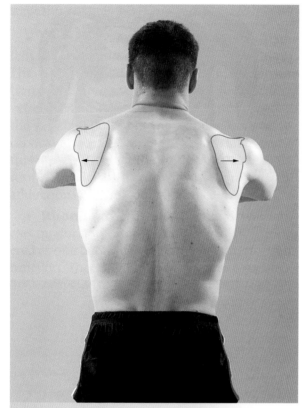

Figure 3.22 Abduction of the scapulae.

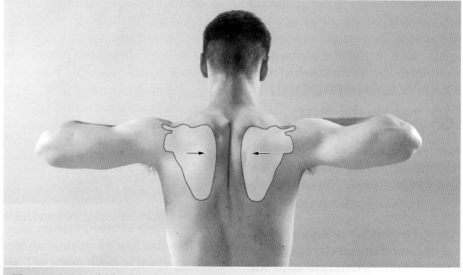

Figure 3.23 Adduction of the scapulae.

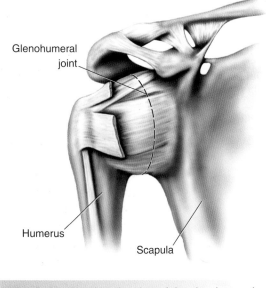

Figure 3.27 Anterior view of the glenohumeral joint.

(anterior movement of the arm) and extension (return from flexion) in the sagittal plane about a frontal horizontal axis, abduction (movement away from the midline of the body) and adduction (movement toward the midline of the body) in the frontal plane about a sagittal horizontal axis, and internal (inward, medial) and external (outward, lateral) rotation in the horizontal plane about a vertical axis. *Hyperextension* is extension beyond the anatomical position. Because the shoulder joint is a triaxial joint, it is capable of combining fundamental movements to produce **circumduction**. Eleven major muscles function to accomplish the six fundamental movements of the shoulder joint: four anterior, two superior, two posterior, and three inferior to the joint.

Anterior Muscles of the Shoulder Joint

The following muscles appear on the anterior aspect of the shoulder joint.

• **Pectoralis major**: The pectoralis major originates on the second to sixth ribs, the sternum, and the medial half of the clavicle and inserts on the anterior area of the surgical neck of the humerus just distal to the greater tuberos-

ity (see figure 3.24). The upper portion of the muscle is often referred to as the clavicular part, and the lower portion is referred to as the sternal part. Contraction of the pectoralis major muscle produces flexion, adduction, and internal rotation of the shoulder joint.

✋ Hands On

Locate your partner's pectoralis major using figure 3.28 as a guide.

Figure 3.28 Locating the pectoralis major.

• **Coracobrachialis**: The coracobrachialis originates on the coracoid process of the scapula (where the tendon of origin is conjoined with the tendon of origin of the short head of the biceps brachii) and inserts on the middle of the medial side of the humerus opposite the deltoid tubercle on the lateral side (figures 3.24 and 3.29). The coracobrachialis flexes the shoulder joint and,

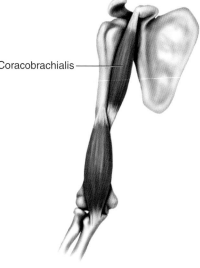

Coracobrachialis

Figure 3.29 Anterior view of the coracobrachialis.

because of its angle of pull, assists with adduction of the joint.

👋 Hands On

Have your partner place her arm at a right angle to the body (abduction) as you palpate the coracobrachialis (figure 3.30).

- **Biceps brachii**: Although the biceps brachii is frequently considered a flexor of the elbow (figure 3.31), both the long-head tendon and short-head tendon of the biceps brachii cross the shoulder joint. The long head originates on the supraglenoid tubercle on the superior edge of the glenoid of the scapula, and the short head originates on the coracoid process of the scapula (and is conjoined with the coracobrachialis tendon of origin). Both heads combine to form the belly of the muscle and insert on the tuberosity of the **radius**, which is one of the two forearm bones. Actions produced by contraction of this muscle at the shoulder joint include flexion and abduction by the long-head tendon and flexion, adduction, and internal rotation by the short-head tendon.

- **Subscapularis**: The subscapularis muscle (figure 3.31) is located on the anterior surface of the scapula between the scapula and the thorax. It originates on the large subscapular fossa on the anterior surface of the scapula and inserts on the lesser tuberosity of the humerus. When it contracts, the subscapularis produces internal rotation and flexion at the shoulder joint. This muscle is one of four shoulder joint muscles that attach to a musculotendinous structure often referred to as the **rotator cuff** (discussed later in this chapter).

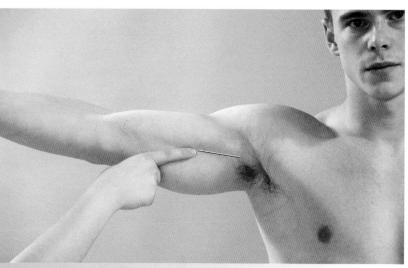

Figure 3.30 Locating the coracobrachialis.

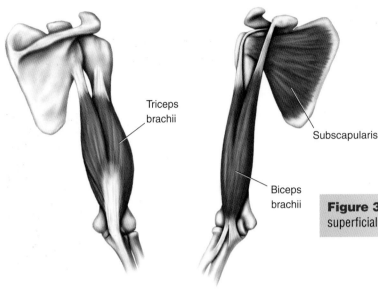

Triceps brachii

Subscapularis

Biceps brachii

Figure 3.31 Anterior and posterior views of two superficial shoulder joint muscles and the subscapularis.

Superior Muscles of the Shoulder Joint

The following muscles appear on the superior aspect of the shoulder joint.

• **Deltoid**: The deltoid is a very large muscle consisting of three parts: anterior, middle, and posterior (see figure 3.24). It covers the shoulder joint, so it is often referred to as the shoulder cap muscle. The anterior (clavicular) fibers originate from the lateral portion of the anterior aspect of the clavicle, the middle (acromial) fibers originate from the lateral edge of the acromion process of the scapula, and the posterior (scapular) fibers originate on the inferior edge of the spine of the scapula. All three portions combine to insert on the deltoid tubercle on the lateral surface of the middle of the humerus. Contraction of the entire deltoid muscle results in abduction of the shoulder joint; contraction of the posterior portion alone results in adduction, extension, and external rotation; and contraction of the anterior fibers alone results in adduction, flexion, and internal rotation. The middle fibers of the deltoid muscle are typically considered to be involved only in shoulder joint abduction. Once the arm has been abducted to the horizontal level from the anatomical position, all three portions of the muscle are considered abductors of the joint.

🖐 Hands On

Have your partner abduct his arm, preferably with some form of resistance, such as holding a book or a weight, as you locate all three portions of the deltoid muscle (figure 3.32).

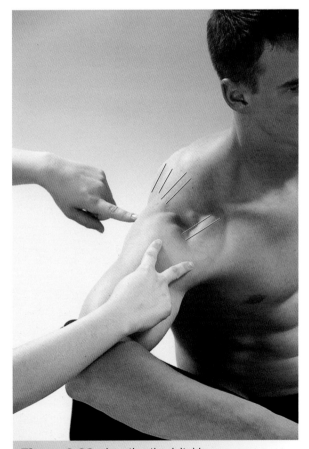

Figure 3.32 Locating the deltoid.

• **Supraspinatus**: Located beneath the deltoid muscle, the supraspinatus muscle (figure 3.33) originates on the supraspinous fossa of the scapula and inserts on the proximal facet of the greater tuberosity of the humerus. The muscle abducts the shoulder joint. Although it contracts throughout the full range of abduction, it is considered the primary initiator of abduction until approximately 30° of abduction, when the deltoid muscle takes over as the major abductor. The supraspinatus muscle is also one of the muscles of the rotator cuff.

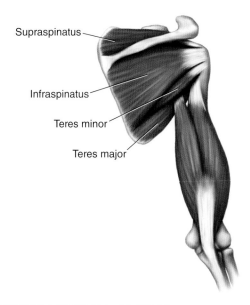

Figure 3.33 Posterior view of the supraspinatus, infraspinatus, teres minor, and teres major muscles.

Posterior Muscles of the Shoulder Joint

The following muscles are found on the posterior aspect of the shoulder joint.

• **Infraspinatus**: The infraspinatus muscle (figure 3.33) gets its name from the anatomical structure where it originates: the infraspinous fossa beneath the inferior surface of the spine of the scapula. The infraspinatus muscle inserts on the middle facet of the greater tuberosity of the humerus. Contraction of the infraspinatus muscle produces external rotation and extension of the shoulder joint. The infraspinatus muscle is also part of the rotator cuff.

👆 **Hands On**

Place your partner's shoulder joint in abduction, external rotation, and extension, and then locate the infraspinatus muscle (figure 3.34).

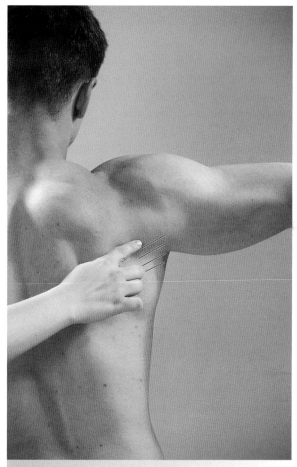

Figure 3.34 Locating the infraspinatus and teres minor.

• **Teres minor**: The teres minor muscle is often considered together with the infraspinatus muscle because they share the same function. The teres minor (figure 3.33) originates on the upper and middle portions of the lateral border of the scapula and inserts on the distal facet of the greater tuberosity of the humerus. Contraction of the teres minor muscle, like the infraspinatus muscle, produces external rotation and extension of the shoulder joint. This muscle is also one of the shoulder joint muscles of the rotator cuff.

Hands On

Place your partner's shoulder joint in abduction, external rotation, and extension, and then locate the teres minor muscle (figure 3.34).

Rotator Cuff

Four of the muscles of the shoulder joint insert on a musculotendinous structure running between the facets located on the lesser and greater tuberosities of the humerus. This structure is commonly referred to as the rotator cuff (figure 3.35). The motions produced at the shoulder joint by these four muscles (subscapularis, supraspinatus, infraspinatus, and teres minor) have been presented, but these four muscles are also responsible for maintaining the stability of the shoulder joint, which is particularly necessary because the socket of this ball-and-socket joint is so shallow and thus provides little stability. In

an action such as throwing, the muscles of the rotator cuff not only generate the force necessary for throwing but also decelerate the force generated. In other words, the actions of the rotator cuff muscles actually prevent the entire upper

FOCUS ON

Rotator Cuff

Understanding the action of the rotator cuff makes it easier to understand why some people involved in repetitive throwing activities (e.g., pitchers, quarterbacks) develop rotator cuff problems. The muscles not only generate the force needed for a throwing activity by contracting concentrically (shortening) but they also apply a braking action through an eccentric (lengthening) contraction to prevent the upper extremity from leaving the body by dissipating the force they generated.

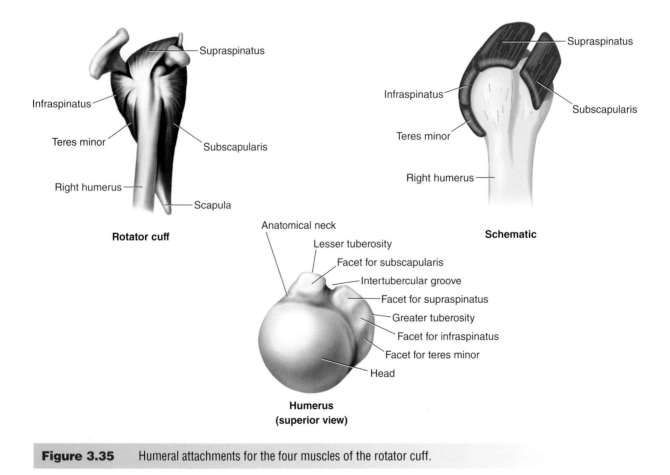

Figure 3.35 Humeral attachments for the four muscles of the rotator cuff.

extremity from following the object thrown by keeping the humeral head in the glenoid fossa.

Inferior Muscles

The following muscles cross the shoulder joint inferiorly (underneath).

• **Latissimus dorsi**: The latissimus dorsi muscle (see figure 3.24), a large muscle of the back, originates on the spinous processes of the lower six thoracic and all five **lumbar vertebrae**, the posterior aspect of the **ilium** (see chapter 8), the lower three ribs, and the inferior angle of the scapula; passes beneath the axilla (armpit); and inserts on the edge of the intertubercular groove on the anterior aspect of the humerus. Contraction of the latissimus dorsi muscle produces internal rotation, extension, and adduction of the shoulder joint.

🖑 Hands On

Place your partner's arm in the position of external rotation and abduction, and locate the latissimus dorsi muscle (figure 3.36).

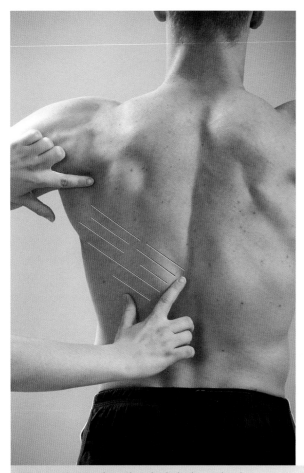
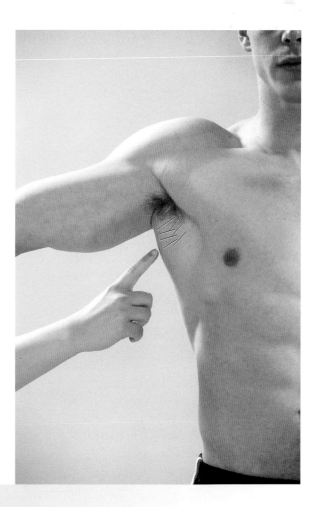

Figure 3.36   Locating the latissimus dorsi.

• **Teres major**: The teres major muscle (figure 3.33) originates on the lower portion of the lateral border of the scapula and its inferior angle, crosses beneath the axilla (armpit), and inserts on the area just inferior to the lesser tuberosity of the humerus. Contraction of the teres major produces the same action as the latissimus dorsi: internal rotation, extension, and adduction of the shoulder joint. Because the action of these two muscles is identical, the teres major is often called the latissimus dorsi's "little helper."

Hands On

Placing your partner's arm in the position of abduction, observe the posterior aspect of the shoulder joint, and locate the teres major muscle (figure 3.37).

• **Triceps brachii**: Although the triceps brachii is more often associated with elbow joint action, one of the three tendinous heads of the triceps brachii does cross the shoulder joint and assist with shoulder joint movement (see figure 3.31). Of the lateral-head, long-head, and medial-head tendons of origin of the triceps brachii, the long head originates on the infraglenoid tubercle of the glenoid lip of the scapula and joins the lateral and medial heads to insert, on a common tendon, on the olecranon process of the **ulna**, one of the two bones of the forearm. Contraction of the long head of the triceps brachii assists with shoulder joint extension and adduction.

Hands On

On yourself or a partner, locate all three portions of the triceps brachii muscle (figure 3.38).

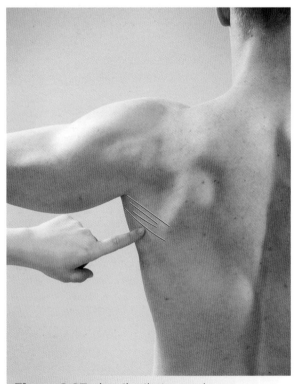

Figure 3.37 Locating the teres major.

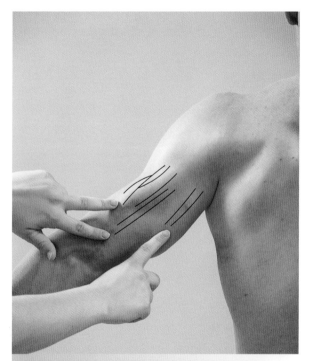

Figure 3.38 Locating the three sections of the triceps brachii.

Glenohumeral Joint

A **dislocation** of any joint usually results in the severe spraining of the ligaments and straining of the muscles that cross the joint. Following are three of the more common dislocations of the glenohumeral joint.

Excessive movement in any joint may result in stress being placed on the ligamentous structures tying the bones of the joint together. As noted earlier regarding the AC joint of the shoulder girdle, a sprain of a ligament is the partial or complete tearing of a ligament. If the ligaments of a joint are disrupted to the extent that the bones of the joint actually displace, this is known as a dislocation. Dislocations of the glenohumeral joint are not uncommon. The three most common forms of glenohumeral dislocations are anterior (subcoracoid), posterior (subspinous), and downward (subglenoid) dislocation (see figure 3.39).

Anterior dislocation is the most common form of shoulder dislocation, typically resulting from excessive abduction and external rotation of the shoulder joint. The humeral head displaces anteriorly from the glenoid of the scapula and rests anterior to the glenoid just beneath the coracoid process of the scapula (thus the name subcoracoid dislocation). A second form of shoulder joint dislocation is posterior (subspinous) dislocation, which can result from excessive internal rotation and adduction of the shoulder joint. The humeral head displaces posteriorly from the glenoid of the scapula and rests posterior to the glenoid just beneath the spine of the scapula (thus the name subspinous dislocation). The third form of shoulder joint dislocation is known as a downward (subglenoid) dislocation, which may result from excessive shoulder joint abduction with the humerus abutting against the acromion process of the scapula and the head of the humerus being forced downward (subglenoid) beneath the lower edge of the glenoid of the scapula.

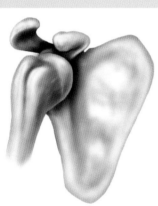 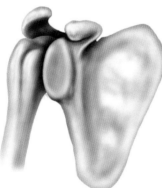 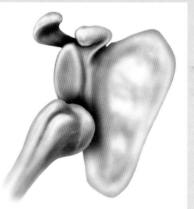

Anterior (subcoracoid)   **Posterior (subspinous)**   **Downward (subglenoid)**

Figure 3.39 Anterior (subcoracoid) glenohumeral dislocation, posterior (subspinous) dislocation, and downward (subglenoid) dislocation.

Combined Actions of the Shoulder Girdle and Shoulder Joint

Each of the shoulder joint actions (flexion, extension, abduction, adduction, internal and external rotation) possesses certain degrees of movement. When any particular movement of the shoulder joint reaches its end point, to move farther in that direction, the glenoid of the scapula must change its position to accommodate additional movement of the humerus. Movement of the shoulder girdle facilitates a greater range of motion in all the fundamental movements of the shoulder joint by changing the position of the glenoid, which is accomplished by movement of the scapula on the thorax (along the **ribs**)

through motion at the AC and SC joints. The prime example of this relationship is referred to as the **scapulohumeral rhythm**. Although initial abduction of the shoulder joint is attributed to GH joint action only (approximately 120° of abduction), the combination of GH joint abduction and rotation of the scapula results in approximately 180° of shoulder joint abduction. The generally accepted ratio of motion between the GH joint and the scapula (scapulothoracic movement) is that for every 2° of GH joint abduction, the scapula rotates 1°.

As it does with abduction of the shoulder joint, the scapula moves to position the glenoid to accommodate all the other fundamental movements of the shoulder joint. Observe this cooperation between shoulder girdle and shoulder joint movements with the suggested learning activities at the end of the chapter.

LEARNING AIDS

REVIEW OF TERMINOLOGY

The following terms are discussed in this chapter. Define or describe each term, and where appropriate, identify the location of the named structure either on your body or in an appropriate illustration.

abduction	downward rotation	pectoralis major
acromial end of the clavicle	elevation	pectoralis minor
acromioclavicular (AC) joint	glenohumeral (GH) joint	protraction
acromioclavicular ligament	glenohumeral ligament	proximal facet
acromion process	glenoid fossa	radius
adduction	glenoid lip (glenoid labrum)	retraction
anatomical neck of the humerus	great scapular notch	rhomboids
articular disc	greater tuberosity	rib
biceps brachii	head of the humerus	rotator cuff
capsular ligament	humerus	scapula
circumduction	ilium	scapular notch
clavicle	inferior angle of the scapula	scapulohumeral rhythm
clavicular notch	inferior glenohumeral ligament	serratus anterior
conoid ligament	inferior transverse scapular ligament	shoulder girdle
conoid tubercle	infraglenoid tubercle	shoulder joint
coracoacromial ligament	infraspinatus	spine of the scapula
coracobrachialis	infraspinous fossa	sternal end of the clavicle
coracoclavicular joint	interclavicular ligament	sternoclavicular (SC) joint
coracoclavicular ligament	intertubercular groove	sternoclavicular ligament (anterior,
coracohumeral ligament	lateral border of the scapula	posterior, and superior fibers)
coracoid process	lateral tilt	sternum
costal tuberosity	latissimus dorsi	subclavian groove
costoclavicular ligament	lesser tuberosity	subclavius
deltoid	levator scapulae	subscapular fossa
deltoid tubercle of the clavicle	lumbar vertebra	subscapularis
deltoid tuberosity of the humerus	medial border of the scapula	superior angle of the scapula
depression	medial tilt	superior border of the scapula
dislocation	middle facet	superior glenohumeral ligament
distal facet	middle glenohumeral ligament	superior transverse scapular ligament

supraglenoid tubercle	teres minor	trapezoid ligament
supraspinatus	thoracic vertebra	trapezoid line
supraspinous fossa	thorax	triceps brachii
surgical neck of the humerus	transverse humeral ligament	ulna
teres major	trapezius	upward rotation

SUGGESTED LEARNING ACTIVITIES

1. Place your hands on a partner's scapula. Ask the partner to slowly abduct both shoulder joints. As the humerus moves away from the body, determine when the scapula starts to move. Did the scapula move throughout abduction of the shoulder joint? When did it start to move? Why did it move? What muscle initiated this action? Repeat this activity during shoulder joint flexion, extension, and hyperextension, and internal and external rotation, and ask yourself these same questions.

2. Ask a partner in anatomical position to abduct both shoulder joints to the point where her hands touch, while you place your hands on both scapulae. What movement did the scapulae perform? Did the partner reach the end point through just abduction, or did the partner have to rotate each shoulder joint? If so, why was this necessary? In what direction?

3. Flex your partner's elbow to 90°. Passively abduct and externally rotate the shoulder joint to the point where you feel resistance. What muscles are providing that resistance? (Do this activity gently. Excessive force can harm the joint structures.)

4. Flex your partner's elbow to 90°. Passively adduct and internally rotate the shoulder joint to the point where you feel resistance. What muscles are providing that resistance? (Do this activity gently. Excessive force can harm the joint structures.)

5. Place your hands on top of your partner's shoulders (hands on clavicle and scapula), and push downward while asking your partner to elevate (shrug) the shoulders. What muscles performed this activity? What muscles are being used to oppose your resistance?

MULTIPLE-CHOICE QUESTIONS

1. In the anatomical position, depression of the shoulder girdle is defined as
    a. active
    b. passive
    c. isometric
    d. impossible

2. The only portion of the triceps brachii muscle that crosses the shoulder joint is the
    a. medial-head tendon
    b. lateral-head tendon
    c. short-head tendon
    d. long-head tendon

3. Which part of the deltoid muscle is involved only in shoulder joint abduction?
    a. middle fibers
    b. posterior fibers
    c. anterior fibers
    d. inferior fibers

4. In addition to elevation of the scapula, the levator scapulae muscle performs what other shoulder girdle action?
    a. upward rotation
    b. flexion
    c. downward rotation
    d. extension

(continued)

MULTIPLE-CHOICE QUESTIONS *(continued)*

5. The articulation formed by the clavicle and the scapula is often referred to as the
   a. AC joint
   b. SC joint
   c. GH joint
   d. SH joint

6. Movement of the scapula away from the midline of the body is defined as
   a. upward rotation
   b. adduction
   c. downward rotation
   d. elevation

7. Which of the following muscles does not have a role in the rotator cuff, which provides both motion and stability at the glenohumeral joint?
   a. supraspinatus
   b. subscapularis
   c. infraspinatus
   d. teres major

8. The glenohumeral joint is which type of joint?
   a. nonaxial
   b. uniaxial
   c. biaxial
   d. triaxial

9. What muscle, known as the latissimus dorsi's "little helper," extends, adducts, and internally rotates the shoulder joint?
   a. supraspinatus
   b. infraspinatus
   c. teres minor
   d. teres major

10. The primary movement of the shoulder girdle produced by the contraction of the rhomboids is
    a. abduction
    b. adduction
    c. upward rotation
    d. depression

11. Which of the following muscles is not considered an anterior muscle of the shoulder joint?
    a. pectoralis major
    b. pectoralis minor
    c. subscapularis
    d. coracobrachialis

12. The short head of the biceps brachii originates on the
    a. acromion process
    b. glenoid fossa
    c. coracoid process
    d. greater tuberosity

13. When the humerus is abducted to the point that the arm is held upright over one's head, the scapula is
    a. abducted
    b. adducted
    c. flexed
    d. extended

14. The inferior angle, the medial border, the lateral border, and the spine are all bony landmarks associated with which of the following bones?
    a. sternum
    b. clavicle
    c. scapula
    d. humerus

15. Which of the following ligaments is not present at the AC joint of the shoulder girdle?
    a. acromioclavicular
    b. trapezoid
    c. conoid
    d. costoclavicular

16. The coracoid process of the scapula serves as the attachment for the conjoined tendon of the coracobrachialis muscle and what other muscle?
    a. long head of the biceps
    b. brachioradialis
    c. brachialis
    d. short head of the biceps

17. Which of the following muscles is not considered a posterior muscle of the shoulder girdle?

    a. rhomboids

    b. latissimus dorsi

    c. trapezius

    d. levator scapulae

18. The primary function of the biceps brachii at the shoulder joint is flexion, but the long-head tendon of the biceps brachii also assists with what other movement of the shoulder joint?

    a. abduction

    b. adduction

    c. extension

    d. external rotation

19. The serratus anterior muscle is primarily involved in what shoulder girdle action?

    a. abduction

    b. adduction

    c. elevation

    d. depression

20. The infraspinatus muscle and which of the following muscles are usually considered the primary external rotators of the shoulder joint?

    a. supraspinatus

    b. subscapularis

    c. teres major

    d. teres minor

21. The rhomboid muscles help elevate the scapula as well as rotate it downward to produce what other scapular movement?

    a. abduction

    b. adduction

    c. flexion

    d. extension

22. The lesser tuberosity of the humerus serves as the source of attachment for which of the following muscles?

    a. infraspinatus

    b. subscapularis

    c. supraspinatus

    d. teres minor

23. Which of the following shoulder girdle actions is performed by the pectoralis major?

    a. upward rotation

    b. adduction

    c. elevation

    d. none

24. The lowest part of the trapezius muscle assists with which of the following movements of the shoulder girdle (when performed from the anatomical position)?

    a. elevation

    b. depression

    c. downward rotation

    d. upward rotation

25. For movement of the shoulder joint to occur in the frontal plane, which of the following joint actions must take place?

    a. internal rotation

    b. circumduction

    c. flexion

    d. abduction

26. The costoclavicular ligament is found in which of the following joints of the shoulder girdle?

    a. sternoclavicular

    b. humeroclavicular

    c. acromioclavicular

    d. thoracoclavicular

27. One of the tendons of which of the following muscles lies in the anatomical structure known as the intertubercular groove?

    a. biceps brachii

    b. coracobrachialis

    c. deltoid

    d. supraspinatus

FILL-IN-THE-BLANK QUESTIONS

1. Two ligaments known as the conoid and the trapezoid ligaments, which join the scapula and the clavicle together, are commonly called the _____ ligament.

2. The muscle that originates from the base of the skull (external occipital protuberance) to the end of the thoracic vertebrae (approximately two-thirds of the way down the back) is known as the _____.

3. A muscle of the rotator cuff that crosses the anterior portion of the shoulder joint and is a major internal rotator of the shoulder joint is the _____ muscle.

4. The muscle known as the initiator of shoulder joint abduction is the _____ muscle.

5. At the lateral end of the spine of the scapula, there is a wide, bony projection of the spine known as the _____.

6. The muscle that runs between the coracoid process of the scapula and the medial surface of the humerus opposite the deltoid tuberosity is known as the _____ muscle.

7. The glenoid fossa (shoulder joint socket) is located on the lateral aspect of the _____.

8. When all three portions of the deltoid muscle contract together, the shoulder joint moves into _____.

9. The pectoralis major muscle has two distinct parts: a lower portion known as the sternal part and a superior portion known as the _____ part.

10. The coracoid process is found on the _____.

11. A broad superficial muscle of the lower back, lateral and inferior to the trapezius, that is a powerful internal rotator of the shoulder joint is known as the _____ muscle.

FUNCTIONAL MOVEMENT EXERCISE

Abduction is defined as movement away from the midline of the body. This photo shows 90° of abduction of the arm (from the anatomical position). For this action to occur, motion happens at both the shoulder joint and the shoulder girdle. List one muscle acting as a prime mover, one as an antagonist, one as a fixator, and one as a synergist for abduction of the shoulder to occur.

	Shoulder joint	Shoulder girdle
Prime mover		
Antagonist		
Fixator		
Synergist		

The Elbow and Forearm

The elbow and the forearm are composed of three bones: the **humerus**, the **ulna**, and the **radius**. Together these three bones form four joints, three at the proximal end of the forearm (**radiohumeral**, **ulnohumeral**, and **proximal radioulnar**) and one at the distal end of the forearm (**distal radioulnar**).

Bones of the Elbow and Forearm

The humerus, the ulna, and the radius meet to form the structure commonly known as the **elbow**, and the ulna and radius form the **forearm** (figure 4.1). The proximal end of the humerus was discussed in chapter 3 on the shoulder joint. The distal end of the humerus provides the bony attachments for the soft tissues that span the upper arm and the forearm to form the

elbow joint. At the distal end of the humerus, the shaft widens out to form two bony prominences: the **lateral** and **medial epicondyles**.

🖐 Hands On

Find these structures on your own elbow or your partner's elbow (figure 4.2). The epicondyles are easily palpated and are visible on most people.

The parts of the humerus between the shaft and these epicondyles, where the bone actually widens, are known as the **lateral** and **medial supracondylar ridges**. There are three fossa (depressions) on the distal humerus that provide areas for other bony structures to move into during elbow joint actions. On the anterior surface, between the two epicondyles, is the **coronoid fossa**. This is where the coronoid process of the ulna is positioned during elbow flexion. Lateral to the coronoid fossa is the **radial fossa**.

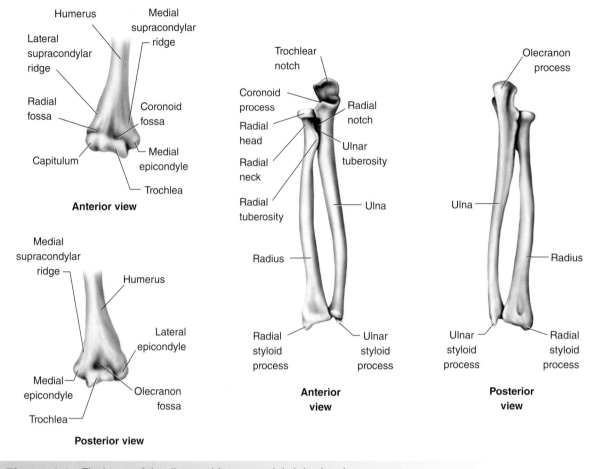

Figure 4.1   The bones of the elbow and forearm and their landmarks.

The head of the radius moves into this fossa during elbow flexion. Just distal to the lateral epicondyle is a smooth, round surface known as the **capitulum**. This surface is where the head of the radius rotates during forearm movement. At the very distal end of the humerus is a spool-like structure known as the **trochlea**. This is the structure on which the olecranon process of the ulna attaches.

A posterior view of the distal end of the humerus also reveals some structures previously discussed: the supracondylar ridges, the epicondyles, and the trochlea. In addition, the **olecranon fossa**, which is formed between the two epicondyles, appears. This is the depression into which the **olecranon process** of the ulna moves when the elbow joint is moved into extension.

Under normal conditions, when the elbow joint is in full extension, you should be able to observe that the lateral epicondyle, the olecranon process, and the medial epicondyle form a straight line. If the normal elbow joint is flexed to 90°, these three structures should form an isosceles triangle with the olecranon process distal to the epicondyles (figure 4.2).

The two bones of the forearm are known as the radius and the ulna. From the anatomical position, the radius is on the lateral aspect and the ulna on the medial aspect of the forearm. In the elbow joint, the ulna has the prominent role of articulating with the humerus, whereas the radius plays the more prominent role of articulating with the bones of the wrist. The proximal end of the radius consists of the **radial head**,

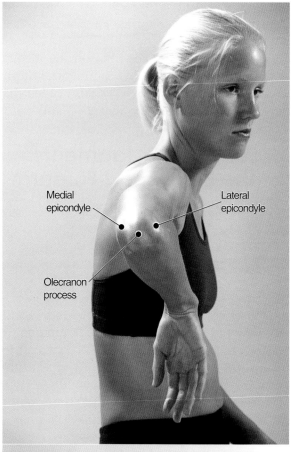

Figure 4.2 Locating the medial and lateral epicondyles.

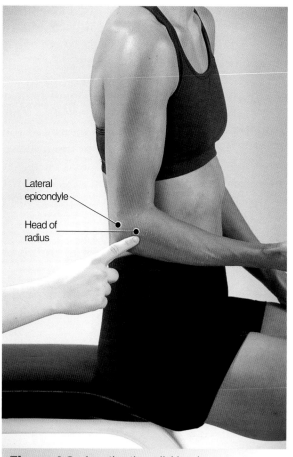

Figure 4.3 Locating the radial head.

the **radial neck**, and a large **radial tuberosity** distal to the neck on the medial aspect of the upper portion of the shaft of the bone.

Hands On

You can easily palpate the head of the radius at the lateral aspect of your elbow (figure 4.3).

At the distal end of the shaft of the radius, on the lateral aspect, is a large prominence known as the **radial styloid process** (also known as the lateral styloid process).

Hands On

This structure can be easily palpated in the area where the hand and forearm come together (wrist) on the thumb side (figure 4.4).

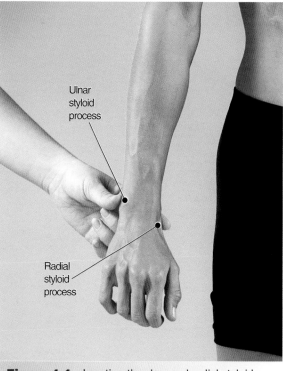

Figure 4.4 Locating the ulnar and radial styloid processes.

On the medial aspect of the distal end of the radius is the **ulnar notch**. On the distal surface of the radius are two distinct facets where bones of the wrist articulate (figure 4.5).

The medial forearm bone, the ulna, has a very large prominence at the proximal end known as the olecranon process (see figure 4.1). This structure has a cuplike surface, known as the trochlear notch, that rotates about the trochlea of the humerus to form the articulation between the humerus and the ulna commonly known as the elbow joint. At the very anterior portion of the trochlear notch is a smaller prominence known as the **coronoid process**. Just lateral to the coronoid process is the **radial notch** of

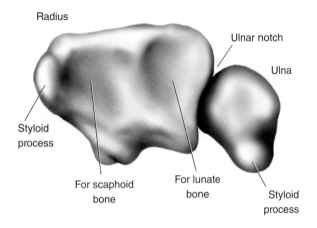

Figure 4.5 Facets on the distal surface of the radius where bones of the wrist articulate.

the ulna. The head of the radius articulates with the ulna (proximal radioulnar joint) at the radial notch of the ulna.

At the distal end of the ulna on the medial aspect is an easily palpated prominence known as the **ulnar styloid process** (also called the medial styloid process; see figures 4.1 and 4.4).

Joints and Ligaments of the Elbow and Forearm

With three bones (humerus, ulna, and radius) coming together to form the elbow joint (figure 4.6), there are actually three joints in the anatomical area between the upper arm and the forearm: the "true" elbow joint between the humerus and the ulna (the ulnohumeral joint), the radiohumeral joint between the radius and the humerus, and the proximal radioulnar joint between the radius and the ulna. The **capsular ligament** surrounds all three of these articulations. It is divided into an anterior and posterior part. The anterior part extends from the anterior surface of the humerus just proximal to the coronoid fossa to the anterior surface of the coronoid process and the **annular ligament**. Laterally, the capsular ligament fuses with the **collateral ligaments**. Posteriorly, the capsular ligament attaches to the tendon of insertion of the triceps brachii muscle, the edge of the olecranon, the lateral epicondyle, and the posterior surface

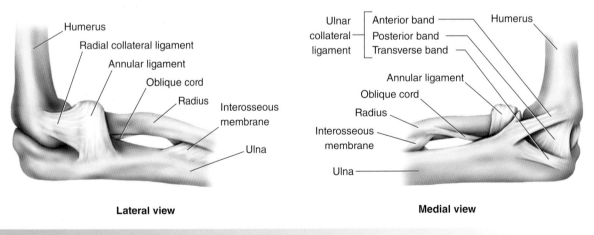

Figure 4.6 Medial and lateral views of the major ligaments of the elbow.

of the humerus in the area of the trochlea and the capitulum. Distally, the capsular ligament attaches to the lateral and superior edges of the olecranon process, the posterior aspect of the annular ligament, and posterior to the radial notch of the ulna.

The ligaments that fuse with the anterior portion of the capsular ligament are known as the **radial** (lateral) **collateral ligament** and the **ulnar** (medial) **collateral ligament**. The radial collateral ligament runs between the inferior border of the lateral epicondyle of the humerus to the annular ligament and the radial notch of the ulna.

The ulnar collateral ligament has three distinct parts (see figure 4.6): an **anterior band** that runs between the anterior inferior area of the medial epicondyle of the humerus and the medial aspect of the coronoid process of the ulna; a **posterior band** running between the medial epicondyle of the humerus and the medial border of the olecranon process of the ulna; and the **transverse band**, which does not cross the elbow joint, running between the anterior band on the coronoid process of the ulna and the posterior band on the olecranon process of the ulna.

The remaining joint of the elbow, the proximal radioulnar joint, is actually a joint between the bones of the forearm. The joint is between the head of the radius and the radial notch of the ulna. The annular ligament runs between the anterior and posterior edge of the radial notch on the ulna and forms a ring completely around the head of the radius.

Running between the shafts of the ulna and radius is a ligamentous band of connective tissue known as the **interosseous ligament** (also referred to as the **interosseus membrane**) that helps distribute pressure between the ulna and radius when force is applied and also serves as a place of attachment for several forearm muscles. At the proximal end of the interosseous ligament (membrane) is a ligament known as the **oblique cord** that prevents separation between the ulna and radius.

At the distal end of the ulna and radius, just proximal to the wrist, is the distal radioulnar joint. In the concave ulnar notch of the radius, the round head of the ulna rotates, forming a pivot joint. The ligaments of this joint are the **dorsal radioulnar** and the **volar radioulnar** (figure 4.7).

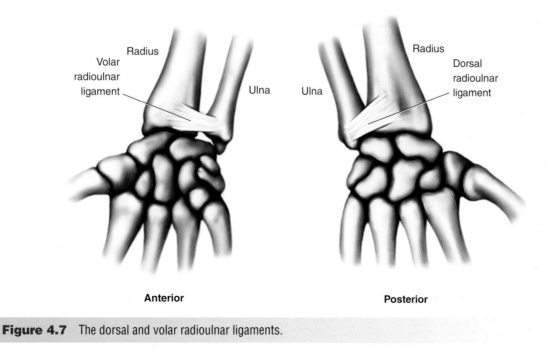

Radius

Volar radioulnar ligament

Ulna

Anterior

Radius

Dorsal radioulnar ligament

Ulna

Posterior

Figure 4.7 The dorsal and volar radioulnar ligaments.

Fundamental Movements and Muscles of the Elbow and Forearm

Remember that the starting position for description of all fundamental movements of any joint is the anatomical position. The elbow joint (articulation between the trochlea of the humerus and the olecranon process of the ulna) is a uniaxial joint capable of flexion and extension in the sagittal plane about a frontal horizontal axis. Five major muscles produce the motions of flexion and extension of the elbow joint. They are the brachialis (flexion), the brachioradialis (flexion), the biceps brachii (flexion), the triceps brachii (extension), and the anconeus (extension). The brachialis, brachioradialis, and biceps brachii muscles are anterior to the elbow joint, and the triceps brachii and anconeus muscles are posterior to the joint. Four muscles are responsible for the movements of supination and pronation of the forearm (figure 4.8). The biceps brachii (supination) has already been mentioned in its other role at the elbow joint (flexion). The other three muscles involved with forearm motion are the supinator (supination), the pronator quadratus (pronation), and the pronator teres (pronation).

Anterior Muscles of the Elbow

The anterior muscles of the elbow are three in number: the brachialis, the brachioradialis, and the biceps brachii. Two of these muscles (the brachialis and brachioradialis) are involved exclusively in one movement of the elbow joint (flexion), whereas the third (the biceps brachii) is involved in elbow joint movement and also movement of the forearm.

- **Brachialis**: The brachialis muscle originates on the middle of the anterior shaft of the humerus and inserts on the coronoid process of the ulna. Because of its origin and insertion, its only function is flexion of the elbow joint. It is located beneath the biceps brachii muscle (figures 4.9 and 4.10).

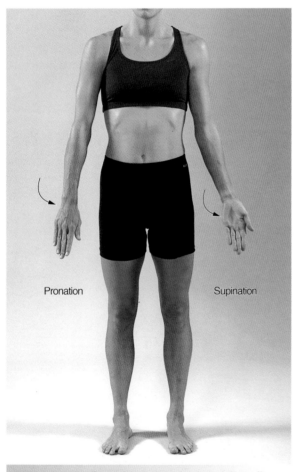

Figure 4.8 Pronation and supination of the radio-ulnar joints.

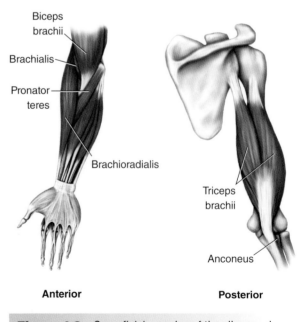

Figure 4.9 Superficial muscles of the elbow and forearm.

👆 Hands On

Have your partner abduct and externally rotate her shoulder joint so you can locate the brachialis muscle (figure 4.11).

• **Brachioradialis**: The brachioradialis muscle originates on the lateral epicondyle of the humerus and inserts on the radial styloid process. The muscle crosses the anterior aspect of the elbow joint and therefore is a flexor of the elbow joint (figure 4.9).

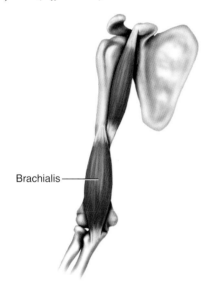

Brachialis

Figure 4.10 Anterior view of the brachialis.

👆 Hands On

Have your partner flex his elbow joint against a slight resistance applied by the opposite hand as you locate the brachioradialis muscle (figure 4.12).

Figure 4.12 Locating the brachioradialis.

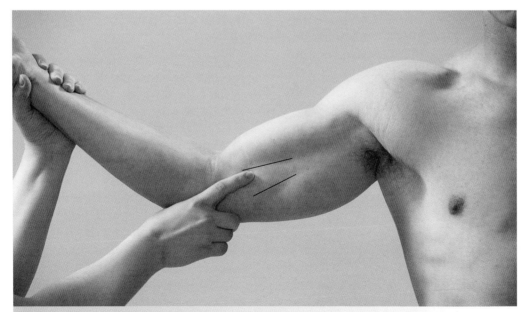

Figure 4.11 Identifying the brachialis.

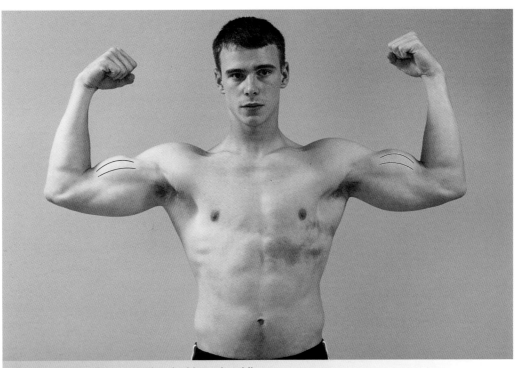

Figure 4.13   Demonstrating the biceps brachii.

- **Biceps brachii**: The biceps brachii is commonly considered a flexor of the elbow (see figure 4.9), but both the long-head tendon and short-head tendon of the biceps brachii also cross the shoulder joint and contribute to shoulder motion (figure 4.13). The long head originates on the supraglenoid tubercle on the superior edge of the glenoid of the scapula, and the short head originates on the coracoid process of the scapula (and is conjoined with the coracobrachialis tendon of origin). Both heads combine into the belly of the muscle, which inserts on the tuberosity of the radius. Actions produced by contraction of this muscle are flexion at the elbow joint and supination at the forearm.

Posterior Muscles of the Elbow

The anconeus and triceps brachii are posterior muscles of the elbow.

- **Triceps brachii**: The triceps brachii is most often associated with elbow joint extension (see figure 4.9), but one of its three tendinous heads crosses the shoulder joint and assists with shoulder joint movement. The long head originates on the infraglenoid tubercle of the glenoid lip of the scapula and joins the lateral and medial heads to insert, on a common tendon, on the olecranon process of the ulna. Contraction of the triceps brachii extends the elbow joint.

Hands On

To palpate the three heads of the triceps brachii (the lateral head, long head, and medial head), have a partner extend the elbow against a resistance while also extending the shoulder joint as you locate all three heads (figure 4.14).

- **Anconeus**: The anconeus originates on the lateral epicondyle of the humerus and inserts on the olecranon process of the ulna. This muscle assists the triceps brachii in elbow extension (see figure 4.9 and figure 4.14). A common error is to assume that the anconeus is also involved in movement of the forearm; on careful inspection, it can be seen that this muscle's origin and insertion points allow only one action when the muscle contracts: elbow extension.

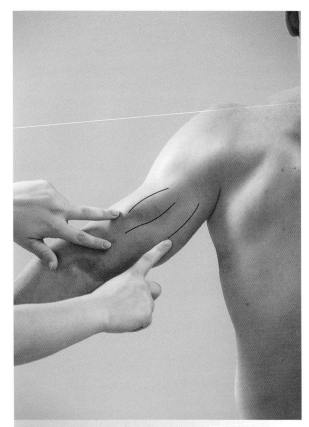

Figure 4.14 Finding the three sections of the triceps brachii.

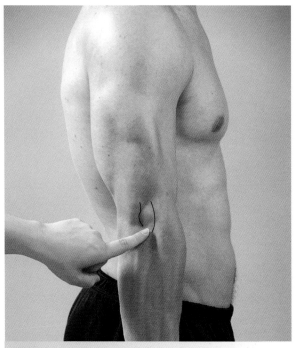

Figure 4.15 Finding the anconeus.

Hands On

Locate the anconeus muscle on the posterior lateral aspect of your or your partner's elbow (figure 4.15).

Muscles of the Forearm

Muscles of the forearm include two pronator and two supinator muscles. All are major producers of forearm movement. Only the biceps brachii has another function, that of elbow flexion, as previously mentioned.

• **Pronator teres**: Originating on the coronoid process of the ulna and inserting on the lateral surface of the radius, the pronator teres, as indicated by its name, is responsible for pronation of the forearm (see figure 4.9).

✋ **Hands On**

Manually resist your partner's attempt to flex his elbow and pronate his wrist while you look for the pronator teres muscle (figure 4.16).

Figure 4.16 Finding the pronator teres muscle.

Throwing Action

The pronator teres muscle and a group of muscles that originate in the common flexor tendon, presented in chapter 5 on the wrist and hand, come under great stress in the overhead throwing motion, particularly at the end of the windup stage of throwing. This pronator–flexor muscle group also is stressed at the release stage of throwing a curveball in baseball pitching. Inflammation of the pronator–flexor group at its origin in the area of the medial epicondyle of the humerus is often referred to as pitcher's elbow or Little League elbow.

A further complication from this throwing action is the spraining and possible complete disruption of the medial or ulnar collateral ligament of the elbow joint. In 1974, Dr. Frank Jobe, a Los Angeles orthopedic surgeon, reconstructed the torn ulnar collateral ligament of Major League Baseball pitcher Tommy John. Dr. Jobe used a section of the tendon of insertion of the palmaris longus, an anterior muscle of the forearm, and wove it through drill holes in both the humerus and the ulna, making the tendon a substitute for the disrupted ligament. Today, the sports world refers to this reconstruction technique as "the Tommy John surgery."

An understanding of proper throwing mechanics is essential to prevent elbow and shoulder joint problems. Studies in kinesiology, biomechanics, athletic training, and sports medicine provide that understanding.

• **Pronator quadratus**: This muscle originates on the radius and inserts on the ulna just proximal to the wrist (figure 4.17). Its name reflects its function and its shape. Because the ulna is the stable bone in the distal radioulnar articulation, when the pronator quadratus muscle contracts, the radius is pulled over the ulna, and forearm pronation takes place.

• **Supinator**: The supinator originates on the ulna and inserts on the radius at the proximal ends of the bones on the posterior aspect (figure 4.17). The name of this muscle indicates its function: forearm supination.

• **Biceps brachii**: Because of the position of the biceps brachii attachment to the radial tuberosity, when the forearm is in pronation, contraction of the biceps brachii causes the radius to rotate externally (laterally), causing the forearm to supinate (see figures 3.31 and 4.9).

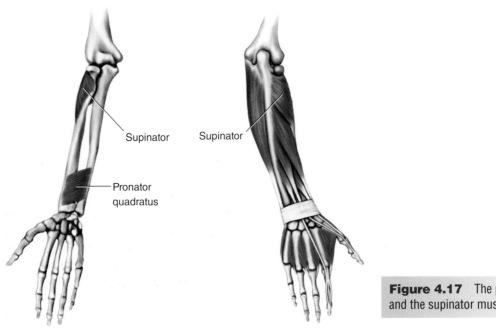

Supinator

Supinator

Pronator
quadratus

Figure 4.17 The pronator quadratus and the supinator muscles.

FOCUS ON

Tennis Elbow

Tennis elbow is the bane of tennis players. In tennis, when you strike the ball using a backhand stroke, particularly if you attempt to put topspin on the ball as you strike it, the forearm moves from pronation to supination, the wrist moves from flexion to extension, and the supinator muscle along with the muscles that originate in the common extensor tendon (presented in chapter 5 on the wrist and hand) contracts. If the tennis ball is hit off center (outside the so-called sweet spot), a torque (turning force) is applied to the racket that opposes the force applied by the supinator–extensor muscle group. This can result in tennis elbow, an inflammation in the area of the supinator–extensor muscle group's origin on the lateral epicondyle of the humerus. An understanding of proper stroke technique is essential to prevent tennis elbow from developing.

As a challenge, think about how the two-hand backhand stroke might help prevent tennis elbow. Hint: The biomechanical action of the forearm is different in the one-hand and two-hand backstrokes. How is the stress reduced?

LEARNING AIDS

REVIEW OF TERMINOLOGY

The following terms are discussed in this chapter. Define or describe each term, and where appropriate, identify the location of the named structure either on your body or in an appropriate illustration.

anconeus	interosseous ligament (membrane)	radial notch
annular ligament	lateral epicondyle	radial styloid process
anterior band	lateral supracondylar ridge	radial tuberosity
biceps brachii	medial epicondyle	radiohumeral joint
brachialis	medial supracondylar ridge	radius
brachioradialis	oblique cord	supinator
capitulum	olecranon fossa	transverse band
capsular ligament	olecranon process	triceps brachii
collateral ligament	posterior band	trochlea
coronoid fossa	pronator quadratus	ulna
coronoid process	pronator teres	ulnar collateral ligament
distal radioulnar joint	proximal radioulnar joint	ulnar notch
dorsal radioulnar ligament	radial collateral ligament	ulnar styloid process
elbow	radial fossa	ulnohumeral joint
forearm	radial head	volar radioulnar ligament
humerus	radial neck	

SUGGESTED LEARNING ACTIVITIES

1. Either with a skeletal model or a partner, observe the relationship between the lateral and medial epicondyles of the humerus and the olecranon process of the ulna, both in the anatomical position and with the elbow flexed to 90°. In which position do these three points form a straight line? In which position do they form an isosceles triangle?

2. Grasp a broom handle, stick, barbell, or similar object with both hands. First with palms up (supination of the forearms) and then with palms down (pronation of the forearms), perform a biceps curl (flex the elbows from the anatomical position of elbow extension to the position of full elbow flexion). Note the difference in the feeling in

(continued)

SUGGESTED LEARNING ACTIVITIES *(continued)*

your forearms, wrists, and hands when the curl is performed in forearm pronation and forearm supination. What causes the difference when the forearms are in pronation at the beginning of the movement?

3. Apply resistance to your partner's attempt to flex her elbow. Place the fingers of your other hand on either side of your partner's biceps brachii muscle. What is the name of the muscle you feel contracting beneath the biceps brachii?

4. Grip a tennis racket or similar object, and slowly perform the backhand stroke. Note the position of your forearm (pronated or supinated) and your wrist (flexed or extended). Stop your stroke at a point where you think the racket will strike the ball. Picture the ball being struck near the upper edge of the racket as opposed to the middle portion of the racket (the so-called sweet spot). Which way will your racket rotate in your grip at the point when the ball is hit? Will your wrist likely be forced to flex or extend on contact with the ball? Will your forearm likely be forced to pronate or supinate on contact? What muscles of the forearm and wrist contract as you perform the backhand stroke? What happens to these muscles if the ball is hit off center, causing a rotation (torque) of the racket and your forearm? What traumatic condition can be caused by this action if it is repeated every time you strike a backhand? Which of the following factors can affect (positively or negatively) this condition: racket head size, racket grip size, stroke mechanics, a two-hand backstroke?

MULTIPLE-CHOICE QUESTIONS

1. Which of the following anterior muscles is involved in supination of the forearm?
   a. pronator teres
   b. brachioradialis
   c. biceps brachii
   d. brachialis

2. Which of the following muscles assists the triceps brachii in extension of the elbow joint?
   a. extensor digitorum
   b. extensor carpi radialis
   c. anconeus
   d. supinator

3. The relationship between the ulnar (medial) styloid process and the radial (lateral) styloid process is such that the radial (lateral) styloid process is more
   a. proximal
   b. distal
   c. anterior
   d. posterior

4. The ligament running between the shafts of the ulna and radius is known as the
   a. ulnar collateral ligament
   b. radial collateral ligament
   c. annular ligament
   d. interosseous ligament (membrane)

5. The radial head rotates on what aspect of the humerus?
   a. coronoid fossa
   b. capitulum
   c. olecranon fossa
   d. radial tubercle

6. The structures known as the supracondylar ridges are located where in anatomical relation to the epicondyles?
   a. proximal to the epicondyles
   b. distal to the epicondyles
   c. lateral to the epicondyles
   d. medial to the epicondyles

7. Of the following muscles involved in pronation–supination of the forearm, which crosses the elbow joint?
   a. supinator
   b. biceps brachii
   c. pronator teres
   d. pronator quadratus

8. The lateral collateral ligament of the elbow joint is also known as the
   a. ulnar collateral ligament
   b. radial collateral ligament
   c. annular ligament
   d. interosseous ligament (membrane)

9. The olecranon process of the elbow joint is located on the posterior aspect of the
   a. distal humerus
   b. proximal humerus
   c. proximal radius
   d. proximal ulna

10. The medial collateral ligament of the elbow joint is also known as the
    a. ulnar collateral ligament
    b. radial collateral ligament
    c. annular ligament
    d. interosseous ligament (membrane)

11. The pronator–flexor group of forearm muscles originates on the
    a. lateral humeral epicondyle
    b. coronoid process
    c. medial humeral epicondyle
    d. olecranon process

12. The supinator–extensor group of forearm muscles originates on the
    a. lateral humeral epicondyle
    b. coronoid process
    c. medial humeral epicondyle
    d. olecranon process

FILL-IN-THE-BLANK QUESTIONS

1. The olecranon fossa is found on the _____.

2. The ligament that holds the radial head to the ulna is known as the _____ ligament.

3. The coronoid process of the elbow joint is located on the anterior aspect of the _____.

4. The function of the brachialis muscle is _____.

5. The spool-like structure at the distal end of the humerus is known as the _____.

6. When the elbow joint is held in full extension, the olecranon process and the epicondyles form a _____.

7. From the anatomical position, internal rotation of the forearm is called _____.

8. The muscle *chiefly* responsible for the extension of the elbow joint is the _____.

9. A muscle found just proximal to the wrist that helps in the action of turning the forearm and the palm of the hand downward is the _____.

FUNCTIONAL MOVEMENT EXERCISE

Pronation of the forearm is defined as movement of the radius over the ulna. This photo shows flexion of the elbow to 90° (from the anatomical position), and pronation of the forearm. For this action to occur, motion happens at both the elbow joint and the forearm. List one muscle acting as a prime mover, one as an antagonist, one as a fixator, and one as a synergist for flexion of the elbow and pronation of the forearm to occur.

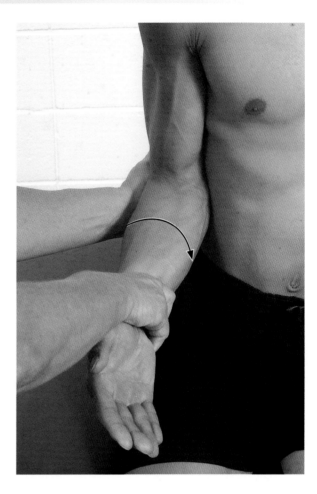

	Elbow joint flexion	Forearm pronation
Prime mover		
Antagonist		
Fixator		
Synergist		

The Wrist and Hand

The wrist and the hand are complicated structures with multiple bones (figure 5.1), ligaments, joints, and muscles. Because of the fine movements performed by the hand and thumb, these areas are very complex and require more time and effort to learn about than areas previously presented. Although the thumb is often considered one of the hand's five fingers, its movements are unique, and it is discussed separately. Humans have **prehensile hands** (i.e., capable of grasping). The capacity to grasp is a direct result of the ability of the thumb to perform opposition, which is discussed under the movements of the thumb. The structure of the thumb joint and its muscles of the thenar eminence (and, to a lesser extent, the hypothenar eminence of the little finger) contribute to the ability of primates (the mammalian order that includes monkeys, apes, and human beings) to grasp things in their hands. These structures are presented in detail later in this chapter.

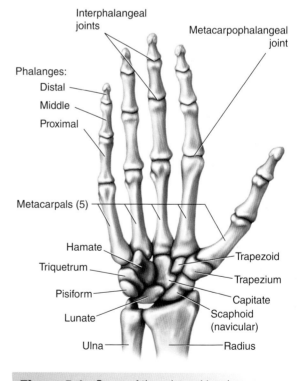

Figure 5.1 Bones of the wrist and hand.

Bones of the Wrist and Hand

The wrist contains eight bones, roughly aligned in two rows, known as the **carpal** bones. The proximal row of carpal bones contains the bones that articulate with the forearm (radius and ulna), and the distal row of carpal bones articulate with the long bones of the hand (the **metacarpals**). The proximal row of carpal bones, from lateral to medial, are identified as the **scaphoid** (also known as the **navicular**), the **lunate**, the **triquetrum**, and the **pisiform**. The pisiform and the scaphoid bones in the proximal row are easy to palpate.

🖐 Hands On

Place your index finger on the spots indicated in figures 5.2 and 5.3, and apply pressure downward to feel these bones.

The scaphoid (navicular) is a peanut-shaped bone that is the most frequently fractured bone

of the wrist, typically when forced into the distal end of the radius from a fall on an extended wrist. The lunate, because of its smooth, dome-shaped proximal end, is the most frequently dislocated bone of the wrist, typically when it is forced into the distal end of the radius from a fall on a flexed wrist. The bones in the distal row of carpals articulate with the bones of the hand. They are, from lateral to medial, the **trapezium** (also known as the **greater multangular** or **multangulus major**), the **trapezoid** (also known as the **lesser multangular** or **multangulus minor**), the **capitate**, and the **hamate**.

🖐 Hands On

Downward pressure on the spot illustrated in figure 5.2 will bring you in contact with the hamate bone in the distal row of carpal bones.

There are five bones of the hand known as the metacarpal bones. The metacarpal bone of the thumb is usually referred to as the first metacarpal. The second metacarpal is that of the index

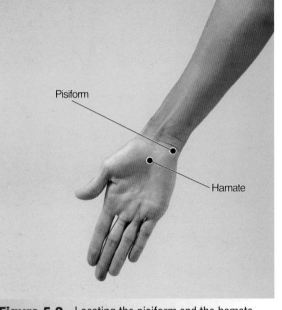

Figure 5.2 Locating the pisiform and the hamate.

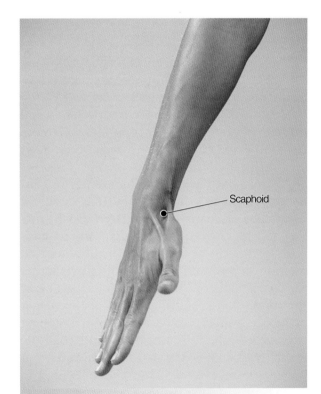

Figure 5.3 Locating the scaphoid.

finger, the third metacarpal is that of the middle finger, the fourth metacarpal is that of the ring finger, and the fifth metacarpal bone is that of the little finger. Distal to each of the metacarpal bones of the hand are the **phalanges** of the fingers. The thumb has two phalanges known as the proximal phalanx and the distal phalanx. The other fingers each consist of three phalanges: a proximal phalanx, a middle phalanx, and a distal phalanx.

Joints and Ligaments of the Wrist and Hand

With the large number of bones composing the wrist (ulna, radius, eight carpals, and five metacarpals), it makes sense that there are many, many joints that make up the structure known as the wrist (figure 5.4). There are joints between the forearm bones and the proximal row of the carpals (**radiocarpals**), joints between the proxi-

mal and distal rows of carpals (**midcarpals**), and joints between the distal row of carpals and the five metacarpal bones of the hand (**carpometacarpals**). In addition to these wrist joints, there are joints between the carpal bones within each row (**intercarpals**).

What is commonly referred to as the wrist joint is the articulation between the distal end of the radius and primarily two bones of the proximal row of carpal bones: the scaphoid (navicular) and the lunate. The movement between these bones produces a gliding type of action as they roll or slide over each other. The radiocarpal joints are classified as condyloid joints because of their movements.

There are five main ligaments of the wrist (figure 5.5). A **capsular ligament** runs between the distal ends of the ulna and radius to the proximal row of carpal bones. There is a **volar** (palmar) **radiocarpal ligament** and a **dorsal radiocarpal ligament**. The volar radiocarpal ligament is found between the anterior surface of the radius and its styloid process and the proximal row of carpal bones (figure 5.5). The dorsal radiocarpal ligament is found between the distal end of the radius and the proximal row of carpal bones (figure 5.5). The two additional ligaments of the wrist are collateral ligaments: the **radial** (lateral) and **ulnar** (medial) **collateral ligaments** (figure 5.5). The radial collateral ligament of the wrist runs between the styloid process of the radius and the scaphoid carpal bone. The ulnar collateral ligament runs between the styloid process of the ulna and the medial portions of the pisiform and triquetrum bones.

The intercarpal joints between the carpal bones of the wrist are connected by three forms of **intercarpal ligaments**: those that connect the four carpal bones in the proximal row, those that connect the four carpal bones in the distal row, and those that connect the carpals of the proximal row to those of the distal row. The intercarpal ligaments can be further divided into volar, dorsal (see figure 5.5), interosseous, radial and ulnar collateral, pisohamate, and pisometacarpal ligaments. All of these ligaments are referred to in this text simply as the intercarpal ligaments. The intercarpal joints move in a gliding motion.

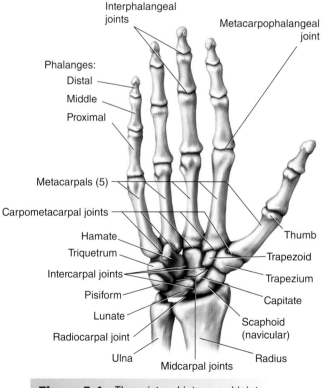

Figure 5.4 The wrist and intercarpal joints.

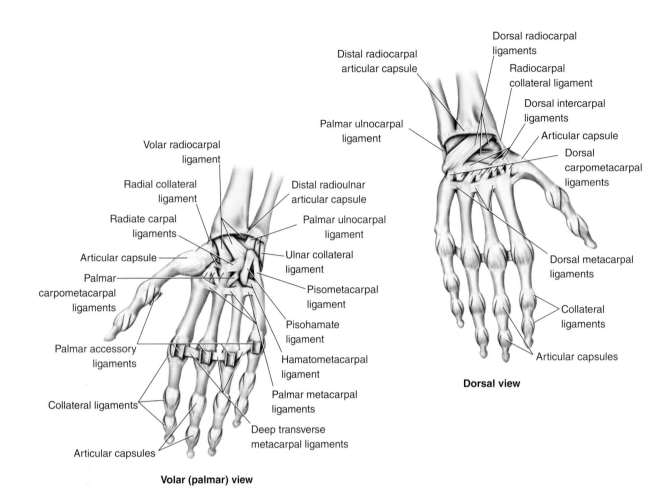

Volar radiocarpal ligament

Radial collateral ligament

Radiate carpal ligaments

Articular capsule

Palmar carpometacarpal ligaments

Palmar accessory ligaments

Collateral ligaments

Articular capsules

Distal radiocarpal articular capsule

Palmar ulnocarpal ligament

Distal radioulnar articular capsule

Palmar ulnocarpal ligament

Ulnar collateral ligament

Pisometacarpal ligament

Pisohamate ligament

Hamatometacarpal ligament

Palmar metacarpal ligaments

Deep transverse metacarpal ligaments

Volar (palmar) view

Dorsal radiocarpal ligaments

Radiocarpal collateral ligament

Dorsal intercarpal ligaments

Articular capsule

Dorsal carpometacarpal ligaments

Dorsal metacarpal ligaments

Collateral ligaments

Articular capsules

Dorsal view

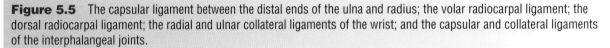

Figure 5.5 The capsular ligament between the distal ends of the ulna and radius; the volar radiocarpal ligament; the dorsal radiocarpal ligament; the radial and ulnar collateral ligaments of the wrist; and the capsular and collateral ligaments of the interphalangeal joints.

The last group of joints considered to be part of the wrist are the carpometacarpal joints. There are five carpometacarpal joints: four between the four carpal bones in the distal carpal row and the bases of the four metacarpal bones of the hand, and one between the trapezium and the base of the first (thumb) metacarpal bone. The motion of these joints is gliding. The ligaments of the four carpometacarpal joints of the hand are the **dorsal**, **volar**, **interosseous**, and **capsular carpometacarpal ligaments**. The dorsal and volar (palmar) carpometacarpal ligaments (see figure 5.5) are found between the dorsal and volar surfaces of the distal row of carpal bones and the bases of the metacarpal bones. The interosseous

ligaments are found between the hamate and capitate bones and the bases of the third and fourth metacarpal bones. The capsular ligaments are located between the distal row of carpals and the bases of the four metacarpals of the hand. The first (thumb) carpometacarpal joint is unique compared with the four other carpometacarpal joints; its ligamentous structure consists of a loose capsular carpometacarpal ligament that is found between the trapezium and the base of the first (thumb) metacarpal bone. The carpometacarpal joint of the thumb is referred to as a saddle joint because of its shape.

Two additional ligamentous structures of the wrist are the flexor (volar) and extensor (dorsal)

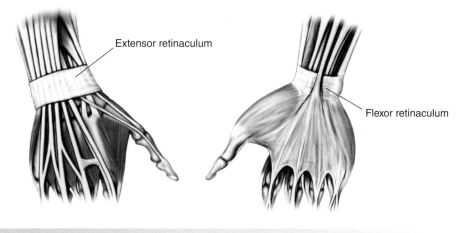

Figure 5.6 The extensor and flexor retinacula.

FOCUS ON

Carpal Tunnel Syndrome

In addition to the various tendons, blood vessels and nerves pass through the carpal tunnel. Anything causing inflammation of these tissues in the tunnel, such as direct trauma or overuse of the muscles that have tendons passing through the tunnel, can cause swelling within the tunnel. Pressure on the tendons, blood vessels, or nerves can result in pain and diminished functioning of any of these tissues. This condition is often referred to as carpal tunnel syndrome.

Repetitive motions that stress the wrist may lead to carpal tunnel syndrome. A common cause of carpal tunnel syndrome is typing at a computer keyboard. In an effort to alleviate this stress, design engineers and biomechanists use an understanding of anatomy to produce new and safer keyboards. The field of study that attempts to improve biomechanical working conditions is known as ergonomics.

retinacula (figure 5.6), which are bands of connective tissue over the volar surface (**flexor retinaculum**) and the dorsal surface (**extensor retinaculum**) of the wrist. The flexor retinaculum forms a bridge over the carpal bones to form the carpal tunnel through which the flexor muscle tendons of the wrist and hand pass. There is a much smaller space between the extensor retinaculum and the carpal bones, through which pass the extensor tendons of the wrist and hand.

The joints of the hand and fingers include five **metacarpophalangeal (MP) joints**, which are articulations between the five long metacarpal bones of the hand and the five proximal phalanges of the fingers. Also the four fingers, having three phalanges each, have **proximal interphalangeal (PIP) joints** and **distal interphalangeal (DIP) joints**. The thumb, having only two phalanges, has only one interphalangeal (IP) joint. All five MP, all four PIP, and all four DIP joints of the fingers and the IP joint of the thumb have capsular ligaments and ulnar and radial collateral ligaments (see figure 5.5).

Fundamental Movements of the Wrist and Hand

Remember that the primary movement among all the numerous joints of the wrist is defined as gliding. A combination of these gliding joint actions results in the wrist having four fundamental movements (figures 5.7 and 5.8): flexion (which takes place primarily in the radiocarpal

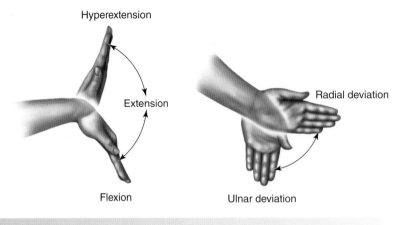

Figure 5.7 Movements of the wrist.

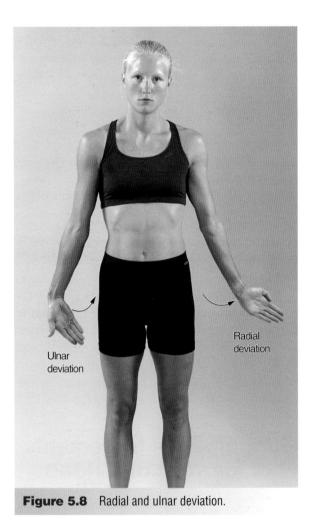

Figure 5.8 Radial and ulnar deviation.

joints), extension (which takes place primarily in the midcarpal joints), **ulnar deviation** (adduction), and **radial deviation** (abduction).

Although movement at the wrist that results in the hand moving toward the midline of the body might be considered adduction, it is known as ulnar deviation. Likewise, moving the hand away from the midline of the body might be considered abduction, but it is known as radial deviation. Keep in mind that the wrist is a biaxial joint (i.e., it can move in two planes about two axes) and is capable of circumduction. Circumduction was defined in chapter 2 as a combination of fundamental movements of a biaxial or triaxial joint. The MP, PIP, DIP, and IP joints of the fingers and thumb all are capable of flexion and extension. The MP joints are also capable of abduction and adduction. The thumb is capable of additional movements that are presented later when the muscles of the thumb are discussed.

Extrinsic Muscles of the Wrist and Hand

Several muscles that are responsible for movement in the wrist actually originate above the elbow joint on either the medial or lateral epicondyle of the humerus. Although these muscles cross the elbow joint, they are not typically considered muscles that create movement in the elbow joint. They are considered muscles of the wrist and hand. Muscles that originate externally to the hand (on the humerus, ulnar, or radius) and insert within the hand are referred to as **extrinsic muscles** of the hand.

Anterior Muscles

There are five major muscles of the hand and wrist that appear on the anterior (volar) surface of the forearm. Four of these muscles (figure 5.9) originate on the medial epicondyle of the humerus on a structure known as the **common flexor tendon**: the flexor carpi radialis, the flexor carpi ulnaris, the flexor digitorum superficialis, and the palmaris longus. The fifth muscle, not part of the common flexor tendon, is the flexor digitorum profundus.

- **Flexor carpi radialis**: Part of the group of muscles originating from the medial epicondyle of the humerus off of the structure known as the common flexor tendon, this muscle inserts on the bases of the second and third metacarpal bones of the hand (figure 5.9). Contraction of the flexor carpi radialis muscle produces flexion and radial deviation of the wrist.

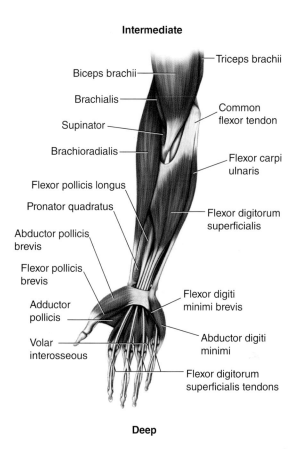

Intermediate

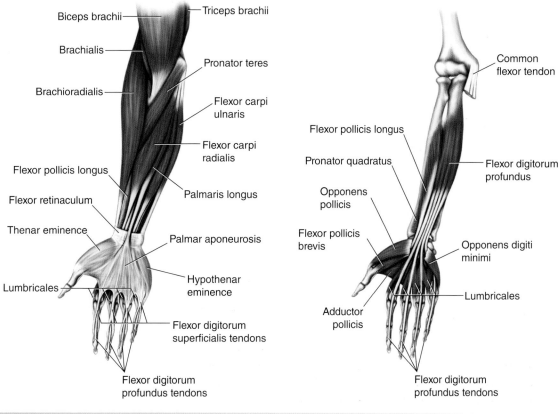

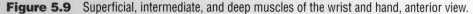

Figure 5.9 Superficial, intermediate, and deep muscles of the wrist and hand, anterior view.

Figure 5.10   Finding the flexor carpi radialis.

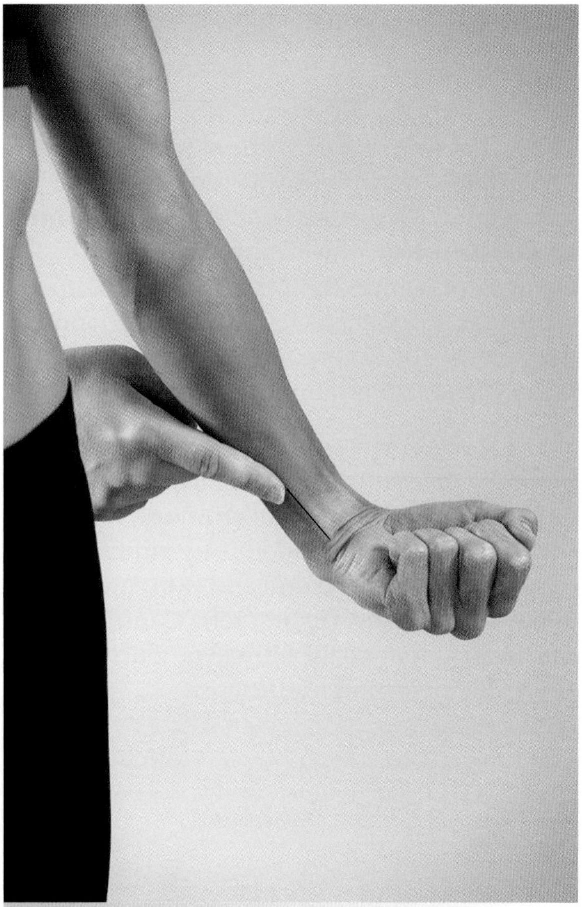

Figure 5.11   Finding the flexor carpi ulnaris.

🖐 Hands On

Make a fist and flex your wrist, find the tendon of the flexor carpi radialis, and trace it upward with your finger until you feel the belly of the muscle (figure 5.10).

- **Flexor carpi ulnaris**: Originating on the common flexor tendon, the flexor carpi ulnaris inserts on the pisiform and hamate carpal bones of the wrist and the base of the fifth metacarpal bone of the hand (figure 5.9). This muscle produces flexion and ulnar deviation of the wrist.

🖐 Hands On

Flex your wrist against a resistance, and note the tendon of the flexor carpi ulnaris (figure 5.11).

- **Flexor digitorum superficialis**: This muscle originates on the common flexor tendon; its tendon of insertion splits into four separate tendons that split again and insert on each side

of the bases of the middle phalanges of the four fingers (figure 5.9). This muscle flexes the wrist, the MP joint, and the PIP joint of the four fingers.

🖐 Hands On

Apply pressure to the area of the forearm containing the tendon of the flexor digitorum superficialis, and flex and extend your fingers. You should feel or see movement. This movement is produced by the flexor digitorum superficialis (figure 5.12).

- **Palmaris longus**: This is the fourth, and last, muscle that has its origin on the common flexor tendon, and it inserts on the palmar aponeurosis (figure 5.9). This muscle's very long tendon passes above the flexor retinaculum, unlike the tendons of the other common flexor tendon muscles, which pass beneath the retinaculum and through the carpal tunnel. Contraction of this muscle results in flexion of the wrist and tightening of the palmar fascia.

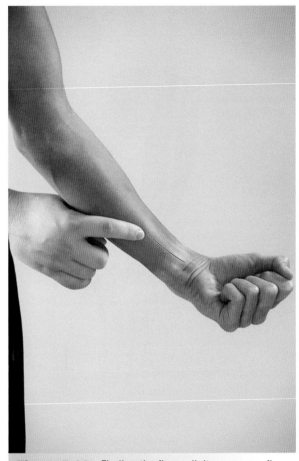

Figure 5.12 Finding the flexor digitorum superficialis.

Figure 5.13 Locating the palmaris longus.

The palmaris longus is absent in 20% of people, but this is not a problem because its function is to assist the other flexor muscles and not to act as a prime flexor. The tendon of this muscle is often used to reinforce elbow ligaments that need surgical reconstruction.

🖐 Hands On

Press your fingers and thumb together, flex your wrist, and feel for movement in the lower third of the anterior aspect of your forearm (figure 5.13). The palmaris longus muscle creates this movement.

- **Flexor digitorum profundus**: This muscle is not part of the common flexor tendon group. It originates on the proximal portion of the volar surface of the ulna and divides into four tendons that pass through the carpal tunnel and split to insert on either side of the bases of the distal phalanges of the four fingers (figure 5.9).

This muscle is involved in flexion of the wrist, the four MP joints, the four PIP joints, and the four DIP joints of the fingers.

Posterior Muscles

Six major muscles of the hand and wrist appear on the posterior (dorsal) surface of the forearm (figure 5.14). Four of these muscles originate on the lateral epicondyle of the humerus on a structure known as the **common extensor tendon**: the extensor carpi radialis brevis, the extensor carpi ulnaris, the extensor digitorum (communis), and the extensor digiti minimi (proprius). The fifth and sixth muscles, not part of the common extensor tendon, are the extensor carpi radialis longus and the extensor indicis.

- **Extensor carpi radialis brevis**: This muscle (see figure 5.14) is one of four muscles originating from the common extensor tendon

Superficial   Deep

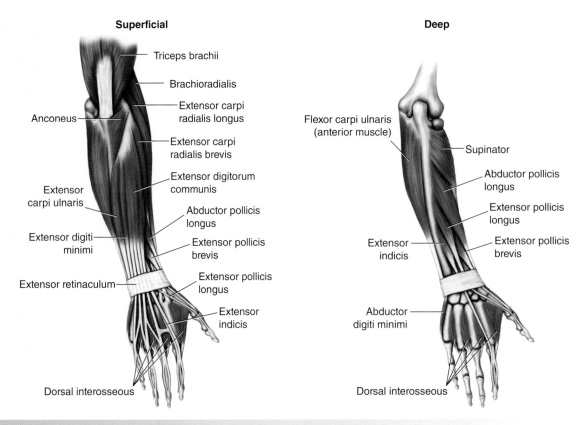

Figure 5.14 Superficial and deep muscles of the wrist and hand, posterior view.

on the lateral epicondyle of the humerus and inserts on the dorsal aspect of the third metacarpal bone. The action produced by this muscle is extension and radial deviation of the wrist.

🤚 Hands On

Against resistance, extend your wrist. On the radial side of the dorsal aspect of the lower third of the forearm, you should see and feel two tendons: the extensor carpi radialis longus and the extensor carpi radialis brevis (figure 5.15).

• **Extensor carpi ulnaris**: Another of the four muscles originating on the common extensor tendon, this muscle (figure 5.14) inserts on the dorsal aspect of the fifth metacarpal bone, and it extends and ulnarly deviates the wrist.

🤚 Hands On

Extend your wrist while it is in ulnar deviation, and you will feel the extensor carpi ulnaris muscle (figure 5.16).

• **Extensor digitorum (communis)**: The third muscle originating from the common extensor tendon, this muscle is typically identi-

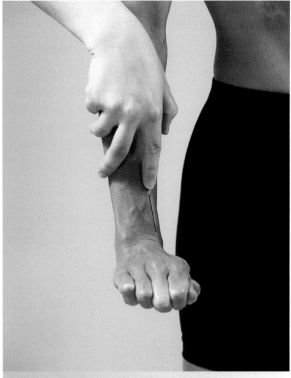

Figure 5.15 Locating the extensor carpi radialis longus and brevis.

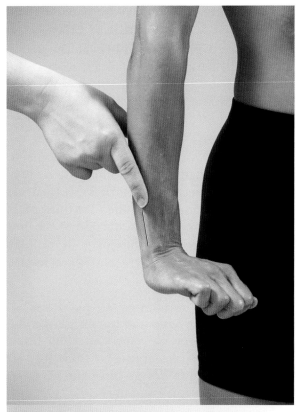

Figure 5.16 Locating the extensor carpi ulnaris.

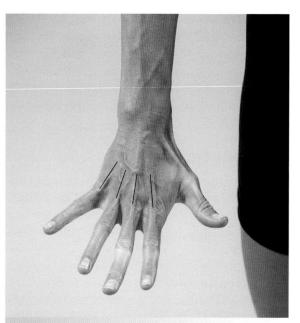

Figure 5.17 Viewing the extensor digitorum communis.

fied as the extensor digitorum, although some texts add the additional term *communis* to the name to distinguish it from similarly named muscles in the foot. The muscle divides into four tendons and inserts on the bases of the distal phalanges of the four fingers (figure 5.14). The muscle extends the wrist, the MP joints, the PIP joints, and the DIP joints of all four fingers.

Hands On

Extend your wrist and flex and extend your fingers. Observe movement of the four tendons of the extensor digitorum on the dorsal aspect of your hand (figure 5.17).

• **Extensor digiti minimi (proprius)**: The fourth muscle originating from the common extensor tendon sometimes includes the term *proprius*. Most texts simply call the muscle the extensor digiti minimi, indicating that the muscle extends the little finger (digiti minimi). The muscle inserts on the base of the proximal phalanx of the fifth finger and extends the wrist and the MP joint of the fifth finger (figure 5.14).

Hands On

Extend your little finger and feel for movement by placing pressure over the fifth metacarpophalangeal joint. This movement is created by the extensor digiti minimi (figure 5.18).

Figure 5.18 Locating the extensor indicis and extensor digiti minimi.

• **Extensor carpi radialis longus**: This is not a muscle of the common extensor tendon. It originates on the lateral supracondylar ridge of the humerus and inserts on the lateral side of the base of the second metacarpal bone. Its actions include extension and radial deviation of the wrist.

🖐 Hands On

Against resistance, extend your wrist. On the radial side of the dorsal aspect of the lower third of the forearm, you should see and feel two tendons: the extensor carpi radialis longus and the extensor carpi radialis brevis (see figure 5.15).

• **Extensor indicis**: This muscle originates on the dorsal surface of the distal portion of the ulna and inserts on the base of the proximal phalanx of the index finger (figure 5.14). Its primary function is to extend the MP joint of the index finger, and it also assists in extension of the wrist.

🖐 Hands On

Place pressure on your second MP joint, and feel for movement as you flex and extend your index finger. The movement is created by the tendons of the extensor digitorum and, just medial to it, the extensor indicis (see figure 5.18).

Intrinsic Muscles of the Hand

Whereas the preceding extrinsic muscles of the hand originate outside the hand and insert within the hand, the following muscles are the intrinsic muscles of the hand, originating and inserting totally within the hand. Three of these muscles (figure 5.9) make up an anatomical structure known as the **hypothenar eminence** (the abductor digiti minimi, the flexor digiti minimi brevis, and the opponens digiti minimi). *Hypo* (from Greek origin, meaning "less than") indicates that these are a group of muscles involved with movement of the fifth (little) finger.

• **Abductor digiti minimi**: The first of three muscles making up the hypothenar eminence, this muscle originates on the pisiform bone of the wrist and inserts on the proximal phalanx of the little finger. It both abducts and assists with flexion of the fifth MP joint.

• **Flexor digiti minimi brevis**: The second muscle of the hypothenar eminence originates on the hamate bone of the wrist and inserts on the proximal phalanx of the fifth finger. It assists in flexion of the fifth MP joint.

• **Opponens digiti minimi**: This is the third and last muscle making up the hypothenar eminence. It originates on the hamate bone of the wrist and inserts on the fifth metacarpal and proximal phalanx bone of the fifth finger, and it assists with flexion and adduction of the fifth MP joint. The ability to bring the little finger toward the thumb to allow grasping is known as **opposition**.

A number of muscles of the hand are not involved with movement of the fifth finger but are involved in the movement of other finger joints. These muscles appear as three distinct muscle groups: the dorsal and palmar (volar) interossei (figure 5.19) and the lumbricales.

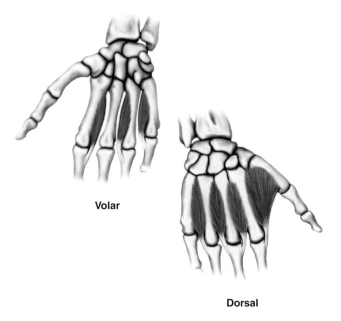

Volar

Dorsal

Figure 5.19 The dorsal and volar interossei muscles of the hand.

• **Dorsal interossei**: There are four dorsal interossei (between the bones) muscles that originate on the four metacarpal bones of the hand (index, middle, ring, and little fingers) and insert on the second (index), third (middle), and fourth (ring) proximal phalanges. Two of these muscles attach to the lateral aspect of the second and third phalanges, whereas the other two attach to the medial aspect of the third and fourth phalanges. The first interosseus can be easily located on the dorsal side of the hand (figure 5.20).

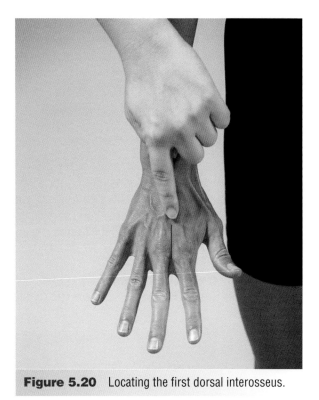

Figure 5.20   Locating the first dorsal interosseus.

Hands On

Spread your fingers and observe what the second, third, and fourth MP joints do. The dorsal interossei muscles cause the second MP joint to abduct (move laterally from the midline of the hand) and the fourth MP joint to adduct (move medially from the midline of the hand), whereas the third MP joint remains stationary. Why? Depending on the angle of the MP joints, these muscles also assist with flexion and extension of the joints.

• **Palmar interossei**: There are three palmar (volar) interossei muscles that originate on the second, fourth, and fifth metacarpal bones of the hand and insert on the second, fourth, and fifth proximal phalanges. Two of these muscles attach to the medial aspects of the fourth and fifth phalanges, whereas the third attaches to the lateral aspect of the second phalanx.

Hands On

Spread your fingers. As you return your fingers toward the middle finger, observe what the second, fourth, and fifth MP joints do. The palmar interossei muscles cause the second MP joint to adduct (move medially toward the midline of the hand) and the fourth and fifth MP joints to abduct (move laterally to the midline of the hand), whereas the third MP joint remains stationary. Why? Depending on the angle of the MP joints, these muscles also assist with flexion and extension of the joints.

• **Lumbricales**: These four muscles are found deep within the hand; they originate on the tendons of the flexor digitorum profundus and insert on the tendons of the extensor digitorum communis in the area of the proximal phalanges (figure 5.9). This group of muscles assists with the flexion of the MP joints and extension of the PIP and DIP joints (figure 5.21).

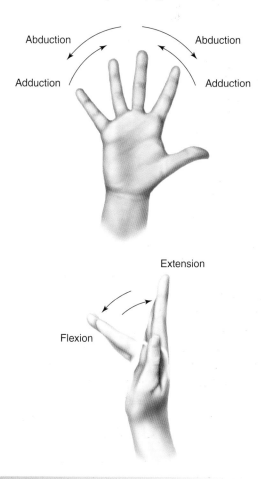

Figure 5.21   Movements of the metacarpophalangeal joints.

Muscles of the Thumb

Muscles involved with movement of the thumb are divided into extrinsic (originating outside of and inserting within the hand) and intrinsic (originating and inserting within the hand) groups.

Extrinsic Muscles

These muscles all originate extrinsically, proximal to the thumb, and insert within the thumb. Note that all contain the word *pollicis* (from the Latin word *pollex*, for thumb) in their names.

• **Extensor pollicis longus**: This muscle originates on the ulna, inserts on the distal phalanx of the thumb, and extends both the IP and MP joints of the thumb (figure 5.14). This muscle also assists with radial deviation of the wrist and supination of the forearm.

• **Extensor pollicis brevis**: This muscle originates on the radius and inserts on the proximal phalanx of the thumb (figure 5.14). It extends the MP joint and assists with radial deviation of the wrist.

The tendons of the extensor pollicis brevis and the **abductor pollicis longus**, as they cross the wrist, form a depression (fossa) directly over the scaphoid carpal bone. This depression is commonly known as the **anatomical snuffbox** (figure 5.22). This term dates to the period in history when tobacco in the form of snuff was used by placing it on this area and then inhaling it through the nose.

• **Abductor pollicis longus**: This muscle originates on the ulna and inserts on the base of the first metacarpal bone (figure 5.14). It abducts the first metacarpal bone and also assists with radial deviation and flexion of the wrist.

 **Hands On**

Abduct your thumb, and apply pressure at the base of the thumb between the two tendons (see figure 5.22). The abductor pollicis longus is difficult to palpate on some people; try several people until you can palpate this muscle as your subject abducts the thumb (figure 5.23).

• **Flexor pollicis longus**: This muscle originates on the radius and inserts on the distal

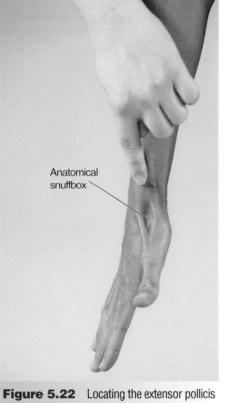

Anatomical snuffbox

Figure 5.22 Locating the extensor pollicis longus and brevis and the anatomical snuffbox.

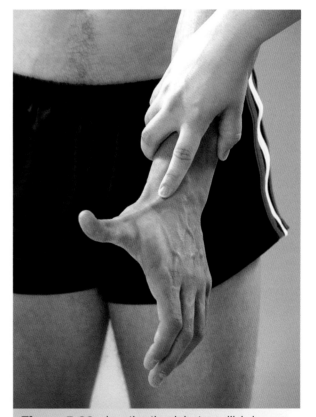

Figure 5.23 Locating the abductor pollicis longus.

phalanx of the thumb (see figure 5.9). It flexes both the IP and MP joints of the thumb and also assists with flexion of the wrist (figure 5.24).

Intrinsic Muscles

The following group of muscles both originate and insert totally within (intrinsically to) the hand. The bellies of these muscles form a thick pad of tissue just proximal to the thumb that is commonly known as the **thenar eminence** (see figure 5.9). Again, note that all four muscles have the term *pollicis* (thumb) in their names.

• **Abductor pollicis brevis**: Originating from the trapezium and scaphoid bones and inserting on the base of the proximal phalanx of the thumb, this muscle assists with abduction of the thumb (figure 5.9).

Hands On

Abduct your thumb and observe or feel the movement on the lateral aspect of the thenar eminence. This movement is produced by the abductor pollicis brevis.

• **Flexor pollicis brevis**: This muscle (figure 5.9) originates on the trapezium bone and inserts

on the proximal phalanx of the thumb. It assists the flexor pollicis longus in flexing the MP joint of the thumb.

Hands On

Flex your thumb toward your second finger and feel the flexor pollicis brevis move (figure 5.25).

• **Opponens pollicis**: Originating on the trapezium and inserting on the first metacarpal bone, this muscle adducts (opposes) the thumb (figure 5.9).

Hands On

Move your thumb toward the four fingers, and palpate along the lateral aspect of the thumb's MP joint. The movement you feel is created by the opponens pollicis (figure 5.26).

• **Adductor pollicis**: This two-headed muscle has one head (oblique) originating on the capitate and second and third metacarpal bones and the other head (transverse) originating on the third metacarpal bone. The two heads combine and insert on the base of the thumb's proximal phalanx (figure 5.9). The action of this muscle is adduction (opposition) of the thumb.

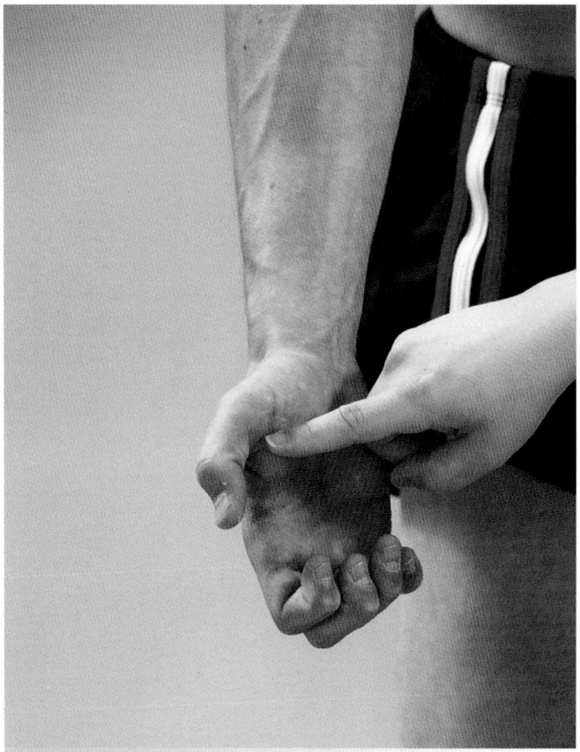

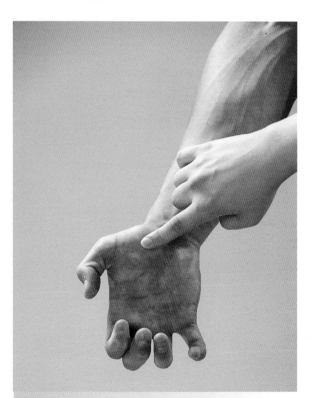

Figure 5.24 Locating the flexor pollicis longus.

Figure 5.25 Finding the flexor pollicis brevis.

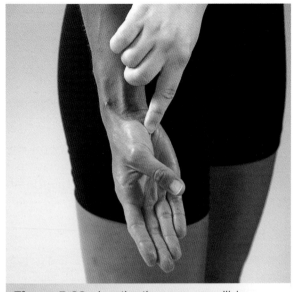

Figure 5.26 Locating the opponens pollicis.

👋 Hands On

Apply the thumb and index finger of your opposite hand to the space between the thumb and index finger, feeling for the tendon of the adductor pollicis.

Figures 5.27 through 5.29 illustrate the movements that the thumb is capable of performing as the result of actions by its extrinsic and intrinsic muscles. The thumb is capable of flexion and extension (figure 5.27), adduction and abduction (figure 5.28), and movement in opposition (figure 5.29). Note that, in describing thumb movements in the anatomical position, there are exceptions to the usual references. Thumb flexion and extension appear to take place in the frontal plane, whereas abduction and adduction appear to take place in the sagittal plane.

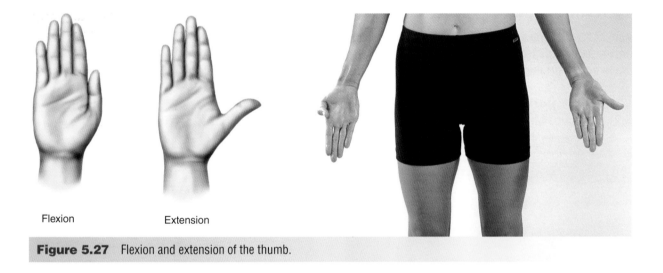

Flexion   Extension

Figure 5.27 Flexion and extension of the thumb.

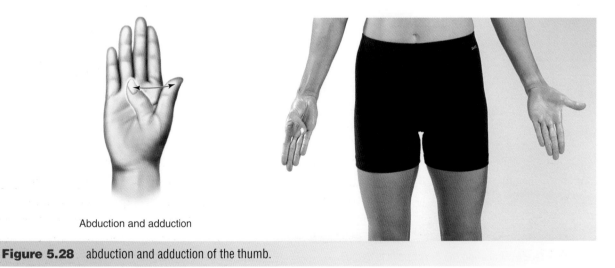

Abduction and adduction

Figure 5.28 abduction and adduction of the thumb.

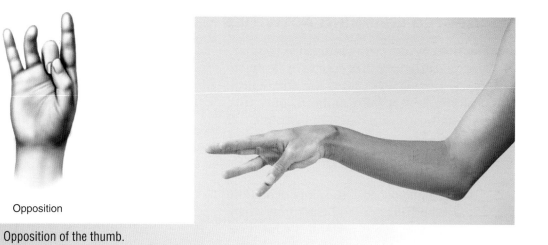

Opposition

Figure 5.29 Opposition of the thumb.

LEARNING AIDS

REVIEW OF TERMINOLOGY

The following terms are discussed in this chapter. Define or describe each term, and where appropriate, identify the location of the named structure either on your body or in an appropriate illustration.

abductor digiti minimi
abductor pollicis brevis
abductor pollicis longus
adductor pollicis
anatomical snuffbox
capitate
capsular carpometacarpal ligament
capsular ligament
carpal
carpometacarpal joint
common extensor tendon
common flexor tendon
distal interphalangeal joint
dorsal carpometacarpal ligament
dorsal interossei
dorsal radiocarpal ligament
extensor carpi radialis brevis
extensor carpi radialis longus
extensor carpi ulnaris
extensor digiti minimi (proprius)
extensor digitorum (communis)
extensor indicis
extensor pollicis brevis
extensor pollicis longus

extensor retinaculum
extrinsic muscle
flexor carpi radialis
flexor carpi ulnaris
flexor digiti minimi brevis
flexor digitorum profundus
flexor digitorum superficialis
flexor pollicis brevis
flexor pollicis longus
flexor retinaculum
greater multangular
hamate
hypothenar eminence
intercarpal joint
intercarpal ligament
interosseous carpometacarpal
ligament
lesser multangular
lumbricales
lunate
metacarpal
metacarpophalangeal joint
midcarpal joint
multangulus major

multangulus minor
navicular
opponens digiti minimi
opponens pollicis
opposition
palmar (volar) interossei
palmaris longus
phalange
pisiform
prehensile hands
proximal interphalangeal joint
radial collateral ligament
radial deviation
radiocarpal joint
scaphoid
thenar eminence
trapezium
trapezoid
triquetrum
ulnar collateral ligament
ulnar deviation
volar carpometacarpal ligament
volar radiocarpal ligament

SUGGESTED LEARNING ACTIVITIES

1. Extend and abduct the thumb on one of your hands. Note the two tendons that rise up at the proximal end of the thumb where it joins the wrist. With your other thumb, apply pressure downward between the two tendons.

a. Name the two tendons.
b. Give the common name for the area where you are applying pressure.
c. What carpal bone lies directly beneath the area where you are applying pressure?

(continued)

SUGGESTED LEARNING ACTIVITIES *(continued)*

2. Flex both of your wrists as much as you can, and then extend both of your wrists as much as you can. Do this from the anatomical position. Repeat both motions, only this time start with your hands in tight fists.

   a. Which motion, from the anatomical position, had greater range: flexion or extension?

   b. What happened to the range of motion for both flexion and extension when the exercise was performed while you made a fist? If there was any change in the range of motion of the wrist between the neutral position and when a fist was made, what anatomical structures were responsible for these changes?

3. Holding your hand in the anatomical position, spread your fingers apart. What movements occurred at the

   a. index finger MP joint (by what muscle)?

   b. middle finger MP joint (by what muscle)?

   c. ring finger MP joint (by what muscle)?

   d. little finger MP joint (by what muscle)?

4. With your fingers spread apart, return them to the anatomical position. What movements occurred at the

   a. index finger MP joint (by what muscle)?

   b. middle finger MP joint (by what muscle)?

   c. ring finger MP joint (by what muscle)?

   d. little finger MP joint (by what muscle)?

5. Touch the tip of your thumb to the tip of your little finger.

   a. What movements occurred in the thumb joints?

   b. What muscles performed these movements?

MULTIPLE-CHOICE QUESTIONS

1. Which of the following joints is not capable of circumduction?

   a. distal interphalangeal

   b. metacarpophalangeal

   c. wrist

   d. glenohumeral

2. Movement of the wrist into abduction is anatomically known as

   a. ulnar deviation

   b. radial deviation

   c. flexion

   d. extension

3. Which one of the following carpal bones is in the distal row of carpals?

   a. hamate

   b. scaphoid

   c. pisiform

   d. lunate

4. The structure known as the anatomical snuffbox is formed by the tendons of the extensor pollicis brevis and the

   a. flexor pollicis longus

   b. abductor pollicis longus

   c. extensor pollicis longus

   d. adductor pollicis longus

5. The lumbricales, located deep within the hand, both extend the interphalangeal joints and flex the

   a. metacarpophalangeal joints

   b. intercarpal joints

   c. carpometacarpal joints

   d. midcarpal joints

6. Which of the following muscles does not originate on the common extensor tendon?

   a. extensor carpi ulnaris

   b. extensor carpi radialis longus

   c. extensor digitorum

   d. extensor carpi radialis brevis

7. The muscles of the forearm that flex and extend only the wrist insert on either the carpal bones or the

a. metacarpals
b. proximal phalanges
c. middle phalanges
d. distal phalanges

8. The hypothenar eminence is formed by muscles that move the

a. thumb
b. index finger
c. little finger
d. ring finger

9. Which of the following carpal bones is also known as the navicular bone?

a. pisiform
b. scaphoid
c. hamate
d. lunate

10. Which of the following muscles does not originate on the common flexor tendon?

a. flexor carpi ulnaris
b. flexor digitorum superficialis
c. flexor digitorum profundus
d. flexor carpi radialis

11. Which of the following muscles is considered an extrinsic muscle of the little finger?

a. extensor digiti minimi
b. flexor digiti minimi brevis
c. abductor digiti minimi
d. opponens digiti minimi

12. Which of the following joints is considered biaxial?

a. interphalangeal
b. proximal interphalangeal
c. metacarpophalangeal
d. distal interphalangeal

13. Which of the following muscles is not considered an intrinsic muscle of the thumb (thenar eminence)?

a. flexor pollicis brevis
b. abductor pollicis brevis
c. flexor pollicis longus
d. adductor pollicis

14. The opponens pollicis muscle moves the thumb

a. toward the little finger
b. toward the radius
c. into flexion
d. into extension

15. Which of the following movements is not considered a fundamental movement of the wrist?

a. flexion
b. ulnar deviation
c. radial deviation
d. circumduction

FILL-IN-THE-BLANK QUESTIONS

1. The only flexor muscle of the wrist whose tendon of insertion does not pass under the flexor (volar) retinaculum is the _____.

2. Extension of the wrist occurs primarily at the _____ joints.

3. The tendons passing through the structure known as the carpal tunnel are considered primarily _____ of the wrist and hand.

4. The thenar eminence is formed by the intrinsic muscles of the _____.

5. Flexion of the wrist primarily occurs at the _____ joints.

6. Ulnar deviation of the wrist in the frontal plane is otherwise also known as _____ of the wrist.

(continued)

FILL-IN-THE-BLANK QUESTIONS *(continued)*

7. The extensor indicis extends the MP, PIP, and DIP joints of the _____ finger.

8. The most lateral carpal bone of the proximal carpal row is the _____.

9. The movement of the thumb that allows grasping of objects is known as _____.

10. The radiocarpal joint is the articulation between the radius, the lunate, and the _____ bone.

11. The band of tissue at the wrist that keeps the wrist and hand flexor muscles from rising up under tension is known as the _____.

12. The wrist joint is stabilized on the medial and lateral sides by _____ ligaments from the radius and ulna.

13. The joints between the bones of the wrist and the bones of the hand are known as the _____ joints.

14. The bone lying directly beneath the anatomical snuffbox is the carpal bone known as the _____.

FUNCTIONAL MOVEMENT EXERCISE

Some degree of extension of the wrist and flexion of the fingers is needed for grasping a tennis racket when performing a backhand stroke of a tennis ball, as illustrated here. For this action to occur, motion happens at both the wrist joints (radiocarpals, midcarpals, carpometacarpals, and intercarpals) and the finger joints (MP, PIP, and DIP). List one muscle acting as a prime mover, one as an antagonist, one as a fixator, and one as a synergist for extension of the wrist and flexion of the fingers to occur.

	Wrist joints	Finger joints (MP, PIP, DIP)
Prime mover		
Antagonist		
Fixator		
Synergist		

Nerves and Blood Vessels of the Upper Extremity

The structures known as nerves, arteries, and veins were presented in chapter 1. These structures are important to movement because they provide the stimulus to contract (nerves), they provide the blood supply that carries nutrients (arteries), and they remove by-products of the muscle's efforts (veins) to move the bones. The nervous system and the vascular system are very complex and have various functions. The nerves and blood vessels pertinent to the musculature of each anatomical section of this text (upper extremity; head, spinal column, and thorax including the heart and lungs; and lower extremity) are presented after the bones, ligaments, and muscles have been discussed.

The **nerves** of the upper extremity originate in the spinal cord from the first cervical to the first thoracic vertebral section and are commonly classified as motor nerves, sensory nerves, or mixed nerves containing both motor and sensory capabilities. Motor nerves innervate muscle and, when stimulated voluntarily or involuntarily, cause muscle fibers to contract and create move-

ment in the joints that these muscle fibers cross. These nerves are presented in the summary table at the end of the upper-extremity part of this text (after this chapter).

The major upper-extremity arteries supply the fuel essential for muscular contraction, and the major veins carry away the waste products of muscular contraction. The major arteries supplying muscles are also listed in the summary table at the end of this part.

It is the intent of this text to examine the anatomy of the nervous system and the vascular system of those structures that are essential to the function of the body's musculature. Figure 6.1 shows some of the relevant nerves that are covered in this text.

Spinal nerves innervating the musculature form **plexuses**, or networks. In the upper extremity, the nerves arise from the **brachial plexus** (C5, C6, C7, C8, and T1 nerve roots and often the nerve roots from C4 and T2) (figure 6.2). Although there are only seven cervical vertebrae, there are eight cervical spinal nerve roots. This is

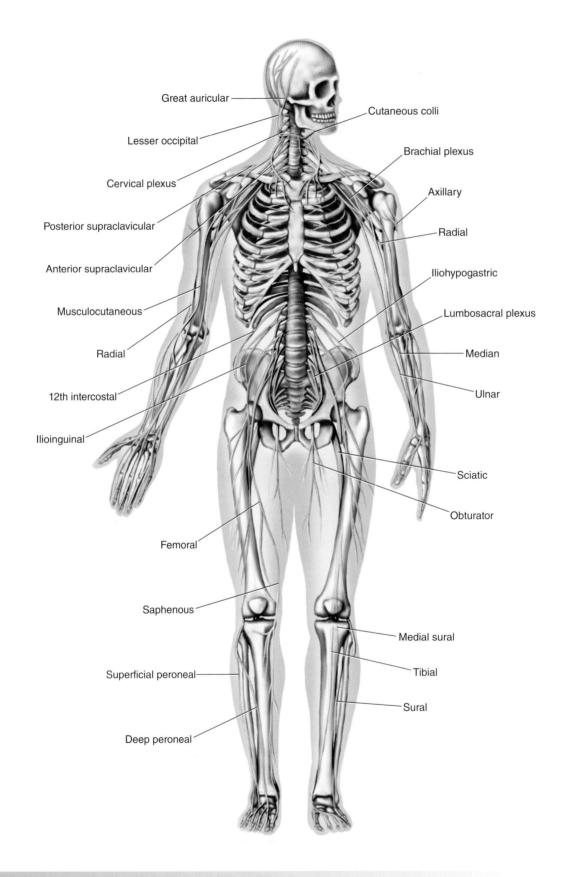

Figure 6.1 Nervous system superimposed over the skeleton.

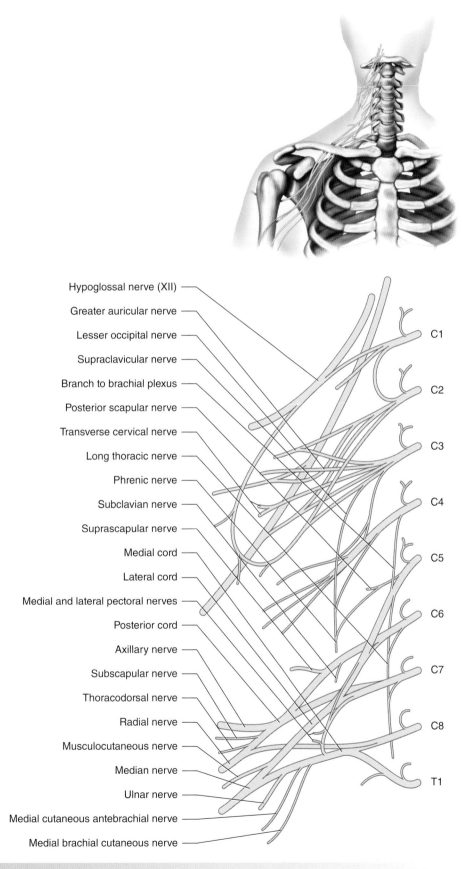

Hypoglossal nerve (XII)

Greater auricular nerve

Lesser occipital nerve

Supraclavicular nerve

Branch to brachial plexus

Posterior scapular nerve

Transverse cervical nerve

Long thoracic nerve

Phrenic nerve

Subclavian nerve

Suprascapular nerve

Medial cord

Lateral cord

Medial and lateral pectoral nerves

Posterior cord

Axillary nerve

Subscapular nerve

Thoracodorsal nerve

Radial nerve

Musculocutaneous nerve

Median nerve

Ulnar nerve

Medial cutaneous antebrachial nerve

Medial brachial cutaneous nerve

C1
C2
C3
C4
C5
C6
C7
C8
T1

Figure 6.2 View of C1 through T1 nerve roots.

easily understood when you realize that the first cervical nerve root originates *above* the first cervical vertebra and the eighth cervical nerve root originates *below* the seventh cervical vertebra.

Nerves of the Brachial Plexus

Spinal nerves have both anterior and posterior rami (branches or arms). The anterior rami form the brachial plexus (figure 6.3), which is typically divided into two parts based on position relative to the clavicle: the **supraclavicular** (above the clavicle) and the **infraclavicular** (below the clavicle) parts.

Supraclavicular Nerves

The **dorsal scapular nerve** (C5 anterior and posterior rami) innervates the levator scapulae and rhomboid major and minor muscles (figure 6.3). The **long thoracic nerve** (C5, C6, and C7 anterior and posterior rami) (figures 6.2 and 6.3) innervates the serratus anterior muscle. Not illustrated are nerves to the scalene and longus colli muscles of the cervical spine (C2–C8) and a communicating nerve (C5) to the phrenic nerve of the cervical plexus, which are discussed in

chapter 10 on the spinal column. Two nerves coming from one of three groups of nerves of the brachial plexus known as the *upper, middle,* and *lower trunks* are the **subclavian nerve** (C4, C5, C6), innervating the subclavian muscle, and the **suprascapular nerve** (C4, C5, C6) (figures 6.2 and 6.3), innervating the supraspinatus and infraspinatus muscles of the rotator cuff.

Infraclavicular Nerves

Three large cords of nerves (lateral, medial, and posterior) are formed in the brachial plexus from divisions of the upper, middle, and lower trunks of spinal nerves (see figure 6.3). The infraclavicular nerves arising from the **lateral cord** are the **lateral anterior thoracic nerve** (C5, C6, C7), innervating the pectoralis major muscle; the **lateral aspect of the median nerve** (C5, C6, C7), innervating most of the anterior forearm muscles and some in the hand; and the **musculocutaneous nerve** (C4, C5, C6) (figure 6.4), innervating the anterior arm muscles.

Hands On

Applying pressure between the triceps brachii and the biceps brachii on the medial aspect of your arm may produce a tingling sensation in your hand because you are compressing the **median nerve** (figure 6.5).

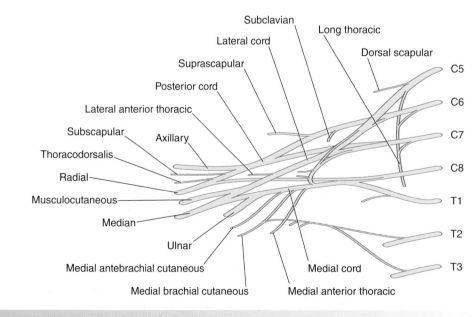

Figure 6.3 Isolated view of the brachial plexus.

There are five major nerves of the brachial plexus that arise from the **medial cord** (see figure 6.3). The **medial anterior thoracic nerve** (C8, T1) innervates the pectoralis major and minor muscles (see figure 6.3). Joining the previously mentioned lateral aspect is the **medial aspect of the median nerve** (C8, T1), which innervates most of the anterior muscles of the forearm and some of the hand (see figures 6.4 and 6.5). The **medial cutaneous nerve of the forearm** (medial antebrachial cutaneous nerve) (T1) is *not* a motor nerve and innervates the skin and fascia on the medial surface of the arm. The **ulnar nerve** (C8, T1) innervates the flexor carpi ulnaris and the muscles of the medial aspect of the hand not innervated by the median nerve (see figure 6.4).

🖐 Hands On

The ulnar nerve is easily located between the medial epicondyle of the humerus and the olecranon process of the ulna (figure 6.6). Sudden pressure on this area often produces a tingling sensation in the forearm and is referred to as "hitting your funny bone."

The fifth nerve, the **medial brachial cutaneous nerve**, is not a motor nerve and innervates the skin and fascia on the medial aspect of the upper arm down to the medial epicondyle and olecranon process areas of the elbow.

Four brachial plexus nerves compose the **posterior cord** (see figure 6.3). The **axillary nerve**

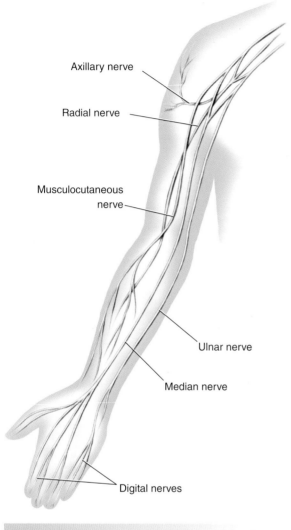

Axillary nerve

Radial nerve

Musculocutaneous nerve

Ulnar nerve

Median nerve

Digital nerves

Figure 6.4 Major nerves of the upper extremity.

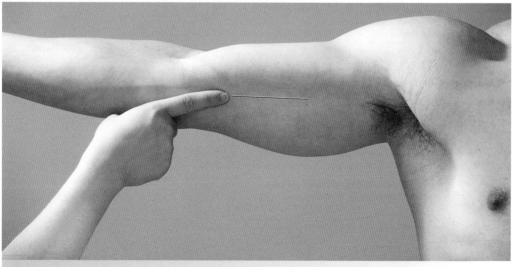

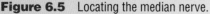

Figure 6.5 Locating the median nerve.

(C5, C6) (figure 6.4) innervates the deltoid and teres minor muscles. The **radial nerve**, also known as the **musculospiral nerve** (C5, C6, C7, C8, T1), innervates the posterior muscles of the arm and forearm (see figure 6.4). The **subscapular nerves** (C5, C6), the **upper sub-** **scapular nerve** and the **lower subscapular nerve**, innervate the subscapularis and teres major muscles (figure 6.3). The **dorsal thoracic nerve**, also known as the **thoracodorsalis** (C6, C7, C8), innervates the **latissimus dorsi** muscle (see figure 6.3).

Figure 6.6   Finding the ulnar nerve.

Vertebral Trauma

FOCUS ON

In a "pinched" nerve or a "stinger" type of trauma in the cervical region or upper extremity, the nerves of the brachial plexus are impinged or stretched. This type of trauma often occurs from the performance of questionable (if not illegal in football) techniques such as butt blocking and spear tackling. Butt blocking is an offensive player's technique that occurs when a blocker makes initial contact with an opponent by striking the opponent's chest ("in the numbers") with the blocker's helmet (face, head, forehead, nose). This can result in the blocker incurring a compression of the cervical vertebrae, causing the impingement of brachial plexus nerves (from a direct head-on blow) or causing the blocker's cervical spine to be excessively laterally flexed. In turn, either or both of the brachial plexus nerves are stretched on one side of the cervical spine while being impinged between cervical vertebrae on the opposite side (when the blow is struck off center, causing the cervical spine to laterally flex). Spear tackling (or spearing, headhunting) is a defensive player's technique in which a ballcarrier is struck by the tackler's head as the initial point of contact with the opponent. The same or similar results that could occur from butt blocking are possible with this tackling technique. Aside from these football examples, any force that may cause compression of the cervical spine or excessive lateral flexion of the cervical spine can result in trauma, impinging or stretching the brachial plexus nerves or even both.

For players to avoid injury to the structures of the brachial plexus, coaches need to learn proper coaching techniques. In the event of injury to the brachial plexus, coaches and athletic trainers must have a sufficient background in first aid to minimize further injury. An understanding of anatomy is vital for the prevention and treatment of pinched nerves.

Major Arteries of the Upper Extremity

The blood vessels of the body are divided into **arteries**, **arterioles**, **capillaries**, **venules**, and **veins**. The arteries (see figure 1.26), which carry blood away from the heart, and some of the larger arterioles are easily observed, whereas the smaller arterioles and capillaries are microscopic. The artery walls are lined with **smooth muscle fibers** that add pumping action; the veins have few or no smooth muscle fibers but have small valves that permit blood flow in only one direction. Any attempt to reverse this direction is blocked by the closing of these valves. The veins, which carry blood to the heart, and some of the larger venules are easily observed, whereas the smaller venules and capillaries are microscopic. Cells receive their nutrients and release their by-products at the capillary level.

The major arteries of the upper extremity include the **subclavian**, **axillary**, **brachial**, **radial**, **deep volar arch**, and **ulnar arteries** (figure 6.7). The subclavian artery originates near the heart, runs posterior to the clavicle, and has three major parts: the costocervical trunk, the internal mammary, and the thyrocervical trunk. Each of these three parts has branches that supply various structures of the thorax and upper extremity.

Hands On

Apply finger pressure to the middle superior aspect of your clavicle (take your pulse) to palpate the subclavian artery (figure 6.8). Note that using your fingers rather than your thumb is the preferred technique for taking a pulse. Some authorities believe that the thumb also has a pulse and could give a false impression when used to palpate for the presence of a pulse in another anatomical area.

The axillary artery is an extension of the subclavian artery that begins at the outside border of the first rib and runs to the lower portion of the teres major muscle, where it becomes known as the brachial artery.

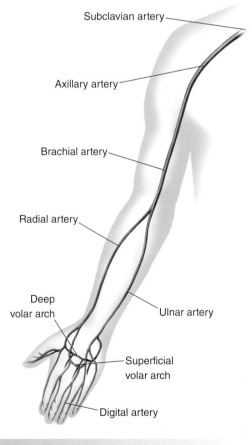

Figure 6.7 The major arteries of the upper extremity.

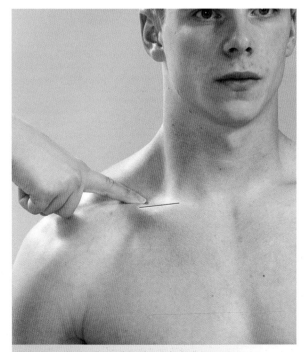

Figure 6.8 Locating the subclavian artery.

✋ Hands On ▰▰▰▰▰

Palpate the axillary area beneath the coracobrachialis muscle for a pulse (figure 6.9). This is a pulse from the axillary artery.

The brachial artery extends from the axilla to the elbow, where it divides into the radial and ulnar arteries. The branches of the brachial artery supply the structures of the upper arm (shoulder to elbow).

✋ Hands On ▰▰▰▰▰

Palpate the area on the medial aspect of the upper arm between the triceps brachii and the biceps brachii muscles (take your pulse) to locate the brachial artery (figure 6.9). This area is often considered a *pressure point* for attempting to reduce blood flow below that site (as in applying pressure in first aid for bleeding in the forearm or hand).

The radial artery (the lateral branch of the brachial artery) and its branches supply the anterior lateral structures from the elbow, through the forearm and wrist, into the hand. This artery is often associated with taking a pulse, because the radial artery is compressed against the distal end of the radius bone (figures 6.7 and 6.10). The medial branch of the brachial artery is the ulnar artery (figures 6.7 and 6.11). The ulnar artery and its branches supply the posterior and anterior medial structures of the forearm and hand. In

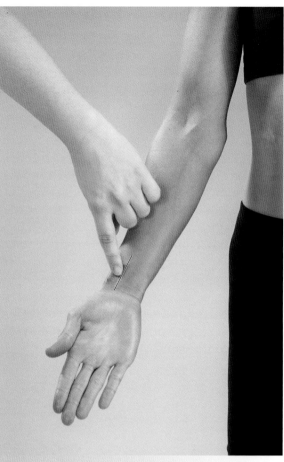

Figure 6.10   Finding the radial artery.

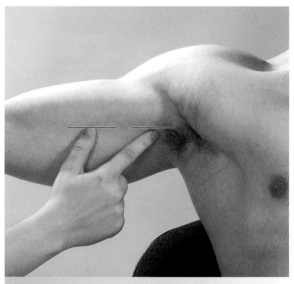

Figure 6.9   Locating the axillary and brachial arteries.

Figure 6.11   Locating the ulnar artery.

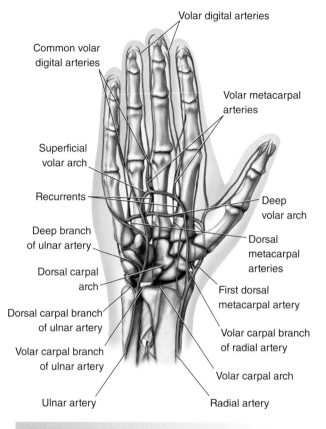

Figure 6.12 The dorsal carpal branch of the ulnar artery, the dorsal carpal arch, and the superficial and deep volar arches.

the wrist, a branch of the ulnar artery (**dorsal ulnar carpal branch**) joins a branch of the radial artery (**dorsal radial carpal branch**) to form the structure known as the **dorsal carpal branch** (figure 6.12). In the palm of the hand (volar surface), the radial artery joins with branches of the ulnar artery to form two structures known as the **superficial volar arch** and the deep volar arch (figures 6.7 and 6.12). Branches from these volar arches, including the **digital arteries** on either side of the fingers, supply the structures of the fingers and thumb (see figures 6.7 and 6.12).

Major Veins of the Upper Extremity

The veins that return blood to the heart (see figure 1.27) are commonly divided into **deep veins** and **superficial veins**. With a few exceptions, the deep veins have the same names as the arteries they parallel, such as the **subclavian vein** and **axillary vein**. The superficial veins have specific names and are located near the skin. Unlike the arteries, veins do not always appear exactly where one might expect, and often they may be absent altogether.

The major deep veins of the upper extremity (figure 6.13) are the subclavian and the axillary. The axillary vein drains the blood from the upper extremity into the subclavian vein, which combines with the **internal jugular vein** to form the **brachiocephalic vein** (figure 6.14). Feeding into the axillary vein are the veins paralleling the major arteries of the upper extremity: the **brachial vein**, **radial vein**, **ulnar vein**, **venous arch**, and **digital veins** (figure 6.13).

The major superficial veins of the upper extremity (figure 6.13) include the **basilic**, the **cephalic**, and the **median veins**. The **median cubital vein** crosses the anterior aspect of the elbow joint (cubital fossa) laterally to medially, and the cephalic vein can be observed superior to

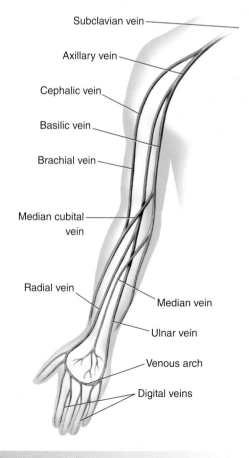

Figure 6.13 The major veins of the upper extremity.

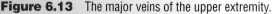

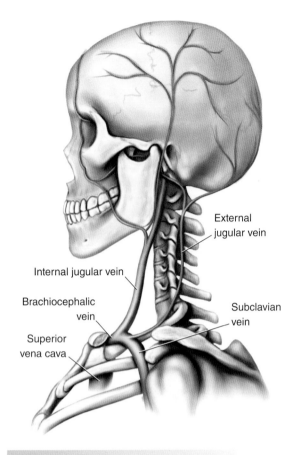

Figure 6.14 The brachiocephalic and internal jugular veins.

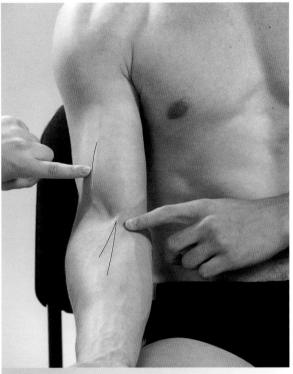

Figure 6.15 Locating the median cubital and cephalic veins.

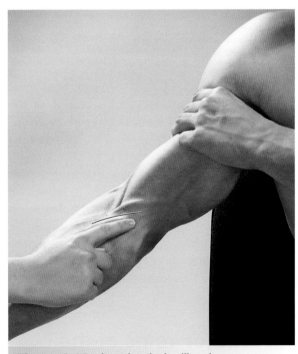

Figure 6.16 Locating the basilic vein.

the brachioradialis muscle and slightly lateral to the biceps brachii muscle (figure 6.15). The basilic vein originates on the ulnar aspect of the venous arch in the hand. At the elbow, the basilic vein receives the median cubital vein and, on reaching the lower portion of the teres major muscle, joins the axillary vein (figure 6.16). It drains the structures on the ulnar side of the volar and dorsal aspects of the hand, forearm, and upper arm. The cephalic vein originates on the radial side of the dorsal venous arch in the hand; at the elbow it is connected to the basilic vein by the median cubital vein and then joins the axillary vein. Also at the elbow, the cephalic vein connects to the **accessory cephalic vein** (figure 6.17). The cephalic vein drains the structures from the lateral aspect of the hand, forearm, and upper arm. The median vein originates in the **palmar venous vessels** of the hand and connects to the **basilic cubital vein**. In the elbow area, the median vein is joined by the median cubital vein, which drains the medial aspect of the forearm

and hand (figure 6.18). The median cubital vein is often used when blood is donated or a sample of blood is needed for testing purposes.

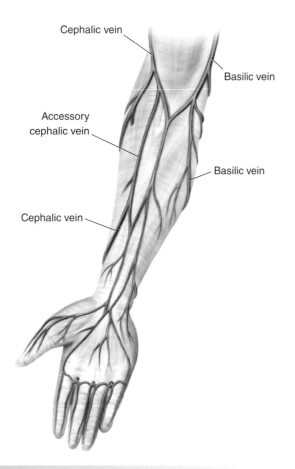

Cephalic vein

Basilic vein

Accessory cephalic vein

Basilic vein

Cephalic vein

Figure 6.17 The cephalic and accessory cephalic veins.

Figure 6.18 Venous network of the median vein of the forearm.

REVIEW OF TERMINOLOGY

The following terms are discussed in this chapter. Define or describe each term, and where appropriate, identify the location of the named structure either on your body or in an appropriate illustration.

accessory cephalic vein
arteriole
artery
axillary artery
axillary nerve
axillary vein
basilic cubital vein
basilic vein
brachial artery
brachial plexus
brachial vein
brachiocephalic vein
capillary
cephalic vein
deep vein
deep volar arch
digital artery
digital vein
dorsal carpal branch
dorsal radial carpal branch
dorsal ulnar carpal branch
dorsal scapular nerve

dorsal thoracic nerve
infraclavicular nerve
internal jugular vein
lateral anterior thoracic nerve
lateral aspect of the median nerve
lateral cord
latissimus dorsi
long thoracic nerve
lower subscapular nerve
medial anterior thoracic nerve
medial aspect of the median nerve
medial brachial cutaneous nerve
medial cord
medial cutaneous nerve of the
forearm
median cubital vein
median nerve
median vein
musculocutaneous nerve
musculospiral nerve
nerve
palmar venous vessel

plexus
posterior cord
radial artery
radial nerve
radial vein
smooth muscle fiber
subclavian artery
subclavian nerve
subclavian vein
subscapular nerve
superficial vein
superficial volar arch
supraclavicular nerve
suprascapular nerve
thoracodorsalis
ulnar artery
ulnar nerve
ulnar vein
upper subscapular nerve
vein
venous arch
venule

SUGGESTED LEARNING ACTIVITIES

1. With your hand in front of your body and your fingers facing upward, make a fist four to six times by opening and closing your hand. After your last effort, keep your hand closed in a fist. With your other hand, apply pressure with your index finger and thumb to either side of your wrist (on the volar surface), as hard as you can. Continue the pressure while opening your hand (releasing the fist formation). Note the color of the palmar surface of your hand.

   a. Release the pressure being applied by your thumb. What happened to the color of your hand? Which side of your hand changed first? What vessels were compressed and released?

   b. Release the pressure being applied by your index finger. What happened to the color of your hand? Which side of your hand changed first? What vessels were compressed and released?

2. Have your partner encircle your upper arm just below the axilla with his hands. As your partner squeezes your arm, flex your elbow and wrist isometrically for a few seconds.

   a. What, if any, superficial veins of the arm and forearm can you identify?

   b. Trading positions with your partner, perform the same maneuver. Were the same veins identifiable, and were they in the same location?

MULTIPLE-CHOICE QUESTIONS

1. The ulnar and radial arteries are branches of which of the following arteries?
   a. dorsal carpal
   b. brachial
   c. volar carpal
   d. median

2. How many cords of nerves make up the brachial plexus?
   a. one
   b. two
   c. three
   d. four

3. The waste material produced by the production of energy within muscle tissue is disposed of through the
   a. arteries
   b. neutralizers
   c. veins
   d. kidneys

4. Which of the following veins drains directly into the subclavian vein?
   a. axillary
   b. ulnar
   c. brachial
   d. radial

FILL-IN-THE-BLANK QUESTIONS

1. When someone either donates blood or has blood drawn for testing purposes, the blood is likely to be drawn from the _____ vein.

2. The levator scapulae and rhomboid muscles are innervated by the _____ nerve.

3. The walls of the blood vessels known as _____ are lined with smooth muscle fibers.

Articulations of the Upper Extremity

Joint	Type	Bones	Ligaments	Movement
Shoulder girdle				
Sternoclavicular	Arthrodial (saddle/ gliding)	Sternum and clavicle	• Sternoclavicular (anterior, superior, and posterior) • Costoclavicular • Interclavicular • (Articular disc)	Elevation, depression, rotation, protraction, retraction
Acromioclavicular	Arthrodial (plane)	Clavicle and scapula	• Acromioclavicular • Coracoclavicular (conoid and trapezoid) • Coracoacromial (for protection of glenohumeral joint) • (Articular disc)	Elevation, depression, rotation, protraction, retraction, winging (posterior movement to keep scapula close to the thorax during scapular abduction), tipping (rotation to keep scapula close to the thorax during scapular elevation)
Shoulder joint				
Glenohumeral	Ball and socket	Humerus and scapula	• Glenohumeral (superior, middle, and inferior) • Coracohumeral • Transverse humeral (keeps long-head biceps tendon in intertubercular groove) • Labrum (deepens socket of the glenoid fossa) • Capsule	Flexion, extension, hyperextension, abduction, adduction, internal and external rotation
Scapulothoracic	(False joint)	Scapula and thorax	• No ligaments (fascia of the serratus anterior muscle and fascia from the thorax)	Elevation, depression, upward rotation and lateral tilt (abduction, protraction) and downward rotation and medial tilt (adduction, retraction) of the scapula on the thorax
Elbow				
Humeroulnar	Hinge	Humerus and ulna	• Capsule • Medial (ulnar) collateral (3 bands: anterior, posterior, and transverse)	Flexion, extension

Joint	Type	Bones	Ligaments	Movement
Radiohumeral	Irregular (gliding)	Humerus and radius	• Lateral (radial) collateral • Annular	Flexion, extension, rotation (pronation, supination)
Radioulnar (proximal)	Pivot	Radius and ulna	• Annular	Pronation, supination
Forearm				
Radioulnar (middle)	Syndesmosis	Radius and ulna	• Oblique cord • Interosseous membrane	Slight (if any)
Radioulnar (distal)	Pivot	Radius and ulna (and articular disc)	• Capsule • Dorsal radioulnar • Volar (palmar) radioulnar	Pronation, supination
Wrist				
Radiocarpal	Condyloid (gliding)	Radius, scaphoid, lunate, triquetrum	• Capsule • Radial collateral • Ulnar collateral • Dorsal radioulnar • Volar (palmar) radioulnar	Flexion, extension, abduction (radial deviation), adduction (ulnar deviation)
Ulnarcarpal (not part of the wrist)	Irregular (gliding)	Ulna, pisiform, triquetrum	• Capsule • Ulnar collateral • Articular disc	Gliding
Wrist: intercarpals				
Proximal row	Arthrodial (gliding)	Radius, scaphoid, lunate, triquetrum	• Capsule • Flexor (volar) retinaculum • Extensor (dorsal) retinaculum • Radial collateral • Ulnar collateral • Volar (palmar) intercarpal • Dorsal intercarpal	Gliding
Middle	Hinge (gliding)	Proximal and distal rows of carpals	• Capsule • Volar (palmar) intercarpal • Dorsal intercarpal • Interosseous intercarpal • Radial collateral • Ulnar collateral • Pisohamate • Pisometacarpal	Gliding

(continued)

PART II Summary Tables *(continued)*

Joint	Type	Bones	Ligaments	Movement
Wrist: intercarpals				
Distal row	Irregular (gliding)	Trapezius, trapezoid, capitate, hamate	• Capsule • Volar (palmar) intercarpal • Dorsal intercarpal • Interosseous intercarpal	Gliding
Hand and fingers				
Carpometacarpal	Modified saddle	Trapezium, trapezoid, capitate, hamate, metacarpals (4)	• Capsule • Volar carpometacarpal • Dorsal carpometacarpal • Interosseous carpometacarpal • (Volar accessory)	Flexion, extension, abduction, adduction, slight opposition of fifth
Metacarpophalangeal	Condyloid	Metacarpals (4) and proximal phalanges (4)	• Transverse metacarpal • Capsule • Radial collateral • Ulnar collateral • (Volar accessory)	Flexion, extension, abduction, adduction
Proximal interphalangeal	Hinge	Proximal phalanges (4) and middle phalanges (4)	• Capsule • Radial collateral • Ulnar collateral • (Volar accessory)	Flexion, extension
Distal interphalangeal	Hinge	Middle phalanges (4) and distal phalanges (4)	• Capsule • Radial collateral • Ulnar collateral • (Volar accessory)	Flexion, extension
Thumb				
Carpometacarpal	Saddle	Trapezium and 1st metacarpal	• Capsule	Flexion, extension, abduction, adduction, opposition
Metacarpophalangeal	Hinge	1st metacarpal and proximal phalanx	• Capsule • Radial collateral • Ulnar collateral • (Dorsal accessory)	Flexion, extension
Interphalangeal	Hinge	Proximal phalanx and distal phalanx	• Capsule • Radial collateral • Ulnar collateral • (Volar accessory)	Flexion, extension

Muscles, Nerves, and Blood Supply of the Upper Extremity

Muscle	Origin	Insertion	Action	Nerve	Blood supply
Shoulder girdle, anterior					
Pectoralis minor	3rd to 5th ribs	Coracoid process of scapula	Downward rotation of scapula	Medial anterior thoracic	Lateral thoracic and thoracoacromial branches of axillary
Serratus anterior	Upper 9 ribs	Vertebral border of scapula	Upward rotation of scapula	Long thoracic	Lateral thoracic and subscapular branches of axillary, transverse cervical branch of thyrocervical branch of subclavian
Subclavian	1st rib	Subclavian groove of clavicle	Ligamentous action at sternoclavicular joint	Medial anterior thoracic	Clavicular branch of thoracoacromial branch of axillary thoracic
Shoulder girdle, posterior					
Levator scapulae	Transverse processes of first 4 cervical vertebrae	Vertical border of scapula	Elevation and downward rotation of scapula	Dorsal scapular	Transverse cervical branch of thyrocervical and intercostals
Rhomboids (major and minor)	Spines of vertebrae C7 to T5	Vertebral border of scapula	Elevation and downward rotation of scapula	Dorsal scapular	Transverse cervical branch of thyrocervical branch of subclavian
Trapezius	External occipital protuberance, all 7 cervical and 12 thoracic spinous processes	Spine of scapula, acromion process, lateral 1/3 of posterior clavicle	Elevation, upward and downward rotation, and adduction of scapula	C2, C3, C4, and spinal accessory (11th cranial)	Transverse cervical branch of thyrocervical branch of subclavian
Shoulder joint, anterior					
Pectoralis major	2nd to 6th ribs, sternum, medial 1/2 of clavicle	Inferior to greater tuberosity of humerus in area of surgical neck	Flexion, adduction, and internal rotation of GH joint	Lateral and medial anterior thoracic	Lateral thoracic and thoracoacromial branches of axillary
Coracobrachialis	Coracoid process of scapula	Middle 1/3 of medial surface of humerus	Flexion and adduction of GH joint	Musculocutaneous	Brachial

(continued)

PART II Summary Tables *(continued)*

Muscle	Origin	Insertion	Action	Nerve	Blood supply
Shoulder joint, anterior *(continued)*					
Biceps brachii (long head)	Supraglenoid tubercle	Radial tuberosity	Abduction and flexion of GH joint, flexion of elbow, supination of forearm	Musculocutaneous	Brachial
Biceps brachii (short head)	Coracoid process of scapula	Radial tuberosity	Adduction and flexion of GH joint, flexion of elbow, supination of forearm	Musculocutaneous	Brachial
Subscapularis	Subscapular fossa	Lesser tuberosity of humerus	Internal rotation and flexion of GH joint, GH joint stability	Subscapular branch from brachial plexus	Subscapular branch of axillary, transverse cervical and transverse scapular branches of thyrocervical
Shoulder joint, superior					
Deltoid (clavicular)	Lateral 1/3 of anterior border of clavicle	Deltoid tubercle	Flexion, adduction, and internal rotation of GH joint (above horizontal, abduction)	Axillary	Anterior and posterior humeral circumflex, brachial, thoracoacromial
Deltoid (acromial)	Lateral border of acromion process	Deltoid tubercle	Abduction of GH joint		
Deltoid (scapular)	Inferior lip of spine of scapula	Deltoid tubercle	Extension, adduction, and external rotation of GH joint (above horizontal, abduction)		
Supraspinatus	Supraspinous fossa of scapula	Proximal facet of greater tuberosity of humerus	Abduction of GH joint, GH joint stability	Suprascapular from C5 through brachial plexus	Transverse cervical and transverse scapular branches of thyrocervical and subscapular
Shoulder joint, posterior					
Infraspinatus	Infraspinatus fossa of spine of scapula	Middle facet of greater tuberosity of humerus	External rotation and extension of GH joint, GH joint stability	Suprascapular through brachial plexus	Transverse cervical and transverse scapular branches of thyrocervical

Muscle	Origin	Insertion	Action	Nerve	Artery
Teres minor	Upper 2/3 of lateral border of scapula	Distal facet of greater tuberosity of humerus	External rotation, extension, and adduction of GH joint; GH joint stability	Axillary through brachial plexus	Posterior humeral circumflex and subscapular
Shoulder joint, inferior					
Latissimus dorsi	Spinous processes of lower 6 thoracic and all lumbar vertebrae, ilium, lower 3 ribs, inferior angle of scapula	Intertubercular groove	Internal rotation, extension, and adduction of GH joint	Thoracodorsal nerve and posterior rami of lower 6 thoracic and lumbar spinal nerves	Subscapular branch of axillary, transverse cervical branch of thyrocervical branch of subclavian
Teres major	Inferior 1/3 of lateral border and inferior angle of scapula	Beneath the lesser tuberosity on anterior humerus	Internal rotation, extension, and adduction of GH joint	Subscapular through brachial plexus	Subscapular branch of axillary
Triceps brachii (long head)	Infraglenoid tubercle of scapula	Olecranon process of ulna	Extension and adduction of GH joint	Radial	Posterior humeral circumflex, profundus branch of brachial
Elbow joint, anterior					
Brachialis	Distal 1/3 of anterior surface of humerus	Coronoid process and tuberosity of ulna	Flexion of elbow	Musculocutaneous	Brachial
Brachioradialis	Proximal aspect of lateral supracondylar ridge of humerus	Lateral aspect of styloid process of radius	Flexion of elbow, supination of forearm	Radial	Radial
Biceps brachii (long head)	Supraglenoid tubercle	Radial tuberosity	Abduction and flexion of GH joint, flexion of elbow, supination of forearm	Musculocutaneous	Brachial
Biceps brachii (short head)	Coracoid process of scapula	Radial tuberosity	Adduction and flexion of GH joint, flexion of elbow, supination of forearm	Musculocutaneous	Brachial
Elbow joint, posterior					
Triceps brachii (lateral head)	Proximal 1/3 of posterolateral aspect of humerus	Olecranon process of ulna	Extension of elbow	Radial	Posterior humeral circumflex, profundus branch of brachial

(continued)

PART II Summary Tables (continued)

Muscle	Origin	Insertion	Action	Nerve	Blood supply
Elbow joint, posterior *(continued)*					
Triceps brachii (long head)	Infraglenoid tubercle of scapula	Olecranon process of ulna	Extension of elbow, extension and adduction of GH joint	Radial	Posterior humeral circumflex, profundus branch of brachial
Triceps brachii (medial head)	Distal 2/3 of posteromedial aspect of humerus	Olecranon process of ulna	Extension of elbow	Radial	Posterior humeral circumflex, profundus branch of brachial
Anconeus	Lateral epicondyle of humerus	Radial aspect of olecranon process and dorsal surface of proximal 1/4 of ulna	Extension of elbow	Radial	Profundus branch of brachial and dorsal interosseous
Forearm, supinators					
Supinator	Lateral epicondyle of humerus and supinator fossa of ulna	Proximal 1/3 of lateral and volar aspect of radius	Supination of forearm	Radial	Dorsal interosseous and radial
Biceps brachii (long head)	Supraglenoid tubercle	Radial tuberosity	Abduction and flexion of GH joint, flexion of elbow, supination of forearm	Musculocutaneous	Brachial
Biceps brachii (short head)	Coracoid process of scapula	Radial tuberosity	Adduction and flexion of GH joint, flexion of elbow, supination of forearm	Musculocutaneous	Brachial
Forearm, pronators					
Pronator teres	Common flexor tendon and coronoid process of ulna	Middle of lateral aspect of radius	Pronation of forearm	Median	Radial branch of brachial
Pronator quadratus	Distal part of volar aspect of ulna	Distal part of volar aspect of radius	Pronation of forearm	Median	Ulnar and radial

Wrist and hand, anterior wrist and extrinsic hand

Muscle	Origin	Insertion	Action	Nerve	Blood supply
Flexor carpi radialis	Common flexor tendon	Bases of 2nd and 3rd metacarpals	Flexion and radial deviation of wrist, elbow flexion, and forearm pronation	Median	Radial
Flexor carpi ulnaris	Common flexor tendon	Pisiform, hamate, and base of 5th metacarpal	Flexion and ulnar deviation of wrist, elbow flexion, and forearm supination	Ulnar	Ulnar branch of brachial
Flexor digitorum superficialis	Common flexor tendon, coronoid process, and proximal 2/3 of radius	Split tendons on each side of base of middle phalanges of 4 fingers	Flexion of PIP, MP, and wrist joints	Median	Ulnar branch of brachial
Palmaris longus	Common flexor tendon	Palmar aponeurosis	Flexion of wrist	Median	Radial branch of brachial
Flexor digitorum profundus	Proximal 3/4 of coronoid process of ulna and interosseous membrane	4 tendons—passing split superficialis tendons—on bases of distal phalanges of 4 fingers	Flexion of DIP, PIP, MP, and wrist joints	Median and ulnar	Ulnar and volar interosseous

Wrist and hand, posterior wrist and extrinsic hand

Muscle	Origin	Insertion	Action	Nerve	Blood supply
Extensor carpi radialis longus	Lateral supracondylar ridge of humerus	Base of 2nd metacarpal	Extension and radial deviation of wrist, supination of forearm	Radial	Radial
Extensor carpi radialis brevis	Common extensor tendon	Base of 3rd metacarpal	Extension and radial deviation of wrist, supination of forearm	Radial	Radial
Extensor carpi ulnaris	Common extensor tendon	Base of 5th metacarpal	Extension and ulnar deviation of wrist	Radial	Ulnar
Extensor digitorum communis	Common extensor tendon	Bases of distal phalanges of 4 fingers	Extension of DIP, PIP, and MP joints of 4 fingers; extension of wrist	Radial	Ulnar
Extensor digiti minimi (proprius)	Common extensor tendon	Base of proximal phalanx of little finger	Extension of 5th MP joint and wrist	Radial	Ulnar

(continued)

PART II Summary Tables (continued)

Muscle	Origin	Insertion	Action	Nerve	Blood supply
Wrist and hand, posterior wrist and extrinsic hand (continued)					
Extensor indicis	Dorsal aspect of distal ulna and interosseous membrane	Base of proximal phalanx of index finger	Extension and radial deviation of 2nd MP joint, extension of wrist	Radial	Ulnar
Wrist and hand, intrinsic hand					
Abductor digiti minimi	Pisiform bone	Base of proximal phalanx of little finger	Abduction and flexion of 5th MP joint	Ulnar	Ulnar
Flexor digiti minimi brevis	Hamate bone	Base of proximal phalanx of little finger	Flexion of 5th MP joint	Ulnar	Ulnar
Opponens digiti minimi	Hamate bone	Medial surface of 5th metacarpal	Flexion and adduction of little finger	Ulnar	Ulnar
Dorsal interossei (4)	Adjacent sides of all 4 metacarpals	Bases of 2nd, 3rd, and 4th proximal phalanges and extensor digitorum communis tendon	Abduction of 2nd and 4th MP joints	Ulnar	Metacarpal branches of radial
Palmar interossei (3)	Ulnar side of 2nd metacarpal and radial side of 4th and 5th metacarpals	Radial side of 4th and 5th phalanges and ulnar side of 2nd proximal phalanx	Adduction and flexion of 2nd, 4th, and 5th MP joints	Ulnar	Metacarpal branches of deep volar arch
Lumbricales	Flexor digitorum profundus tendons	Extensor digitorum communis tendons	Flexion of 4 MP joints, extension of 4 PIP and DIP joints	Median and ulnar	Volar metacarpal branch of deep volar
Thumb, extrinsic					
Extensor pollicis longus	Middle 1/3 of ulna and interosseous membrane	Base of distal phalanx of thumb	Extension of IP and MP joints of thumb, extension and radial deviation of wrist	Radial	Ulnar through the dorsal interosseous
Extensor pollicis brevis	Middle dorsal aspect of radius and interosseous membrane	Base of proximal phalanx of thumb	Extension of MP joint of thumb, abduction of 1st metacarpal, radial deviation of wrist	Radial	Ulnar through the dorsal interosseous

Muscle	Origin	Insertion	Action	Nerve	Blood supply
Abductor pollicis longus	Middle dorsolateral aspect of ulna and interosseous membrane	Base of 1st metacarpal	Abduction of 1st metacarpal, flexion of wrist	Radial	Dorsal interosseous branch of ulnar
Flexor pollicis longus	Middle 1/2 of volar surface of radius	Base of distal phalanx of thumb	Flexion of IP and MP joints, adduction of thumb	Median	Radial, volar interosseous
Thumb, intrinsic					
Abductor pollicis brevis	Trapezius and scaphoid bones	Base of proximal phalanx of thumb	Abduction of MP joint of thumb	Median	Radial
Flexor pollicis brevis	Trapezium	Base of proximal phalanx of thumb	Flexion and adduction of MP joint of thumb	Median	Radial
Opponens pollicis	Trapezium	Lateral surface of 1st metacarpal	Flexion and adduction of MP joint of thumb	Median	Radial
Adductor pollicis	Oblique head: capitate and 2nd and 3rd metacarpal bones; transverse head: 3rd metacarpal	Volar surface of base of proximal phalanx of thumb	Adduction of MP joint of thumb	Median	Deep volar arch

GH = glenohumeral; PIP = proximal interphalangeal; MP = metacarpophalangeal; DIP = distal interphalangeal; IP = interphalangeal.

PART III

The Head, Spinal Column, Thorax, and Pelvis

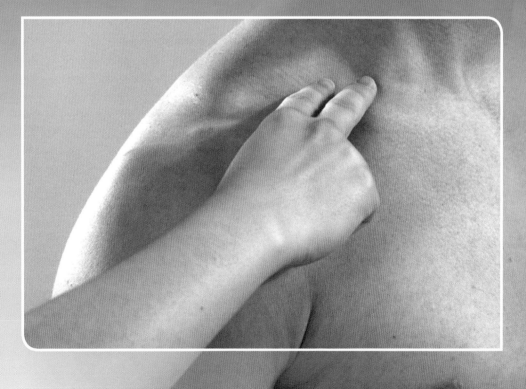

The Head

Although the structures of the skull are not usually associated with a great degree of movement in the human body (the only movable joint in the head is the temporomandibular joint, or TMJ), some of the structures within the head are infinitely involved with all movement. When studying human anatomy, particularly the structures involved in movement, it is also appropriate to take a brief look into the extensive number of bones, ligaments, muscles, and blood vessels of the head as well as the contents within the skull related to the nervous system. This chapter concentrates on the bones, ligaments, and muscles of the skull, while chapter 10 presents the blood vessels and nerves of the head and spinal column.

Bones of the Head

The head (or skull) consists of 28 bones divided into the 8 bones of the skull (**cranium**), 14 of the face, and 6 within the ears. (A 29th bone, the hyoid, is considered a head bone by some, but since it has no bony attachment to any of the other bones, it is not considered a skull bone.) The bones of the skull consist of one **ethmoid** bone, one **frontal** bone, two **temporal** bones, two **parietal** bones, one **sphenoid** bone (figure 7.1), and one **occipital** bone (figure 7.2). The face bones include two **lacrimal** bones, one **mandible** bone, two **maxilla** bones, two **nasal** bones, one **vomer** bone, two **zygomatic** bones, and two **palatine** bones (figure 7.3). Additionally, there are two **nasal concha** bones with superior, middle, and inferior portions that form the lateral wall of the nasal cavity.

The bony markings of the skull are listed here. Most are named based on their purpose. Many of these markings are very small and are typically of greater concern in advanced studies of human anatomy. However, this listing is presented to illustrate the functions of common

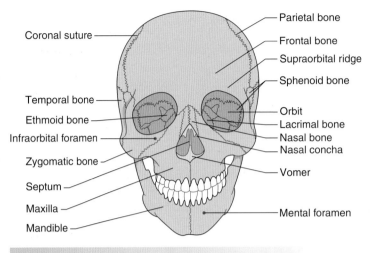

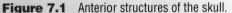

Figure 7.1 Anterior structures of the skull.

Adapted by permission from Watkins 2010.

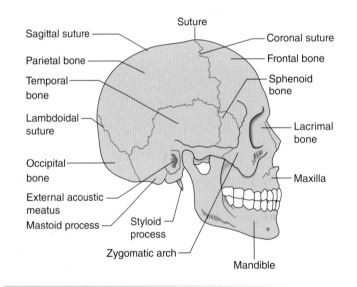

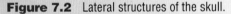

Figure 7.2 Lateral structures of the skull.

Adapted by permission from Cartwright and Pitney 2011.

bony markings of the skull and to facilitate the use of the part III summary table, where bony markings are noted as the origins and insertions of the 84 muscles of the head.

Alveolus: a socket that holds a tooth

Anterior cranial fossa: area that supports the brain's frontal lobe

Body of the mandible: source of attachments for the buccinator, digastric, triangularis, mentalis, platysma, mylohyoid, quadratus labii inferior, geniohyoid, and genioglossus muscles

Canine fossa: source of muscular attachments for the caninus

Carotid canal: temporal bone structure for the internal carotid artery

Cerebellar fossa: surface for the cerebellum

Condyloid fossa: area for a vein from the transverse sinus

Coronoid process: source of the temporalis muscle attachment

Cribriform: plate area for the olfactory nerve between the nasal cavity and the olfactory bulb (which is responsible for the sense of smell)

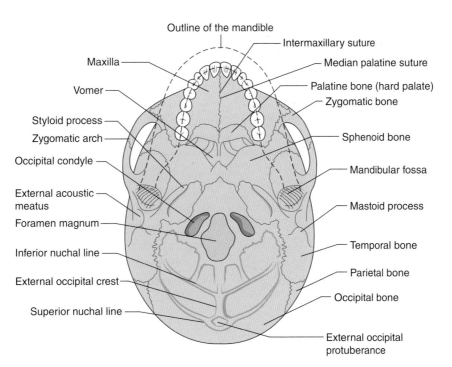

Figure 7.3 Inferior structures of the skull.

Adapted by permission from Watkins 2010.

Crista galli: source of attachment for the falx cerebri

External acoustic meatus: the canal that conducts sound through the temporal bone to the tympanic membrane

External occipital protuberance: source of attachment for the trapezius muscles and the ligamentum nuchae

Foramen lacerum: passageway for the superficial petrossal nerve and the internal carotid artery

Foramen magnum: the large opening in the occipital bone where the medulla oblongata, the occipitoaxial ligaments, the vertebral arteries, the anterior and posterior spinal nerves, and the accessory nerves pass through

Foramen ovale: the opening in the sphenoid bone through which the mandibular and lesser petrossal nerves, accessory meningeal artery, and several small veins and lymph vessels pass

Foramen rotundum: passageway for the maxillary nerve

Foramen spinosum: passageway for the middle meningeal blood vessels and a branch of the mandibular nerve

Hard palate (roof of the mouth): formed by the horizontal portion of the palatine bones and the palatine processes of the maxillary bones

Hypoglossal canal: area on the occipital bone where a meningeal artery and the hypoglossal nerve leave the skull

Hypophyseal fossa: the area where the pituitary gland is located

Incisive canal: passageway for the nasopalatine nerves and the palatine artery

Inferior nuchal line: site of attachment for the semispinalis capitis muscle

Inferior orbital fissure: passageway for the infraorbital blood vessels, the zygomatic branch of the maxillary nerve, and branches of the sphenopalatine ganglion

Inferior tympanic canaliculus: area for passage of the tympanic branches of the glossopharyngeal nerves

Infraorbital foramen: opening in the maxillary bone for the infraorbital branch of the maxillary nerve and the infraorbital branch of the internal maxillary artery

Internal occipital crest: source of attachment for a portion of the inner layer of the dura mater (one of the coverings of the brain)

Jugular foramen: opening in the temporal bone for passage of the inferior petrosal sinus; the transverse sinus (internal jugular vein); meningeal branches of the pharyngeal and occipital arteries; and the 9th, 10th, and 11th cranial nerves

Jugular process: source of attachment for the lateral atlantooccipital ligament and the rectus capitis lateralis

Lacrimal fossa/groove: location of the lacrimal sac

Mandibular foramen: opening in the mandibular canal through which pass the inferior alveolar artery and nerve

Mandibular fossa: area where the head of the mandible articulates with the temporal bone

Mandibular head: forms the temporomandibular joint (TMJ) with the mandibular fossa on the temporal bone

Mastoid canal: passageway for the auricular branch of the vagus nerve

Mastoid foramen: passageway for the emissary vein and artery

Mastoid process: source of attachment for the splenius capitis and the sternocleidomastoid muscles

Mental foramen: passageway through the mandible for the mental nerve and blood vessels

Mylohyoid groove: area on the mandible for passage of the mylohyoid nerves and artery

Neck of the mandible: area for the attachment of the external pterygoid muscle

Occipital condyle: point of articulation between the skull and the superior articular process of the atlas

Optic foramen: passageway for the optic nerve

Orbit: eye socket housing the eye

Ramus of the mandible: bony attachment for the internal pterygoid and masseter muscles

Styloid process: bony attachment for the stylohyoid, styloglossus, and stylopharyngeus muscles and the stylomandibular ligament

Stylomastoid foramen: hole where the facial nerve and the stylomastoid artery pass through the temporal bone

Superior nuchal line: area of attachment for the trapezius, sternocleidomastoid, occipitalis, and splenius capitis muscles

Superior orbital fissure: passageway for the trochlear, oculomotor, and ophthalmic portions of the trigeminal and abducens nerves

Supraorbital notch: passageway in the frontal bone for the supraorbital branches of both the ophthalmic artery and ophthalmic nerve

Temporal fossa: area of attachment for the temporalis muscle

Zygomatic arch: articulation between the temporal process of the zygomatic bone and the zygomatic process of the temporal bone that provides the source of attachment for the masseter muscle and provides space for the temporalis muscle to attach to the coronoid process

The bones of the ear include the **malleus** (hammer), the **incus** (anvil), and the **stapes** (stirrup) (figure 7.4). The malleus is attached to the tympanic membrane (eardrum) and has a head, neck, handle, and both an anterior and lateral process. The handle and lateral process attach to the tympanic membrane, and the anterior process serves as an attachment site for ligamentous fibers. The malleus head connects to the body of the incus. The incus consists of a body and a long and short crus. The long crus attaches to the head of the stapes. The stapes consists of a head, neck, body, and both an anterior and posterior crus. These three bones are suspended, by ligaments, in the tympanic cavity of the middle ear; they transmit sound from the tympanic membrane to the cochlea of the inner ear.

The ear consists of three parts: the **external ear**, the **middle ear**, and the **internal ear**. The external ear is divided into the **pinna** or **auricle**, the **external meatus** (auditory canal), and the **tympanic membrane**. The middle ear contains the three ear bones, the upper end of the **eustachian tube**, and the round window (**fenestra rotunda**) and oval window (**fenestra ovales**) that separate the middle ear from the inner ear. The three ear bones do not articulate with any bones of the skull. The inner ear (figure 7.5) contains the **cochlea** and the **vestibule** (consisting of the **utriculus**, the **sacculus**, and the **semicircular canals**, which are integral parts of the **equilibrium mechanism**). The following

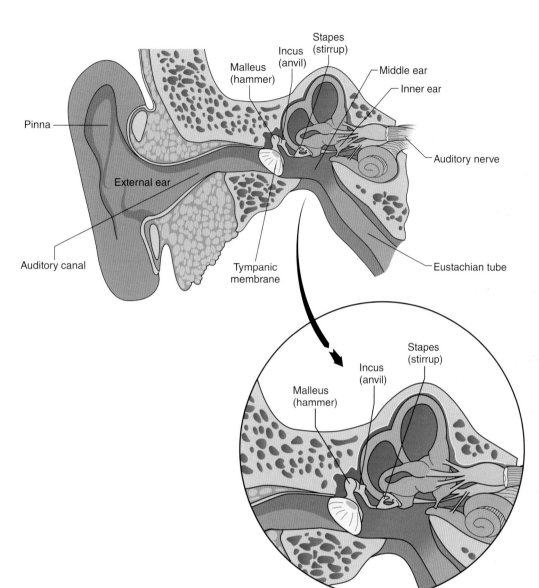

Figure 7.4   Structures and bones of the ear.

Adapted by permission from Cartwright and Pitney 2011.

sequence is the basic explanation of how sound is heard: Sound waves are collected by the outer ear and conducted through the auditory canal to the tympanic membrane. The tympanic membrane transfers the waves to the middle ear structures (the malleus, incus, and stapes), which conduct the waves to the cochlea in the inner ear via the middle ear windows. Cochlear nerves along with nerves from the vestibule combine to form the **vestibulocochlear nerve** (or eighth cranial nerve) and transmit impulses to the medulla oblongata (see chapter 10).

The **hyoid bone**, having no articulations with other skull bones, is suspended under the tongue between the larynx and the mandible. It serves as a point of attachment for a number of muscles of the neck and tongue (figure 7.6).

The mandible forms the lower portion of the jaw and articulates with the temporal bone of the skull (figure 7.7). Initially it consists of two halves before birth, and these two halves are united after birth at an area known as the **mandibular symphysis**.

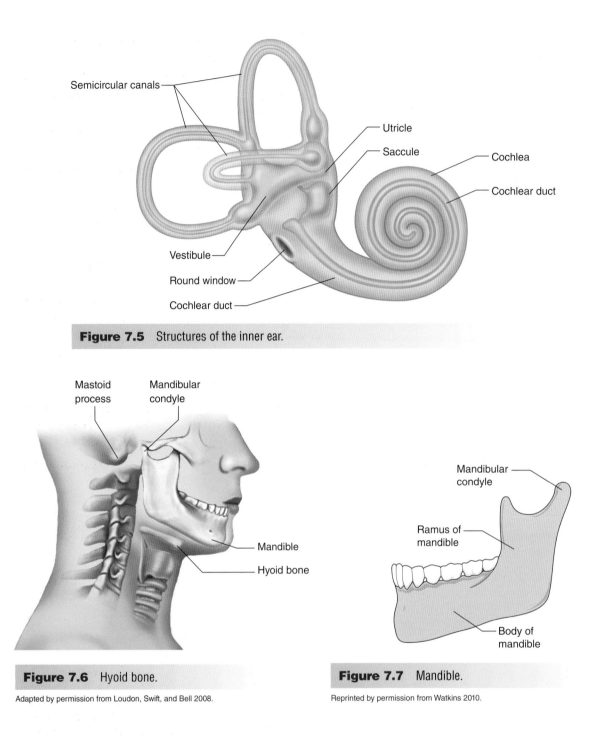

Figure 7.5 Structures of the inner ear.

Figure 7.6 Hyoid bone.

Adapted by permission from Loudon, Swift, and Bell 2008.

Figure 7.7 Mandible.

Reprinted by permission from Watkins 2010.

Joints of the Head

Joints of the skull are classified as **intercranial** (between cranial bones), **interfacial** (between facial bones), and **craniofacial** (between facial and cranial bones). There also are **craniovertebral** joints between the skull and the cervical vertebrae, which are presented in chapter

8. Bones of the skull articulate with each other through immovable joints known as **sutures**. These sutures are easily seen in the skulls of young babies (see the sidebar Focus on Fontanels), but once the bones mature, the sutures (seams) are fused. Most skull sutures are named from the bones they connect. The skull has 34 such sutures, as outlined in the following list. The

suture is listed first, followed by the bones it connects. Items marked with an asterisk (*) indicate a suture that is visible in figure 7.1, 7.2, or 7.3.

Coronal: between the frontal and parietal bones

Ethmoideomaxillary: between the ethmoid and maxillary bones

Frontal: between the frontal bones until about age 6 to 10

Frontoethmoidal: between the frontal and ethmoid bones

Frontomaxillary: between the frontal and maxillary bones

Intermaxillary: between the maxillary bones

Internasal: between the nasal bones

Lacrimoethmoidal: between the lacrimal and ethmoid bones

Lacrimofrontal: between the lacrimal and frontal bones

Lacrimomaxillary: between the lacrimal and maxillary bones

*Lambdoidal: between the occipital and parietal bones

Mastosquamosal: between the mastoid process and the squamous aspect of the temporal bone

Maxillozygomatic: between the maxillary and zygomatic bones

Median palatine: between the palatine process of the maxilla and the palatine bone

Nasofrontal: between the nasal and frontal bones

Nasomaxillary: between the nasal and maxillary bones

Occipitomastoid: between the occipital bone and the mastoid process

Palatoethmoidal: between the orbit of the palatine bone and the ethmoid bone

Palatomaxillary: between the orbit of the palatine bone and the maxillary bone

Parietomastoid: between the parietal bone and the mastoid process

Petrooccipital: between the petrous aspect of the temporal bone and the occipital bone

Petrosquamosal: between the petrous and squamous aspects of the temporal bone

*Sagittal: between the parietal bones

Sphenoethmoidal: between the wing of the sphenoid bone and the ethmoid bone

Sphenofrontal: between the wing of the sphenoid bone and the frontal bone

Sphenoorbital: between the body of the sphenoid bone and the orbital process of the palatine bone

Sphenoparietal: between the wing of the sphenoid bone and the parietal bone

Sphenopetrosal: between the sphenoid bone and the petrous aspect of the temporal bone

Sphenosquamosal: between the wing of the sphenoid bone and the squamous aspect of the temporal bone

Sphenozygomatic: between the wing of the sphenoid bone and the zygomatic bone

Squamosal: between the lower portion of the parietal bone and the squamous aspect of the temporal bone

Temporozygomatic: between the zygomatic process of the temporal bone and the temporal process of the zygomatic bone

Transverse palatine: between the palatine processes of the maxillary bones and the palatine bones

Zygomaticofrontal: between the zygomatic and frontal bones

FOCUS ON

Fontanels

In an infant there are four prominent areas where bones of the skull have not yet grown together. These soft spots between the bones are known as fontanels. The largest fontanel is the anterior (**bregma**), which is the point between the coronal and sagittal sutures. An infant's pulse can be felt and often visibly observed in this area. The posterior (**lambda**) fontanel is the point where the lambdoidal and sagittal sutures meet. The lateral or sphenoidal (**pterion**) fontanel is the point where the greater wing of the sphenoid bone meets the sphenoidal angle of the parietal bone. The mastoid (**asterion**) fontanel is the point where the lambdoidal, occipitomastoid, and parietomastoid sutures meet. Under normal growth conditions these fontanels "close," or the sutures are completely formed, by the age of two.

All of the identified sutures (joints) are considered nonmoving joints, but the skull does contain one movable joint: the **temporomandibular joint** (often referred to as the TMJ). The articulation between the head of the mandible and the **mandibular fossa** of the temporal bone form a diarthrodial joint.

Ligaments of the Head

The mandible joint consists of an **articular disc**, a joint capsule, and a **temporomandibular ligament** (figure 7.8a). Support for the joint also is provided by the **stylomandibular** and **sphenomandibular ligaments** (figure 7.8b). Movement in the joint is a rolling and gliding motion that results in elevation, depression, protrusion, and medial and lateral motion of the mandible.

Sinuses

Within the bones and bony markings there exist several air-filled cavities known as **sinuses** (figure 7.9). These spaces are lined with mucous membranes. All are either directly or indirectly connected to the nasal cavity and are thought to act as resonating cavities for the voice. Inflammation of the mucosa of any of the sinuses is known as sinusitis. There are four major sinuses:

- The **ethmoid sinuses** are located in the ethmoid bone between the eyes and the middle **turbinate** (the bony walls of the nasal cavity form four pairs of projections known as turbinates: supreme, superior, middle, and inferior), and they communicate with the nasal cavity. A paper-thin bone (**lamina papyracea**) separates these sinuses from the eye socket contents.

- The **frontal** sinuses are two triangular-shaped cavities (left and right) located at the base of the frontal bone (forehead); they communicate both directly and indirectly with the nasal cavity.

- Located in the cheeks, the large **maxillary sinuses** are cavities within the maxillary bone,

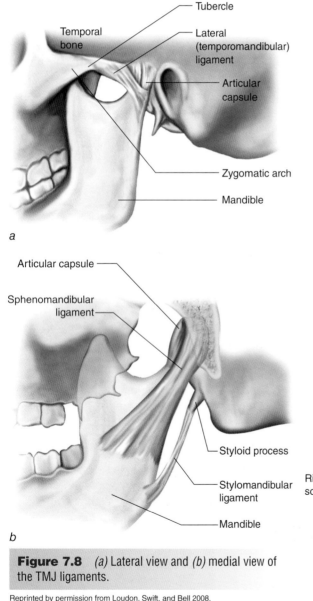

a

b

Figure 7.8 *(a)* Lateral view and *(b)* medial view of the TMJ ligaments.

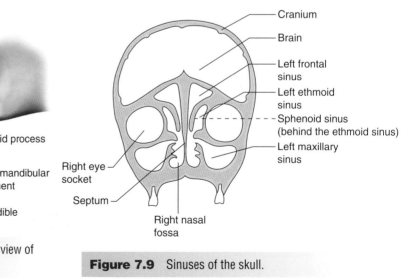

Figure 7.9 Sinuses of the skull.

communicating with the middle meatus of the nasal cavity.

• The **sphenoid sinus** cavities are located in the anterior body of the sphenoid bone behind the ethmoid sinuses, opening into the nasal cavity above the superior nasal concha. The carotid arteries and the optic nerve lie in this area.

Fundamental Movements and Muscles of the Head

You might think that with the temporomandibular joint being the only movable joint of the head, there would be a minimal number of muscles. However, there are numerous muscles specific to movements within the skull as well as many muscles attached to the skull that facilitate movement of the cervical spine (neck). The muscles of the head are divided into six groups: cranial, facial, eyeball, tongue, soft palate, and pharynx (figures 7.10 and 7.11).

• Cranial muscles: The cranial muscles are divided into three groups: (1) **epicranius**, consisting of the **frontalis** and **occipitalis** muscles; (2) external ear, consisting of the anterior, posterior, and superior **auricularis** muscles; and (3) middle ear, consisting of the **stapedius** and **tensor tympani** muscles.

• Facial muscles: The facial muscles (muscles of expression) are divided into four groups: (1) eyelid muscles, consisting of the **corrugator supercilii**, **levator palpebrae superioris**, and **orbicularis oculi** muscles; (2) nose muscles, consisting of the **caput angulare**, **depressor alaeque nasalis**, **dilator naris**, **nasalis**, and **procerus** muscles; (3) mouth muscles, consisting of the **buccinator**, **caninus**, **mentalis**, **orbicularis oris**, **quadratus labii** (**inferior** and **superior**), caput angulare, **caput infraorbitale**, **caput zygomaticum**, **risorius**, **triangularis**, and **zygomaticus** muscles; and (4) mastication muscles, consisting of the **pterygoid** (**internal** and **external**), **masseter**, and **temporalis** muscles.

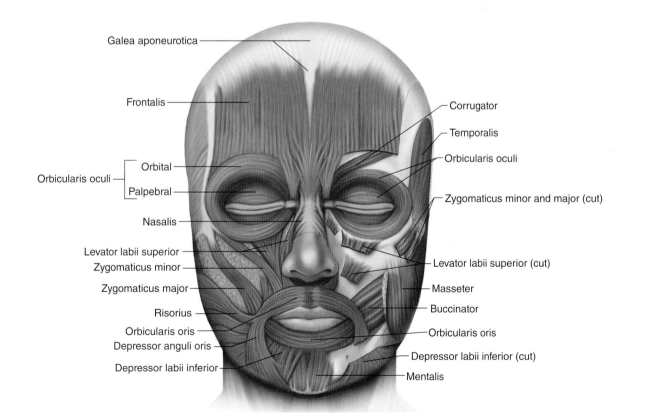

Galea aponeurotica

Frontalis

Orbicularis oculi — Orbital
Palpebral

Nasalis

Levator labii superior

Zygomaticus minor

Zygomaticus major

Risorius

Orbicularis oris

Depressor anguli oris

Depressor labii inferior

Corrugator

Temporalis

Orbicularis oculi

Zygomaticus minor and major (cut)

Levator labii superior (cut)

Masseter

Buccinator

Orbicularis oris

Depressor labii inferior (cut)

Mentalis

Figure 7.10 Superficial and deep muscles of the head.

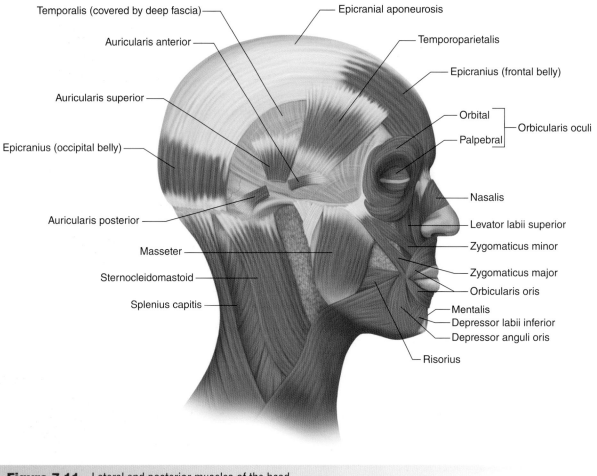

Figure 7.11 Lateral and posterior muscles of the head.

Hands On

Either facing a partner or using a mirror, perform the following activities, considering which of the facial muscles are used: (1) Raise your eyebrows, (2) wrinkle your nose, and (3) form a kiss with your lips. Place your fingers over the area of the TMJ, open and close your mouth, and feel the muscles of mastication contracting. What facial muscles contribute to this action?

• Eyeball muscles: A cross-sectional view of the eye shows the following structures (figure 7.12). The **cornea** is the transparent layer just under the eyelid, and the **iris** is an elastic membrane located between the cornea and the **lens**. The iris is the portion of the eye that is colored (e.g., blue, green, brown); it has an opening in its center known as the **pupil**. The space between the iris and the cornea is known as the **anterior chamber**, and the space between the iris and the lens is known as the **posterior chamber**. Both chambers are filled with a fluid known as **aqueous humor**. A **suspensory ligament** suspends the lens in the eyeball. Posterior to the lens lies a three-layered wall consisting of the outer layer (**sclera**), an inner layer (**retina**), and a layer between these two (**choroid**). The area between the lens and the retina is filled with a fluid known as the **vitreous humor**. The muscles of the eyeball are divided into two groups: (1) extrinsic, consisting of the **obliquus** (**superior** and **inferior**), **orbitalis**, and four **recti** muscles (**superior**, **medialis**, **lateralis**, **inferior**); and (2) intrinsic, consisting of the **ciliary** and two muscles of the iris, the **dilator pupillae** and the **sphincter pupillae**.

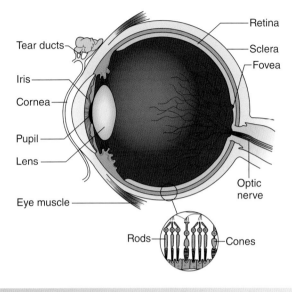

Figure 7.12   Muscles and structures of the eyeball.

Reprinted by permission from Cartwright and Pitney 2011.

👋 Hands On

Either facing a partner or using a mirror, perform the following activities, considering which of the eyeball muscles are used: (1) Look upward, (2) look downward, and (3) look side to side. Cover one eye with your hand, and then quickly uncover it and observe the pupil. Did it increase or decrease in size?

• Tongue muscles: The muscles of the tongue (figure 7.13) are divided into two groups: (1) extrinsic, consisting of the **chondroglossus**, **genioglossus**, **hyoglossus**, and **styloglossus** muscles; and (2) intrinsic, consisting of the **longitudinalis** (**superior** and **inferior**) muscles, **transversus linguae**, and **verticalis linguae**.

👋 Hands on

Either facing a partner or using a mirror, perform the following activities, considering which of the tongue muscles are used: (1) Stick out your tongue, and (2) wiggle your tongue left and right.

• Soft palate muscles: The muscles of the soft palate consist of the **glossopalatinus**, **uvulae**, **levator veli palatini**, **tensor veli palatini**, **pharyngopalatinus**, and **salpingopharyngeus** muscles.

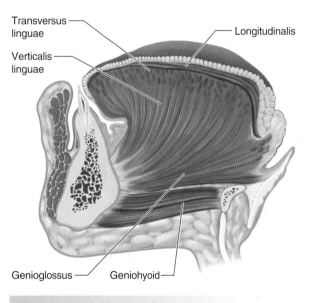

Figure 7.13   Muscles of the tongue.

• Pharynx and larynx muscles: The muscles of the pharynx consist of the **stylopharyngeus** and the **constrictor pharyngis** muscles (**superior**, **inferior**, and **medius**). While typically considered muscles of the neck, the muscles of the larynx are included in this grouping of muscles of the head since their actions relate only to the larynx and not to movement of the cervical spine. The muscles of the larynx are divided into two groups: extrinsic and intrinsic. The extrinsic muscles are considered muscles of the hyoid and are also divided into two groups: (1) the **infrahyoid** group, consisting of the **levator glandulae thyroideae, sternohyoid, sternothyroid, thyrohyoid,** and **omohyoid (anterior** and **posterior bellies);** and (2) the **suprahyoid** group, consisting of the **geniohyoid, mylohyoid, stylohyoid,** and **digastric (anterior** and **posterior bellies)** (figure 7.14). The intrinsic muscles of the larynx are the **aryepiglotticus, arytenoid oblique** and **transverse, cricoarytenoid posterior** and **lateralis, cricothyroid, thyroarytenoid, thyroepiglotticus, ventricularis,** and **vocal muscle.**

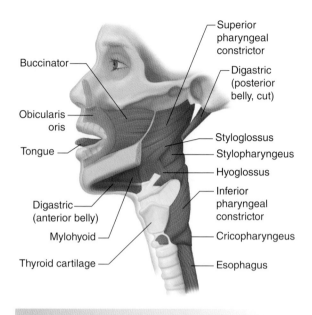

Figure 7.14 Superficial muscles of the pharynx and larynx.

LEARNING AIDS

REVIEW OF TERMINOLOGY

The following terms are discussed in this chapter. Define or describe each term, and where appropriate, identify the location of the named structure either on your body or in an appropriate illustration.

anterior chamber	cricothyroid	lacrimal bone
articular disc	depressor alaeque nasalis	lambda
aqueous humor	digastric (anterior and posterior)	lamina papyracea
aryepiglotticus	dilator naris	lens
arytenoid oblique	dilator pupillae	levator glandulae thyroideae
arytenoid transverse	epicranius	levator veli palatini
asterion	equilibrium mechanism	levator palpebrae superioris
auricle	ethmoid bone	longitudinalis (superior and inferior)
auricularis	ethmoid sinus	malleus
bregma	Eustachian tube	mandible
buccinator	external ear	mandibular fossa
caninus	external meatus	mandibular symphysis
caput angulare	fenestra ovales	masseter
caput infraorbitale	fenestra rotunda	maxilla
caput zygomaticum	frontal bone	maxillary sinus
chondroglossus	frontal sinus	mentalis
choroid	frontalis	middle ear
ciliary	genioglossus	mylohyoid
cochlea	geniohyoid	nasal bone
constrictor pharyngis (superior, inferior, and medius)	glossopalatinus	nasal concha bone
	hyoglossus	nasalis
cornea	hyoid bone	obliquus (superior and inferior)
corrugator supercilii	incus	occipital bone
craniofacial joint	infrahyoid	occipitalis
craniovertebral joint	intercranial joint	omohyoid (anterior and posterior)
cranium	interfacial joint	orbicularis oculi
cricoarytenoid lateralis	internal ear	orbicularis oris
cricoarytenoid posterior	iris	orbitalis

palatine bone	sphenoid bone	tensor veli palatini
parietal bone	sphenoid sinus	thyroarytenoid
pharyngopalatinus	sphenomandibular ligament	thyroepiglotticus
pinna	sphincter pupillae	thyrohyoid
posterior chamber	stapedius	transversus linguae
procerus	stapes	triangularis
pterion	sternohyoid	turbinate
pterygoid (internal and external)	sternothyroid	tympanic membrane
pupil	styloglossus	utriculus
quadratus labii (inferior and	stylohyoid	uvulae
superior)	stylomandibular ligament	ventricularis
recti (superior, inferior, lateralis, and	stylopharyngeus	verticalis linguae
medialis)	suprahyoid	vestibule
retina	suspensory ligament	vestibulocochlear nerve
risorius	suture joint	vitreous humor
sacculus	temporal bone	vocal muscle
salpingopharyngeus	temporalis	vomer
sclera	temporomandibular joint (TMJ)	zygomatic bone
semicircular canal	temporomandibular ligament	zygomaticus
sinus	tensor tympani	

SUGGESTED LEARNING ACTIVITIES

1. With a partner, identify and palpate as many of the external bony markings from pages 130-132 as possible.

2. While your partner opens and closes his mouth, palpate the area of the TMJ, and identify the muscles you feel and what specific TMJ movements are being performed.

3. Ask your partner to follow your index finger as you raise it, lower it, and move it from side to side. Closely observe your partner's eye movements, and try to identify what muscles she is using to perform the movements.

MULTIPLE-CHOICE QUESTIONS

1. The largest fontanel in an infant's skull is the
   a. anterior fontanel
   b. lateral fontanel
   c. posterior fontanel
   d. medial fontanel

2. Joints between the bones of the face and the skull are known as _____ joints.
   a. intercranial
   b. interfacial
   c. craniofacial
   d. craniovertebral

3. Which of the following ligamentous structures is *not* involved with the temporomandibular joint?

   a. temporomandibular ligament
   b. articular disc
   c. zygomaticomandibular ligament
   d. joint capsule

4. Which of the following cranial muscles is *not* considered a muscle of the external ear?
   a. superior auricularis
   b. inferior auricularis
   c. anterior auricularis
   d. posterior auricularis

5. The opening in the center of the iris is known as the
   a. retina
   b. pupil
   c. lens
   d. cornea

(continued)

MULTIPLE-CHOICE QUESTIONS *(continued)*

6. A very prominent source of attachment for both the trapezius muscle and the ligamentum nuchae is the
    a. foramen magnum
    b. mastoid process
    c. external occipital protuberance
    d. styloid process

7. Which of the following muscles is a muscle of mastication?
    a. frontalis
    b. uvulae

    c. mentalis
    d. masseter

8. Which of the following bones connects to the tympanic membrane?
    a. incus
    b. stapes
    c. middle meatus
    d. malleus

FILL-IN-THE-BLANK QUESTIONS

1. The malleus, incus, and stapes are located in the _____ ear.

2. The joints (sutures) between the bones of the skull are classified as _____ joints except for the joint between the mandible and the temporal bone.

3. The _____ sinus is located in the cheekbones of the face.

4. The superior and inferior obliquus muscles are considered _____ muscles of the eyeball.

5. The _____ bone, found in the head, has no attachments to any other skull bones and thus is *not* considered a bone of the skull.

6. A cavity within a bone, typically containing air and lined with a mucous membrane, is known as a _____.

7. The anatomical opening in the occipital bone where the medulla oblongata and spinal nerves pass through is known as the _____.

8. The fluid found in the eye lens and retina is known as the _____.

The Spinal Column and Pelvis

The spinal column (figure 8.1) is a stack of 33 bones called vertebrae held together by ligaments and muscles, with cartilaginous discs (primarily water and protein) between the bones. The 33 vertebrae are divided into five distinct sections. All the vertebrae have many common characteristics, but each group has unique features designed for specific purposes. The most superior group is known as the cervical (neck) spine and contains 7 vertebrae. The next group is known as the thoracic (chest) spine and contains 12 vertebrae. The next group is known as the lumbar (low back) spine and contains 5 vertebrae. The next group is known as the sacral spine and contains 5 vertebrae fused together into one structure known as the sacrum. The last, or most distal, group is known as the coccygeal spine and contains 4 vertebrae fused together into one structure known as the coccyx.

The lateral view of the spinal column reveals four curvatures: anterior (convex) curves in the cervical and lumbar spine and posterior (concave) curves in the thoracic and sacrococcygeal spine. These curvatures may increase or decrease as the body's center of gravity shifts (e.g., with pregnancy, weight gain, weight loss, or trauma). This is a result of one of the spinal column's functions: to maintain, in the upright body position, the brain over the body's center of gravity. Over- or underdevelopment of the musculature on either side of the spinal column, structural deformities, or other causes can result in excessive curvatures of the spinal column. Three of the more common conditions resulting from excessive curvature are **kyphosis** (figure 8.2), excessive posterior curvature of the thoracic spine (hunchback, round shoulders); **lordosis** (figure 8.3), excessive anterior curvature of the lumbar spine (swayback); and **scoliosis** (figure 8.4), excessive lateral curvature of the spinal column, usually in the thoracic spine but sometimes to a lesser extent in the cervical and lumbar spine.

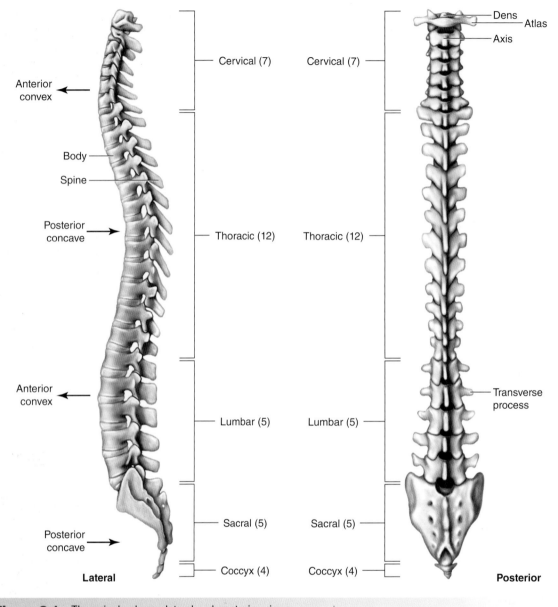

Figure 8.1 The spinal column, lateral and posterior views.

Bones of the Spinal Column

Although all five sections of the spinal column have common structures, the vertebrae of each are spaced uniquely and may be shaped slightly differently, depending on their functions in a particular anatomical area. All vertebrae have a body, two **transverse processes** laterally (serving as sources for ligamentous and muscular attachments), a **spinous process** (serving as another source for ligamentous and muscular attachments), and a **vertebral foramen** (where the spinal cord and nerve roots pass) (figure 8.5). Additionally, each vertebra has the following features. **Superior and inferior articulating facets** are where a vertebra articulates with the vertebrae above and below it. A **lamina** forms the posterior aspect of the vertebral foramen. **Pedicles** form the lateral sides of the vertebral foramen. The **intervertebral foramen** between the vertebrae allows the nerve branches from the spinal cord to pass through. The **isthmus** (also called the **pars interarticularis** or

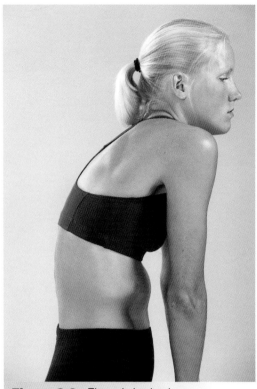

Figure 8.2   Thoracic kyphosis.

Figure 8.3   Lumbar lordosis.

neck) is the bony area between the superior and inferior articulating facets.

The seven **cervical vertebrae** are numbered from the most superior to most inferior as C1, C2, and so on. The first cervical vertebra, the **atlas** (C1), and the second cervical vertebra, the **axis** (C2), are shaped differently from the other five cervical vertebrae (C3–C7) to permit the head to rotate.

The atlas has no significant body but has two large articular facets that provide the surface where the skull and the spinal column articulate (figure 8.6). This is a very firm articulation between the skull and the first cervical vertebrae creating a situation where, if the skull moves, the atlas goes with it. The atlas slides over the axis and rests on top of the two large superior articulating surfaces of the axis, between which lies a rather large bony process from the axis body known as the **dens**, or **odontoid process** (figure 8.6). It is the joint between the atlas (C1), attached to the skull, and the axis (C2) where rotation of the head occurs. Note also that the cervical vertebrae have a **bifid** (split) spinous

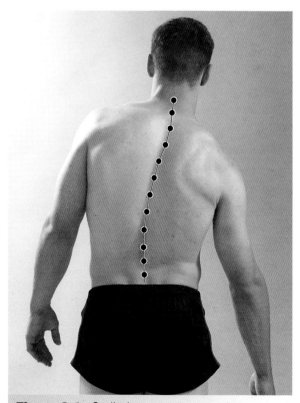

Figure 8.4   Scoliosis.

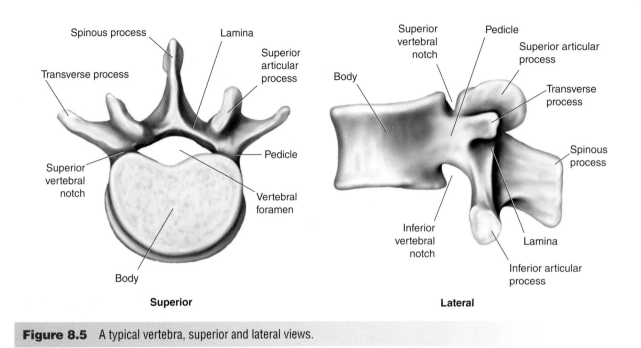

Superior **Lateral**

Figure 8.5 A typical vertebra, superior and lateral views.

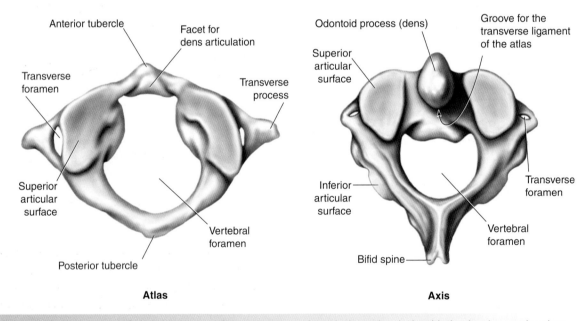

Atlas **Axis**

Figure 8.6 The first cervical vertebra (atlas) and the second cervical vertebra (axis with the dens), superior view.

process and a foramen in each transverse process to provide for the passage of blood vessels through the cervical spine (figure 8.7). These two features are unique to the cervical vertebrae. One other difference to note is the rather long and prominent spinous process of the seventh cervical vertebra (C7). This prominence is easily palpated.

 **Hands on**

Run your finger down the posterior aspect of your cervical spine until you find a large protrusion. This is the C7 spinous process, which serves as an anatomical landmark for determining the spinous processes of the other cervical vertebrae and determining where the thoracic spine begins.

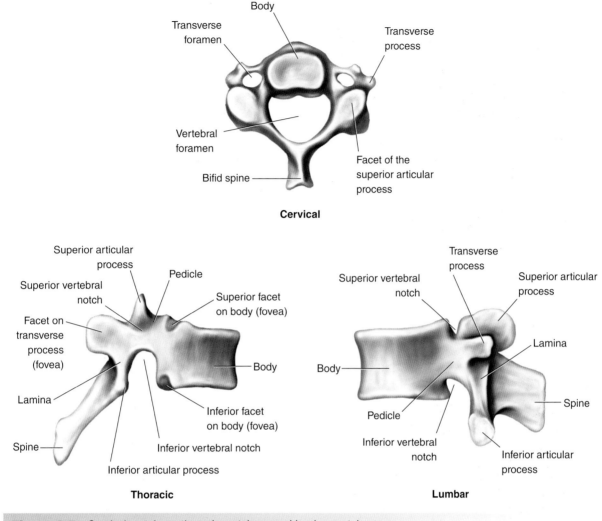

Body

Transverse foramen

Transverse process

Vertebral foramen

Facet of the superior articular process

Bifid spine

Cervical

Superior articular process

Pedicle

Superior vertebral notch

Superior facet on body (fovea)

Facet on transverse process (fovea)

Body

Lamina

Inferior facet on body (fovea)

Spine

Inferior vertebral notch

Inferior articular process

Thoracic

Transverse process

Superior vertebral notch

Superior articular process

Lamina

Body

Spine

Pedicle

Inferior vertebral notch

Inferior articular process

Lumbar

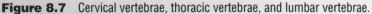

Figure 8.7 Cervical vertebrae, thoracic vertebrae, and lumbar vertebrae.

The 12 **thoracic vertebrae** have bony features similar to all other vertebrae, with a few unique features. Note the longer and more vertical spinous processes (figure 8.7). Also note the articulating surfaces (**fovea**) on the anterior lateral aspects of the transverse processes and on the superior and inferior portions of the posterior lateral aspects of the vertebral bodies. These notches provide the articulation site of the 12 pairs of ribs with the 12 thoracic vertebrae.

The five **lumbar vertebrae** are the largest vertebrae (figure 8.7). They have no foramen through their transverse processes, nor are there any articular facets (fovea) on their bodies. Although separate at birth, the five **sacral vertebrae** (S1–S5) fuse together to form a large triangular-shaped bone known as the **sacrum** (figures 8.8 and 8.9) during the growth process. The two large articular surfaces formed on the lateral aspects of the sacrum are where the spinal column articulates with the bones of the pelvis, forming the pelvic girdle.

The **coccyx** (the final four vertebrae), like the sacrum, is originally four vertebrae that, during the growth process, fuse to form one structure (figure 8.10). It serves as a source of attachment for ligamentous and muscular structures. Some people refer to the coccyx as humans' "vestigial tail," in reference to the evolution of humans from species that had tails.

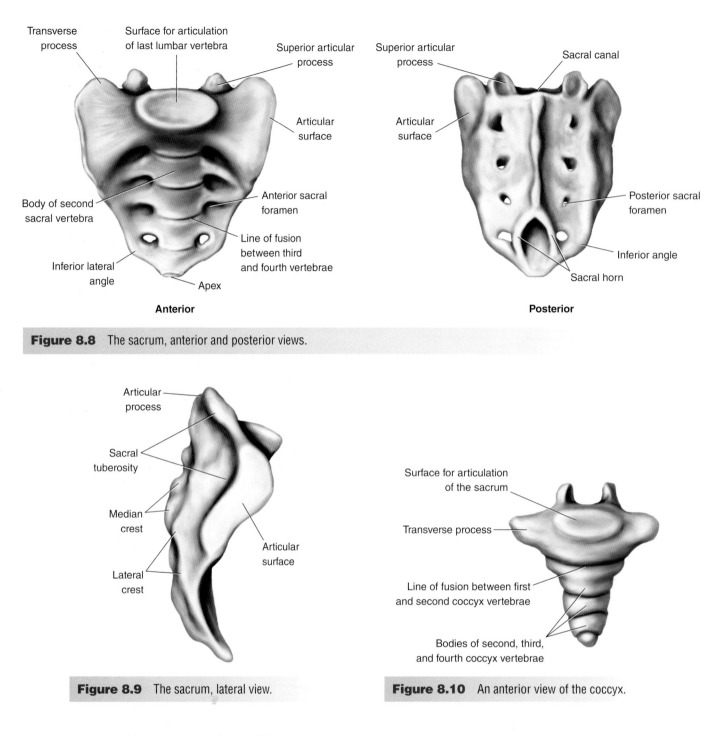

Figure 8.8 The sacrum, anterior and posterior views.

Figure 8.9 The sacrum, lateral view.

Figure 8.10 An anterior view of the coccyx.

Ligaments of the Spinal Column

There is only one movable joint in the skull, the temporomandibular joint, which enables the opening and closing of the mouth; any movement of the head is the result of movement at the joints between the occipital bone of the skull and the first and second cervical vertebrae. There are approximately 25 ligaments between the skull and the cervical spine (figures 8.11 and 8.12). With the exception of one ligament discussed later, we will not devote time to the individual ligaments but simply note the groups they belong to. **Atlantooccipital ligaments** attach the occipital bone of the skull to the first

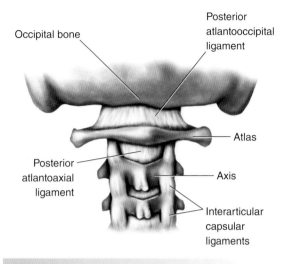

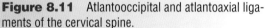

Figure 8.11 Atlantooccipital and atlantoaxial ligaments of the cervical spine.

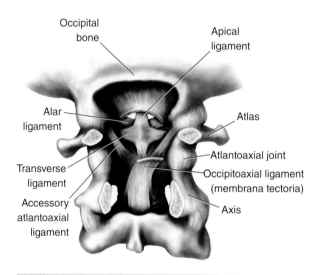

Figure 8.12 Occipitoaxial and atlantoaxial ligaments of the cervical spine.

cervical vertebra (atlas, or C1). **Occipitoaxial ligaments** (also known as the **membrana tectoria)** attach the occipital bone of the skull to the dens (odontoid process) of the second cervical vertebra (axis, or C2), and **atlantoaxial ligaments** attach the atlas and the axis. A final group of ligaments is not involved in movement of the spinal column: the costovertebral ligaments (six per rib) articulate the 12 pairs of ribs of the thorax with the 12 thoracic vertebrae.

One of the atlantoaxial ligaments, the **transverse ligament** (figure 8.13), runs from one

transverse process of the atlas (C1), across the vertebral foramen, to the other transverse process and holds the dens (odontoid process) of the axis (C2) in place. Failure of this ligament to restrain movement of the dens of C2 beneath C1 could cause the dens to do irreparable damage to structures in the vertebral foramen at the C1–C2 level of the spinal column. Many authors nickname the transverse ligament the "hangman's" ligament, because the skull and C1 of a person being hanged are restricted from downward movement by a hangman's noose, while the rest of the body from C2 and below moves downward through the pull of gravity. The dens of C2 tears the transverse ligament of C1 and compresses the spinal cord against C1, disrupting nerves to the heart and lungs.

From the most superior to the most inferior aspects of the spinal column are a number of ligaments that play a major role in the movement of the joints between the vertebrae (figure 8.14). Two ligaments known as the **interbody ligaments** run the entire length of the spinal column. The **anterior longitudinal ligament** runs along the anterior aspect of the bodies of all 33 vertebrae. This ligament is structurally the weakest of all the spinal column ligaments. The **posterior longitudinal ligament** runs along the posterior aspect of the bodies of all 33 vertebrae. The posterior longitudinal ligament forms the anterior wall of the spinal canal.

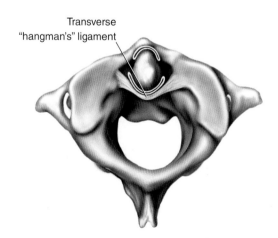

Figure 8.13 The transverse ligament ("hangman's" ligament) of the atlas (C1).

The **ligamentum flavum** runs between the laminae of successive vertebrae (figure 8.14). The **interspinous ligament** runs between the spinous processes of successive vertebrae. Running between the dorsal tips of each vertebra's spinous process, from the coccyx to the external occipital protuberance of the occipital bone, is the **supraspinous ligament**. Between the external occipital protuberance and the spinous process of the seventh cervical vertebra, the supraspinous ligament is known as the **ligamentum nuchae** (figure 8.15).

One additional ligament of the spinal column is the **iliolumbar ligament**, which runs between the transverse processes of the fifth lumbar vertebra to the ilium of the pelvis (figure 8.16). The joints and ligaments between the spinal column and the pelvis are presented later in this chapter.

One other structure of the spinal column is the **intervertebral disc** (figure 8.17). These cartilaginous (primarily water and protein) discs lie on the bodies of each vertebra and serve both as spacers (to help separate the vertebrae and allow nerve roots to pass from the spinal canal to other structures of the body) and as shock absorbers for the spinal column. The disc has two distinct portions: The inner portion is known as the **nucleus pulposus**, and the outer portion is known as the **annulus fibrosus**. The annulus fibrosus

consists of fibrous tissue, whereas the nucleus pulposus consists of soft, pulpy, elastic tissue.

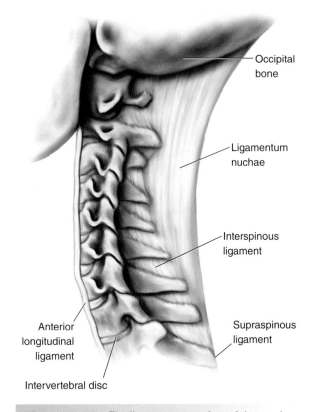

Figure 8.15 The ligamentum nuchae of the cervical spine.

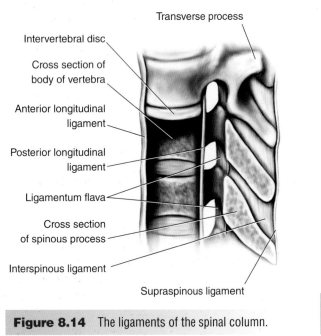

Figure 8.14 The ligaments of the spinal column.

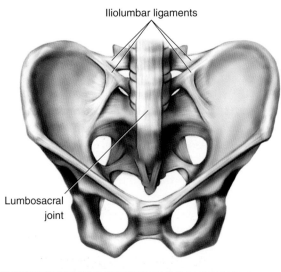

Figure 8.16 The lumbosacral joint and iliolumbar ligament, anterior view.

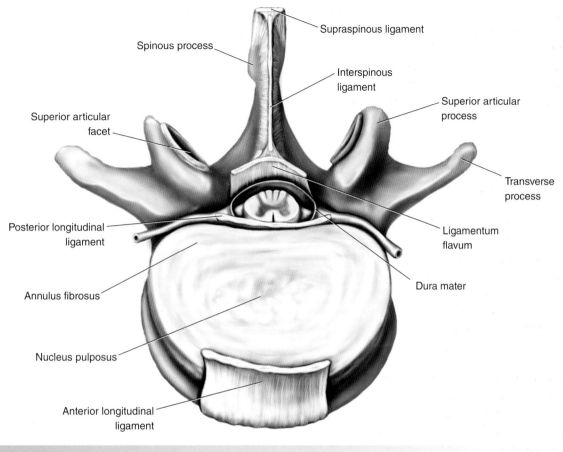

Figure 8.17 The intervertebral disc.

FOCUS ON

Intervertebral Disc

The terms *slipped disc* and *herniated disc* are frequently heard in reference to someone having back pain. If other complications are ruled out (e.g., muscle strain, ligamentous sprain), a slipped disc means the intervertebral disc has protruded into the space occupied by either the spinal cord or its nerve roots, compressing the nerve and causing pain and disability. A more technical term would be *protruded disc*. The term *herniated disc* refers to the actual tearing of the disc structure that allows a portion of the disc material to escape and cause pressure on the spinal cord or its nerve roots. Both the protruded disc and herniated disc can occur as a result of excessive motion of the spinal column (hyperflexion, hyperextension, lateral flexion, or rotation).

A herniated disc in the lumbar spine can be detected from a spinal tap, or lumbar puncture, a procedure in which fluid is removed from the space beneath the arachnoid membrane (which surrounds the spinal cord) of the lumbar region of the spinal cord. Elevated levels of a telltale protein indicate a herniated disc. Disc problems need to be treated by neurosurgeons and orthopedic surgeons.

Fundamental Movements and Muscles of the Spinal Column

Movements of the joints between the vertebrae of the spinal column occur in all three planes about all three axes. The fundamental movements of the spinal column (the cumulative actions of all the joints among the 33 vertebrae) are flexion, extension, lateral flexion (as opposed to abduction and adduction in the extremities), and rotation (figures 8.18 and 8.19). The greatest amount of movement of the

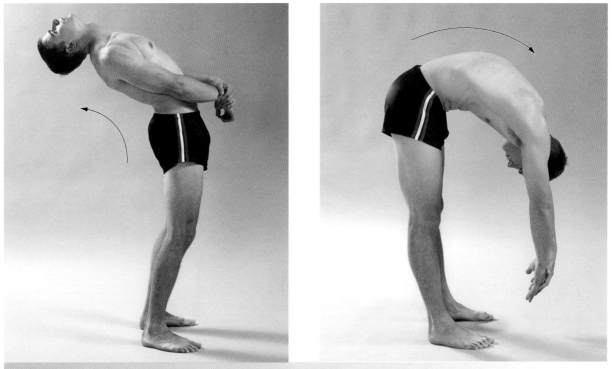

Figure 8.18 Extension and flexion of the spine.

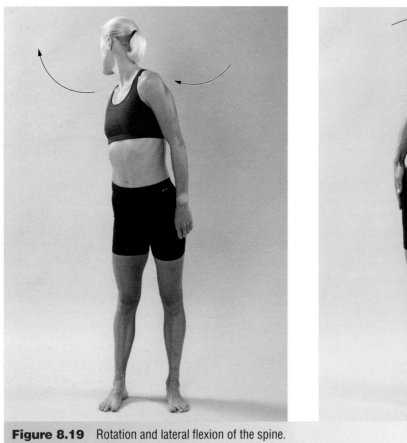

Figure 8.19 Rotation and lateral flexion of the spine.

spinal column takes place in the cervical and lumbar regions. Movement is more restricted in the thoracic region because of the attached ribs. There is no movement in the sacral and coccygeal regions because the vertebrae in these regions are fused together into the five-bone sacrum and the four-bone coccyx.

The muscles responsible for the movements of the spinal column are, for the most part, either anterior or posterior to the spinal column. Those anterior to the spinal column are muscles that flex, laterally flex, or rotate the spine. Those posterior to the spinal column are muscles that extend, laterally flex, or rotate the spine. Some spinal column muscles are specific to a particular area of the spinal column (cervical, lumbar), whereas others have branches in multiple areas of the spinal column.

In the cervical region, the major anterior muscles and muscle groups are the **sterno-cleidomastoid**, the **prevertebrals** (**rectus capitis anterior**, **rectus capitis lateralis**, **longus capitis**, **longus colli**), and the sca-leni (**scalenus anterior**, **scalenus medius**, **scalenus posterior**) (figures 8.20 and 8.21). The sternocleidomastoid muscle, as its name indicates, attaches to the sternum, the clavicle, and the mastoid process of the skull just posterior to the ear.

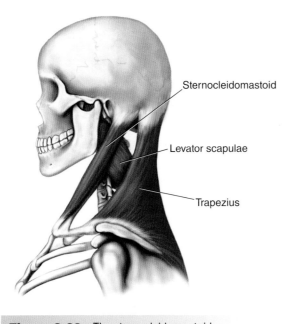

Figure 8.20 The sternocleidomastoid.

Sternocleidomastoid

Levator scapulae

Trapezius

✋ Hands On

Referring to figure 8.22, palpate the sternal and cla-vicular portions of the sternocleidomastoid muscle. Using simple reasoning, and considering there is both a right and left sternocleidomastoid muscle, answer the following question: If this muscle lies anterior and slightly lateral to the cervical spine, what actions is it capable of performing? (Answer: flexion, lateral flexion, and rotation of the cervical spine)

The recti muscles (rectus capitis anterior, rectus capitis lateralis), the longus capitis, and the longus colli flex and rotate the cervical spine. The three scalene muscles (scalenus anterior, scalenus medius, scalenus posterior) run from the cervical vertebrae to the first three ribs. The scaleni not only laterally flex the cervical spine but, during forced respiration, also elevate the first three ribs to allow the lungs to expand.

✋ Hands On

Isolating all three of the scalene muscles is difficult, but you should be able to locate the group by having your partner laterally flex her neck against your resistance (figure 8.23).

Restricted to the posterior aspect of the cervical spine are the rectus (**rectus capitis posterior major** and **minor**), obliquus (**obliquus capitis superior** and **inferior**), and splenius (**splenius capitis** and **cervicis**) muscle groups (see figure 8.21). These muscles are involved in the extension, lateral flexion, and rotation of the cervical spine.

Muscles of the shoulder girdle (presented in chapter 3) also are involved in cervical spine movement. The **levator scapulae**, the **trapezius**, and the most superior portion of the **rhomboids** all have attachments to the cervical spine and assist with the extension, lateral flexion, and rotation of the cervical spine (figure 8.24). Other muscles also are involved in cervical spine movement, but they are not specific to the cervical spine and are presented later in this chapter.

The muscles anterior to the thoracic and lumbar regions of the spinal column are often classified as the abdominal muscles: the **rectus abdominis**, the **internal and obliques**, and the **transversus abdominis** (figure 8.25).

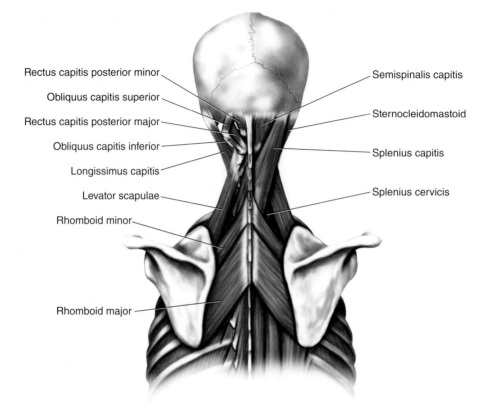

Posterior

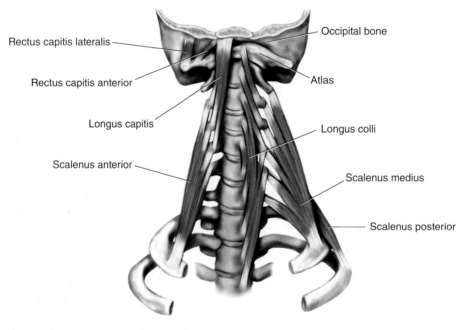

Anterior

Figure 8.21 The posterior and anterior muscles of the cervical spine.

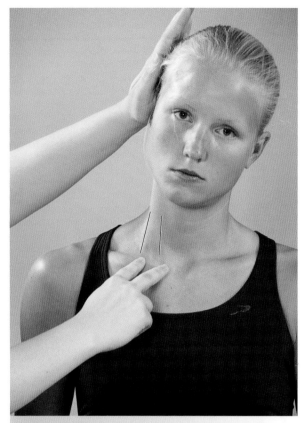

Figure 8.22 Locating the sternal and clavicular parts of the sternocleidomastoid.

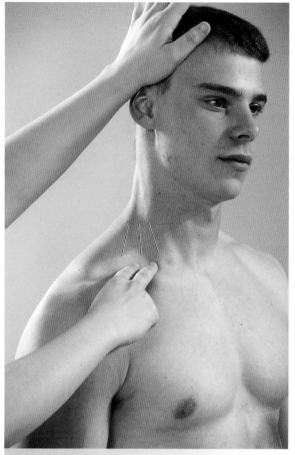

Figure 8.23 Identifying the scalenes.

✋ Hands On

Having your partner perform and hold an abdominal curl should reveal both the rectus abdominis and external oblique muscles (figures 8.26 and 8.27).

The rectus abdominis originates on the pubic symphysis (see later in this chapter) (figure 8.25) and inserts on the fifth, sixth, and seventh ribs and the xiphoid process of the sternum (see chapter 9). It flexes the lumbar and thoracic spine. The external oblique runs between the inferior edges of the last eight ribs and the outer edge of the middle half of the crest of the ilium. It flexes and contralaterally (to the opposite side) rotates the lumbar and thoracic spine. The internal oblique runs between the outer edge of the middle two-thirds of the crest of the ilium and the seventh, eighth, and ninth ribs. It flexes and ipsilaterally (to the same side) rotates the lumbar and thoracic spine. The transversus abdominis runs between the lower six ribs and the middle half of the internal edge of the crest of the ilium,

the **linea alba** (formed by overlapping fascia of the abdominal muscles), and the pubic bone. Its primary function involving the spinal column is to assist with ipsilateral rotation of the thoracic and lumbar spine.

The posterior muscles of the spinal column are actually groups of muscles that cover two or more areas of the spinal column (figure 8.28). The spinalis group contains the **spinalis dorsi** (also called spinalis thoracis), the **spinalis cervicis**, and the **spinalis capitis**. The spinalis group of muscles originates from the second lumbar vertebra through the seventh cervical vertebra and inserts on the thoracic and cervical vertebrae and the occipital bone of the skull. Because their position is so closely aligned with the spinal column, their function is primarily extension. Another group of posterior spinal column muscles is the semispinalis, consisting of the **semispinalis dorsi** (or semispinalis thoracis), **the semispinalis cervicis**, and the **semispinalis capitis**. This

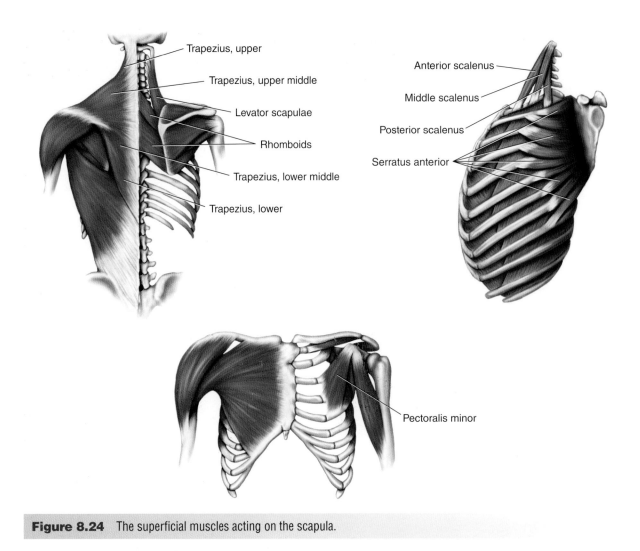

Figure 8.24 The superficial muscles acting on the scapula.

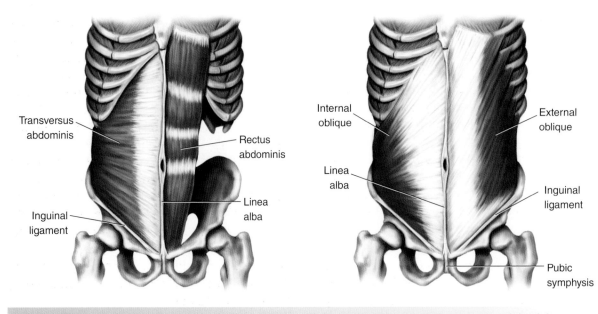

Figure 8.25 The linea alba, rectus abdominis, internal oblique, external oblique, and transversus abdominis.

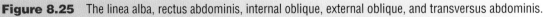

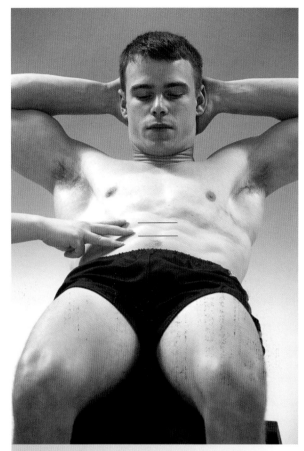

Figure 8.26　Locating the rectus abdominis.

Figure 8.27　Identifying the external oblique.

group originates from the transverse processes of the seventh thoracic vertebra through the last four cervical vertebrae (C4–C7) and inserts on the spinous processes of the upper thoracic, all cervical vertebrae, and the occipital bone of the skull. The semispinalis group extends the cervical and thoracic spine and, particularly in the cervical section, is also involved in lateral flexion and rotation. A third group of posterior spinal column muscles is the iliocostalis group, which includes the **iliocostalis lumborum**, the **iliocostalis dorsi** (or iliocostalis thoracis), and the **iliocostalis cervicis**. This group runs from the posterior aspect of the crest of the ilium and the 3rd through 12th ribs (lumborum and dorsi portions) and the transverse processes of the fourth, fifth, sixth, and seventh cervical vertebrae. All three sections of this group extend the spine, with the upper two sections also involved in lateral flexion and rotation. The fourth group

of posterior spinal column muscles is the longissimus group, which includes the **longissimus capitis**, **longissimus cervicis**, and **longissimus dorsi** (or longissimus thoracis). This muscle group runs from the posterior aspect of the crest of the ilium and the transverse processes of the lumbar, thoracic, and lower cervical vertebrae to all thoracic vertebrae and cervical vertebrae up to the second. This group extends, laterally flexes, and rotates the lumbar, thoracic, and cervical sections of the spinal column.

Three additional muscles of the posterior aspect of the spinal column are the **sacrospinalis** (**erector spinae**), the **quadratus lumborum**, and the **multifidus**. The sacrospinalis originates on the spines of the lumbar and sacral vertebrae and splits into three branches that are part of other muscle groups previously described: the iliocostalis lumborum, the longissimus dorsi, and the spinalis dorsi.

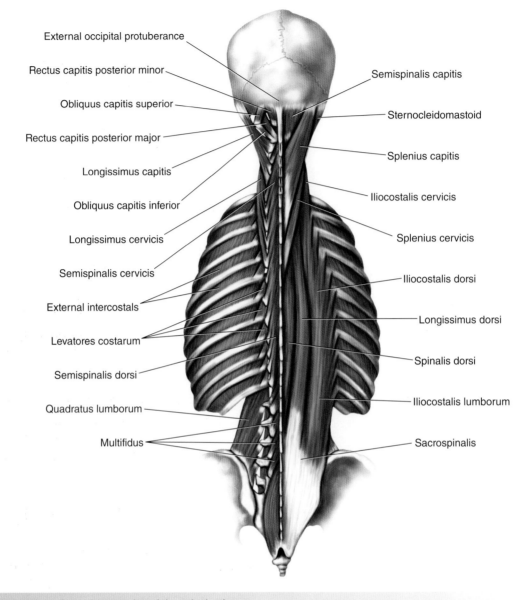

Figure 8.28 Posterior muscles of the spinal column.

External occipital protuberance

Rectus capitis posterior minor

Obliquus capitis superior

Rectus capitis posterior major

Longissimus capitis

Obliquus capitis inferior

Longissimus cervicis

Semispinalis cervicis

External intercostals

Levatores costarum

Semispinalis dorsi

Quadratus lumborum

Multifidus

Semispinalis capitis

Sternocleidomastoid

Splenius capitis

Iliocostalis cervicis

Splenius cervicis

Iliocostalis dorsi

Longissimus dorsi

Spinalis dorsi

Iliocostalis lumborum

Sacrospinalis

Hands On

Ask your partner to flex his trunk at the waist and hold this position. The erector spinae muscle group can be observed running parallel to the spinal column (figures 8.29 and 8.30).

These branches (and the sacrospinalis) extend and laterally flex the lumbar and thoracic spine. The quadratus lumborum runs from the posterior aspect of the crest of the ilium and transverse processes of the lower lumbar vertebrae to the transverse processes of the upper four lumbar vertebrae and the lower aspect of the 12th rib (figure 8.29). Although this muscle does contribute to the extension of the lumbar spine, its primary function is lateral flexion of the lumbar spine. The multifidus is a posterior muscle of the spinal column that covers all movable sections of the column. The multifidus originates from the sacrum; the posterior superior aspect of the iliac spine; and the lumbar, thoracic, and lower four cervical vertebrae and inserts on the spines of the lumbar, thoracic, and all but the first cervical vertebrae (see figure 8.28). This muscle extends and rotates the spinal column.

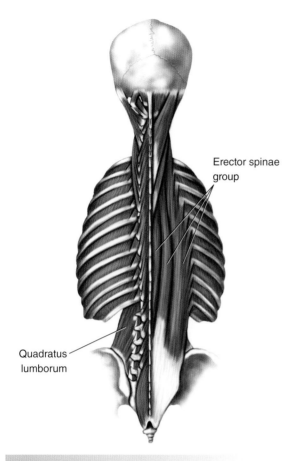

Figure 8.29 The erector spinae group, posterior view.

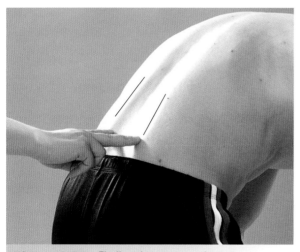

Figure 8.30 Finding the erector spinae group.

As did the cervical region of the spinal column, the thoracic and lumbar spine also have muscles with primary functions in the shoulder joint and shoulder girdle (figure 8.31). The trapezius (with

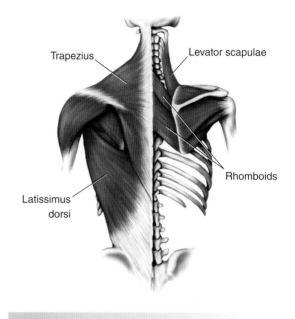

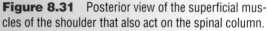

Figure 8.31 Posterior view of the superficial muscles of the shoulder that also act on the spinal column.

attachments to T1–T12), the rhomboids (with attachments to T1–T5), and the **latissimus dorsi** (with attachments to T1–T12 and L1–L5) also contribute to lateral flexion, rotation, and extension of the spinal column.

Bones of the Pelvis

Depending on your point of view, the pelvis could consist of 3, 7, or 11 bones (figure 8.32). The posterior aspect of the pelvis is the sacrum, the five sacral vertebrae fused together (figure 8.33). On both sides of the sacrum are two large bones often referred to as the **innominate bones**. Actually, each innominate bone is three separate bones that have fused together: the **ilium**, the **ischium**, and the **pubic bone**. Note the lines of fusion illustrated on the lateral view of the innominate bone in figure 8.32. The ilium bones articulate with the sacrum posteriorly, and the pubic bones articulate with each other at the **pubic symphysis** to form the pelvis. The male and female pelvises have differences to allow for childbearing by the female (figure 8.34). The female pelvis is proportionally wider and flatter and is tilted forward to a greater degree to allow for this function.

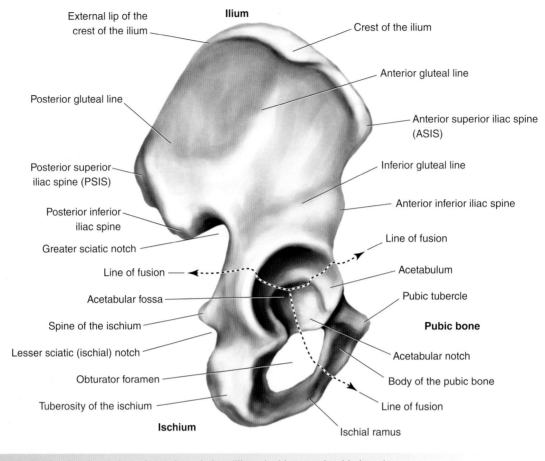

Figure 8.32 The innominate bone, lateral view (ilium, ischium, and pubic bone).

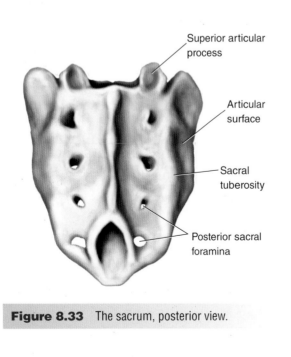

Figure 8.33 The sacrum, posterior view.

The ilium, a wide, winglike structure, is the most superior part of the innominate bone and is the bone that articulates the pelvis with the spinal column through the sacrum.

At the most anterior and most posterior aspects of the ilium are bony prominences known as the **anterior superior iliac spine (ASIS)** and **posterior superior iliac spine (PSIS)**. Running between the ASIS and PSIS is a large ridge of bone serving as a major source of muscular attachments known as the **crest of the ilium** (figure 8.32).

Hands On

Place your hands on your hips and feel your iliac crests (figure 8.35). While you are feeling your iliac crests, follow the crest to its anterior and posterior ends. These end points are your ASIS and PSIS (figure 8.36).

Beneath both the ASIS and PSIS are smaller bony prominences known as the **anterior infe-**

Male   Female

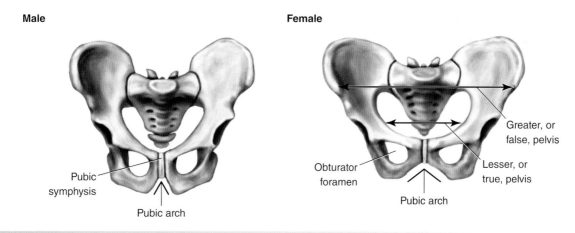

Pubic symphysis

Pubic arch

Obturator foramen

Greater, or false, pelvis

Lesser, or true, pelvis

Pubic arch

Figure 8.34   The differences between the male and female pelvis, anterior views.

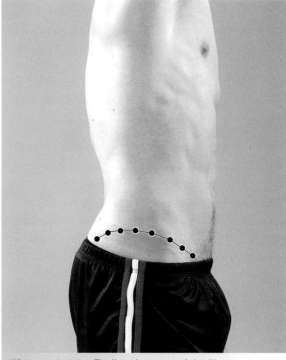

Figure 8.35   Finding the crest of the ilium.

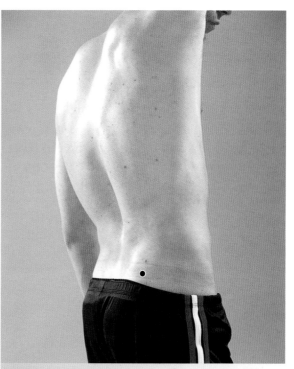

Figure 8.36   Locating the posterior superior iliac spine.

rior iliac spine and the **posterior inferior iliac spine**. Just inferior to the posterior inferior iliac spine is a large notch known as the **greater sciatic notch**.

The ischium is the most posterior of the three bones of the innominate bone and is distal to the ilium. Most posterior on the ischium is a bony prominence known as the **spine of the ischium**. Beneath the spine of the ischium is a notch in the bone known as the **lesser sciatic (ischial) notch**. At the very distal aspect of the

ischium is a large bony prominence known as the **ischial tuberosity**. The ischial tuberosity serves as the source of attachment for the lower-extremity muscle group commonly known as the hamstrings.

Hands On

You sit on your ischial tuberosities. While standing, you can apply pressure to your gluteal areas and palpate your ischial tuberosities.

FOCUS ON

Crest of the Ilium

The crest of the ilium is the source of many muscle attachments, both origins and insertions. When a person contuses this area, it is very painful to attempt to move the lower spinal column and the hips and even to breathe heavily because of the muscular attachments. This type of trauma to the soft tissue attaching to the crest of the ilium is often referred to as a "hip pointer." Injuries such as a hip pointer can sometimes be the result of improperly fitted equipment or failure to use equipment. All coaches should have instruction in the proper use and fitting of equipment. Knowledge of specific anatomical structures and how equipment protects those structures can reduce the incidence and severity of some injuries.

The pubic bone is the most anterior of the three bones making up the innominate bone. The pubic bones articulate with the ischium (**inferior pubic ramus**), the ilium (**superior pubic ramus**), and each other (**body of the pubic bone**). The articulation between the two pubic bones is known as the **pubic symphysis** (see figure 8.34). Just lateral to the pubic symphysis, each pubic bone has a bony prominence on its superior surface known as the **pubic tubercle**.

✋ Hands On

Apply pressure at the most distal aspect of your abdomen, and then move your hand distally to palpate your pubic symphysis.

On the lateral aspect of the innominate bone, the three bones (ilium, ischium, pubic bone) form a deep socket known as the **acetabulum**. This depression is the socket for the triaxial ball-and-socket hip joint. Beneath the acetabulum is a large foramen formed by the ischium and the pubic bone. This foramen is known as the **obturator foramen**.

Ligaments of the Pelvis

The **iliolumbar ligament** runs between the fifth lumbar vertebra and the crest of the

ilium (figure 8.37). Two ligaments articulate the sacrum with the ilium: the **anterior sacroiliac** and the **posterior sacroiliac**. The anterior sacroiliac ligament runs between the anterior surface of the sacrum and the anterior surface of the ilium (figure 8.37). The posterior sacroiliac ligament (figure 8.37) has three sections: the **short sacroiliac**, the **long sacroiliac**, and the **interosseous**. The short sacroiliac ligament runs between the posterior ilium and the lower portions of the sacrum. The long sacroiliac ligament runs between the posterior superior spine of the ilium and the third and fourth vertebrae of the sacrum. The interosseous ligament is made of short fibers that connect the posterior aspects of the sacroiliac joint.

The ligaments of the pubic symphysis are the **anterior pubic**, the **inferior (arcuate) pubic**, the **posterior pubic**, and the **superior pubic**. The function of each of these ligaments is to articulate the anterior, posterior, superior, and inferior aspects of the two pubic bones to form the pubic symphysis. Additionally, there is an **interpubic fibrocartilage disc** between the pubic bones (figure 8.37).

Running between the anterior superior spine of the ilium to the pubic tubercle is a long ligament known as the **inguinal ligament**, which serves as a major source of muscular attachments (figure 8.37). Two ligaments that stabilize the pelvis are the **sacrospinous** and the **sacrotu-**

FOCUS ON

Inguinal Hernia

Excessive stress on the muscles that attach to the inguinal ligament can result in an *inguinal hernia*. The term *hernia* indicates an abnormal protrusion of an organ. Straining the musculature that attaches to the inguinal ligament can cause a portion of the abdominal contents to protrude through the strained muscle tissue, resulting in an inguinal hernia.

Improper lifting techniques can cause hernias and are also a leading contributor to low-back muscular strain. Proper lifting techniques can be taught in weight-training programs and in academic courses in kinesiology and biomechanics.

berous ligaments (figure 8.37). The sacrospinous ligament runs from the sacrum and coccyx to the spine of the ischium. The sacrotuberous ligament runs between the posterior inferior spine of the ilium, the sacrum and coccyx, and the ischial tuberosity.

Fundamental Movements and Muscles of the Pelvis

Movement is not normal in the joints of the pelvic bones. The sacroiliac joint and the joints between the ilium, the ischium, and the pubic bone are essentially fused. The pubic symphysis has minimal movement of a gliding nature.

A combination of movements of the spinal column and the hip joint results in pelvic motion: backward tilt, forward tilt, lateral tilt, and rota-

tion. From the anatomical position (figure 8.38a), backward tilt of the pelvis involves the pubic symphysis moving upward and the sacrum moving downward. This is accomplished by flexion of the lumbar spinal column and extension of the hip joints (figure 8.38b). Forward tilt of the pelvis involves the pubic symphysis moving downward and the sacrum moving upward. Forward pelvic tilt is accomplished by extension of the lumbar spinal column and flexion of the hip joints (figure 8.38c). Lateral tilt of the pelvis results when one ilium is elevated above the other ilium. This is accomplished by lateral flexion of the lumbar spinal column, abduction of one hip joint, and adduction of the other hip joint (figure 8.39). Rotation of the pelvis involves rotation of the lumbar spine, with internal rotation of one hip joint and external rotation of the other hip joint (figure 8.40).

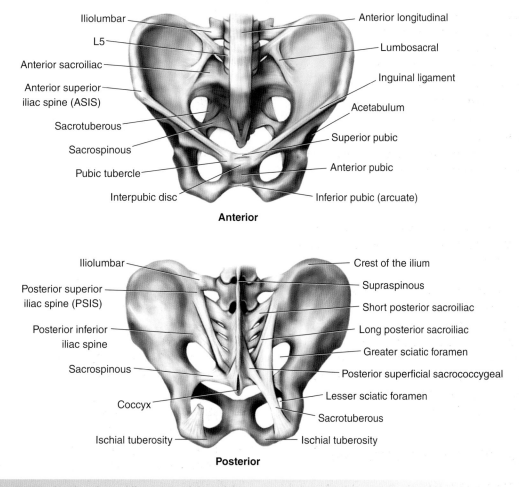

Anterior

Posterior

Figure 8.37 Landmarks and ligaments of the pelvis.

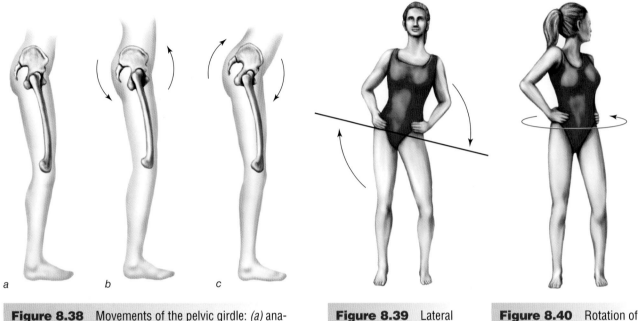

Figure 8.38　Movements of the pelvic girdle: *(a)* anatomical position, *(b)* backward tilt, and *(c)* forward tilt.

Figure 8.39　Lateral tilt of the pelvic girdle.

Figure 8.40　Rotation of the pelvic girdle.

The muscles of the spinal column are presented in this chapter, and the muscles of the hip joint are presented in chapter 11. The spinal column muscles produce the flexion, extension, lateral flexion, and rotation necessary for the pelvic movements, along with the muscles that produce flexion, extension, abduction, adduction, and internal and external rotation of the hip joint. From this information you should be able to understand which muscles produce which movements of the pelvis through spinal column and hip joint movements.

LEARNING AIDS

REVIEW OF TERMINOLOGY

The following terms are discussed in this chapter. Define or describe each term, and where appropriate, identify the location of the named structure either on your body or in an appropriate illustration.

acetabulum	erector spinae	interpubic fibrocartilage disc
annulus fibrosus	external oblique	interspinous ligament
anterior inferior iliac spine	fovea	intervertebral disc
anterior longitudinal ligament	greater sciatic notch	intervertebral foramen
anterior pubic ligament	iliocostalis cervicis	ischial tuberosity
anterior sacroiliac ligament	iliocostalis dorsi	ischium
anterior superior iliac spine (ASIS)	iliocostalis lumborum	isthmus
atlantoaxial ligament	iliolumbar ligament	kyphosis
atlantooccipital ligament	ilium	lamina
atlas	inferior articular facet	latissimus dorsi
axis	inferior (arcuate) pubic ligament	lesser sciatic (ischial) notch
bifid	inferior pubic ramus	levator scapulae
body of the pubic bone	inguinal ligament	ligamentum flavum
cervical vertebra	innominate bone	ligamentum nuchae
coccyx	interbody ligament	linea alba
crest of the ilium	interosseous ligament	long sacroiliac ligament
dens		longissimus capitis

longissimus cervicis
longissimus dorsi
longus capitis
longus colli
lordosis
lumbar vertebra
membrane tectoria
multifidus
nucleus pulposus
obliquus capitis superior and inferior
obturator foramen
occipitoaxial ligament
odontoid process
pars interarticularis
pedicle
posterior inferior iliac spine
posterior longitudinal ligament
posterior pubic ligament
posterior sacroiliac ligament
posterior superior iliac spine (PSIS)
prevertebral muscle

pubic bone
pubic symphysis
pubic tubercle
quadratus lumborum
rectus abdominis
rectus capitis anterior
rectus capitis lateralis
rectus capitis posterior major and
minor
rhomboids
sacral vertebra
sacrospinalis
sacrospinous ligament
sacrotuberous ligament
sacrum
scalenus anterior
scalenus medius
scalenus posterior
scoliosis
semispinalis capitis
semispinalis cervicis

semispinalis dorsi
short sacroiliac ligament
spinalis capitis
spinalis cervicis
spinalis dorsi
spine of the ischium
spinous process
splenius capitis
splenius cervicis
sternocleidomastoid
superior articular facet
superior pubic ligament
superior pubic ramus
supraspinous ligament
thoracic vertebra
transverse ligament
transverse process
transversus abdominis
trapezius
vertebral foramen

SUGGESTED LEARNING ACTIVITIES

1. Imagine you are wearing a jacket with slash-type pockets on both sides of the front of the jacket, just above your waist. Place your hands inside those pockets.

   a. What direction (horizontally, vertically, diagonally) are your fingers pointing?

   b. What abdominal muscle fibers are parallel to the direction your fingers are pointing?

2. In a tackling technique known in football as "clotheslining," a player's head is stopped while the rest of his body continues in motion. This could cause paralysis or even death. Describe anatomically what happens in the cervical spine and how this might have catastrophic results.

3. Excessive weight gain, possibly a pregnancy, or other causes may shift one's center of gravity forward. This could cause the lumbar spine to develop an excessive curvature to counteract the shift.

   a. In which direction would the lumbar spine curve?

   b. What would this curvature be called?

4. Imagine yourself as a baseball pitcher, a football quarterback, a mail carrier with a heavy pouch, a golfer, a tennis player, or any person who participates in an activity that requires use of the musculature on one side of the body more than the other. What type of abnormal curvature of the spine might such activity develop?

5. Name the muscles responsible for turning your head to the right from the anatomical position.

6. Lying supine (on your back), draw your thighs up to your chest.

   a. What movement occurred in your lumbar spine?

   b. What movement occurred in your hip joints?

   c. What movement occurred in your pelvis?

MULTIPLE-CHOICE QUESTIONS

1. How many vertebrae are in the cervical spine?

   a. 4
   b. 5
   c. 7
   d. 12

2. How many vertebrae are in the thoracic spine?

   a. 4
   b. 5
   c. 7
   d. 12

3. How many vertebrae are in the lumbar spine?

   a. 4
   b. 5
   c. 7
   d. 12

4. How many vertebrae are in the coccygeal spine?

   a. 4
   b. 5
   c. 7
   d. 12

5. When a person shakes his head to indicate *no* as an answer to a question, the primary movement takes place between

   a. the skull and C1
   b. C1 and C2
   c. C1 and C7
   d. C2 and C7

6. Which of the following bones of the pelvic girdle is not part of the structure known as the acetabulum?

   a. ilium
   b. ischium
   c. sacrum
   d. pubic bone

7. A forward tilt of the pelvic girdle requires the pubic symphysis to

   a. move laterally
   b. move downward
   c. rotate
   d. move upward

8. A forward tilt of the pelvic girdle requires the sacrum to

   a. move laterally
   b. move downward
   c. rotate
   d. move upward

9. A backward tilt of the pelvic girdle requires the pubic symphysis to

   a. move laterally
   b. move downward
   c. rotate
   d. move upward

10. A backward tilt of the pelvic girdle requires the sacrum to

   a. move laterally
   b. move downward
   c. rotate
   d. move upward

11. Which of the following is not considered a fundamental movement of the vertebral column?

   a. flexion
   b. abduction
   c. extension
   d. rotation

12. Which of the following bones does not provide an attachment for the sternocleidomastoid muscle?

   a. sternum
   b. clavicle
   c. mastoid
   d. humerus

13. Considering that there are right- and left-side scalene muscles, under normal conditions, how many scalene muscles are there in the cervical spine?

   a. 2
   b. 3
   c. 4
   d. 6

14. The erector spinae muscle group supports various segments of the spinal column and is located in what direction relative to the spinal column?

    a. anterior

    b. posterior

    c. superior

    d. inferior

15. Running from the pubic bone to the ribs, which of the following muscles is the most likely to be exclusively a flexor of the lumbar spine?

    a. rectus abdominis

    b. transversus abdominis

    c. obliquus externus

    d. obliquus internus

16. Normally, the cervical spine has what type of curvature?

    a. anterior

    b. posterior

    c. lateral

    d. medial

FILL-IN-THE-BLANK QUESTIONS

1. Under normal conditions, the spinal column has anterior curves in the cervical spine and the _____ spine.

2. Under normal conditions, the spinal column has posterior curves in the coccygeal spine and the _____ spine.

3. The ligament of the spinal column that is considered to be the anterior wall of the spinal canal is the _____ ligament.

4. The cervical vertebra attached to the skull is known as the _____.

5. The anatomical structure that serves as a shock absorber between bodies of the spinal vertebrae is known as the _____.

6. The anterior connection of the two innominate bones is known as the _____.

7. The hamstring muscles originate on the large tuberosity of the _____.

8. An excessive anterior curvature of the lumbar spine is known as _____.

9. An excessive posterior curvature of the thoracic spine is known as _____.

10. An excessive lateral curvature of the spinal column is known as _____.

FUNCTIONAL MOVEMENT EXERCISE

The spinal column has 33 vertebrae divided into the cervical, thoracic, lumbar, sacral, and coccygeal sections. Movements of the spine consist of flexion, extension, lateral flexion, and rotation. Muscle action in the spine is required when you bend over, lean back, and rotate the trunk to the left or right. List one muscle acting as a prime mover, one as an antagonist, one as a fixator, and one as a synergist for returning to the anatomical position after touching the toes (extension) and turning the spine to the left or right position from the anatomical position (rotation).

	Spinal extension	Spinal rotation
Prime mover		
Antagonist		
Fixator		
Synergist		

The Thorax

The thorax (figure 9.1) is a structure formed by bones that create a large compartment known as the thoracic cavity (chest cavity), which houses the lungs and heart. Linings of the lungs and the inner walls of the cavity create a vacuum, and movement of the thorax creates changes in pressure within the thoracic cavity, which causes air to enter into or be expelled from the lungs. Normal inspiration and expiration of air by the lungs are commonly known as quiet respiration, whereas movement of air in and out of the lungs during physical exertion is known as forced respiration. Movement at the joints of the thorax allows for the expansion and contraction of the thoracic cavity, and muscles attached to the bones of the thorax create that movement. For muscles to create movement, oxygen must be supplied to the tissues, and carbon dioxide needs to be removed. The heart and lungs are responsible for providing the oxygen and removing the carbon dioxide. Although respiration and circulation are more suitable discussions for human physiology, a brief presentation of the anatomy of these two major structures within the thorax (heart and lungs) is appropriate in this chapter.

Bones of the Thorax

The anterior bone of the thorax is known as the **sternum** (figure 9.2). The sternum protects the structures beneath it; serves as a source of muscular attachment for muscles of the thorax, the neck, and the abdomen; and provides attachment for costal cartilage. The sternum consists of three parts: the **manubrium**, the **body**, and the **xiphoid process**. The manubrium has a **suprasternal** (jugular) **notch** at its superior edge (figure 9.3). The line where the inferior edge of the manubrium attaches to the body of the sternum is known as the angle of the sternum. The manubrium also has two **clavicular notches** and two **first costal notches** where the sternum articulates with the

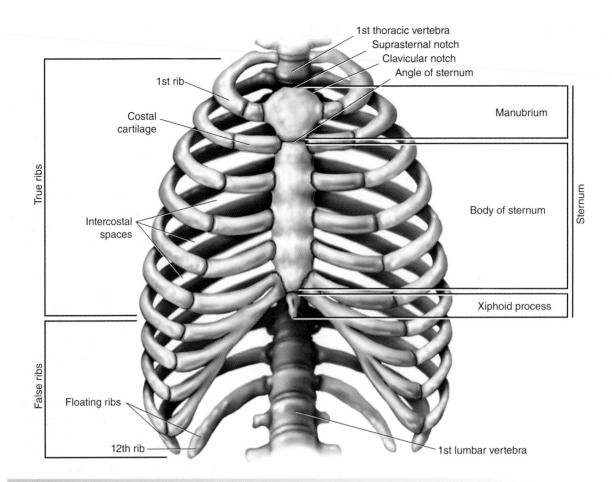

Figure 9.1 Bones of the thorax and their landmarks, anterior view.

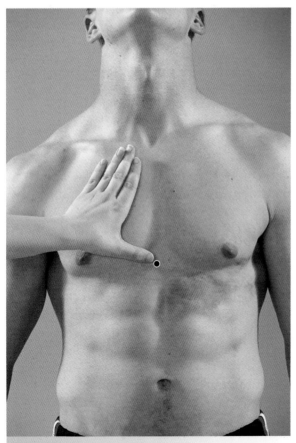

Figure 9.2 Locating the sternum.

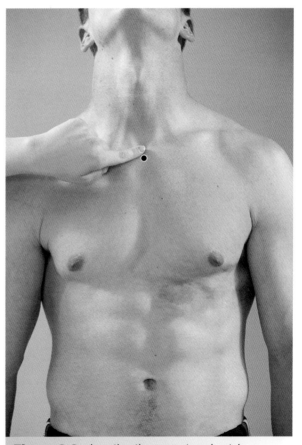

Figure 9.3 Locating the suprasternal notch.

clavicles (sternoclavicular joints) and the first ribs of the thorax. The body of the sternum has costal notches along the lateral sides where the ribs articulate. The most inferior portion of the sternum, the xiphoid process, is cartilaginous in early life and becomes bone in adulthood.

The thorax has 12 pairs of **ribs**: The first (superior) seven pairs are known as true ribs, and the last (inferior) five pairs are known as false ribs. The first seven ribs attach to the sternum (the first to the manubrium and the second through seventh to the sternal body) through costal cartilage, which extends from 1 to 3 inches (2.5–7.6 cm), depending on the length of the rib, between the sternal end of the rib and the sternum. The first three false ribs (ribs 8–10) are attached indirectly to the sternum by costal cartilage that attaches to the costal cartilage of the rib above it. The last two pairs of false ribs are also referred to as floating ribs because they have no anterior attachment. The **head** and the **neck** of a rib (figure 9.4), with an articular facet and tubercle, appear at the end of the rib that articulates with a thoracic vertebra. Lateral to the head of the rib is the angle. This creates the broad back of the human body and gives humans the ability to lie down in a supine position without rolling to one side or the other.

In addition to the sternum and the 12 ribs, the 12 thoracic vertebrae of the spinal column are also considered bones of the thorax. The anatomy of the thoracic vertebrae is covered in chapter 8.

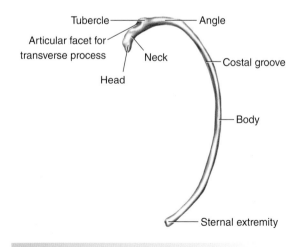

Figure 9.4 Anatomical landmarks of a typical rib.

CPR and the Xiphoid Process

Before the introduction of newer CPR techniques utilizing only chest compressions, cardiopulmonary resuscitation (CPR) also included breathing techniques. The rescuer is still taught to use the xiphoid process as a landmark to establish hand placement on the body of the sternum for compressions. Pressure applied to the xiphoid process not only produces ineffective compressions but also creates the possibility of fracture of the xiphoid process. First-aid courses; CPR courses; emergency medical technician and paramedic courses; and programs in medicine, nursing, physical therapy, athletic training, and other allied health sciences require the acquisition of CPR skills. Knowledge of the anatomy of the thorax helps you perform these CPR skills properly and not expose the victim to further trauma.

Joints and Ligaments of the Thorax

The anterior articulations of the thorax (figure 9.5) include the **chondrosternal joints**, the **costochondral joints**, and the **interchondral joints**. The chondrosternal joints articulate the costal cartilages of the upper seven (true) ribs with the costal notches of the manubrium (first rib) and the body of the sternum (ribs 2–7). The ligaments include the **costochondral (capsular)**, the **costosternal**, and the **interchondral (interarticular)**. A gliding, rotating motion occurs at these joints, which allows external rotation and elevation of the rib. The costochondral joints are the articulations between the sternal end of the rib and the cartilage attaching the rib to the sternum (ribs 1–7) or to the cartilage of the rib above (ribs 8–10). There is no actual ligament associated with the costochondral articulation because the periosteum of the rib (a covering of connective tissue) in this case is actually continuous with the perichondrium of the cartilage (a thin, fibrous tissue that covers cartilage and provides it with nutrition). In athletics, the costochondral joint often comes under stress when the thorax is compressed beyond the normal range of motion of this joint. Slight rotation of the costochondral joint allows upward and outward movement of the rib.

All 12 ribs articulate with the 12 thoracic vertebrae through two articulations for each rib. The **costovertebral articulations** are known as the **corpocapitate** and **costotransverse** **capsular articulations**. The corpocapitate articulation is between the head of the rib and the body of the adjacent thoracic vertebra, and the ligaments involved are the **costotransverse capsular** and the **costovertebral radiate** (figure 9.6). A gliding, rotational movement allows elevation and depression of the ribs. The costotransverse articulations are the joints between the tubercles of the first 10 ribs with

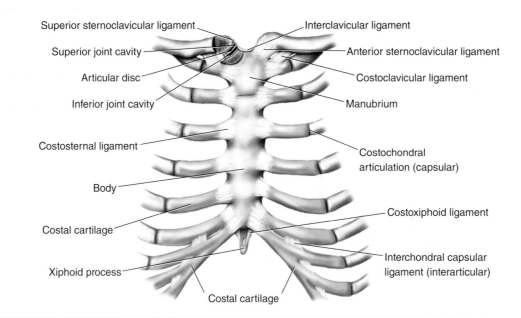
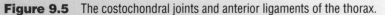

Figure 9.5   The costochondral joints and anterior ligaments of the thorax.

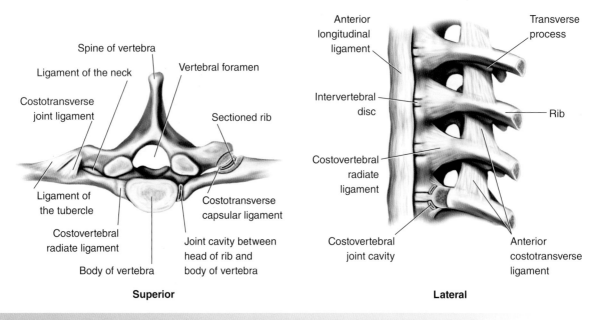

Superior   **Lateral**

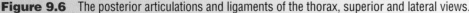

Figure 9.6   The posterior articulations and ligaments of the thorax, superior and lateral views.

Costochondral Sprain

The costochondral joint, like other joints, has a normal range of motion. When that range of motion is exceeded, a costochondral sprain is possible. The thorax of a football player being tackled may be compressed between the tackler and the playing surface, forcing the costochondral joints beyond their normal limits. Likewise, the thorax of a wrestler taken down from a standing position may be compressed between the opponent and the wrestling mat, causing injury to the costochondral joints. Some old-time wrestling coaches refer to this injury as a "rib-out." If the costochondral articulation is sprained, the rib and the cartilage may separate, allowing the bone to move more than normal and resulting in the appearance of a rib attempting to come through the skin at the site of the sprain. The mechanism that causes a costochondral sprain can result in other forms of trauma also (e.g., rib fractures, damage to internal organs of the thoracic cavity).

Recognizing a costochondral sprain requires study of human anatomy, athletic training, sports medicine, and coaching. Care of such a sprain is essential for proper healing. Although it would be ideal to completely immobilize the joint for several weeks (i.e., prevent breathing), this is obviously not possible. Other means of restricting movement need to be used, such as rib belts or taping or wrapping, which will reduce movement of the costochondral articulation during respiration.

the adjacent vertebra's transverse process. The 11th and 12th ribs do not have an articulation with the adjacent transverse processes. The only ligament of these costotransverse articulations is the capsular ligament.

Fundamental Movements and Muscles of the Thorax

Respiration depends on movements of the thorax. As mentioned earlier, normal respiration is known as quiet respiration. The muscles of quiet respiration are considered specific to respiration (figure 9.7). Muscles discussed in other chapters become involved in forced respiration if they have any attachment to the bones of the thorax.

The muscles of the thorax include the **diaphragm**. This large, domelike muscle runs from the xiphoid process, cartilage of the last seven ribs, and lateral aspect of the first four lumbar vertebrae to a structure known as the **central tendon** (figure 9.8). When the diaphragm

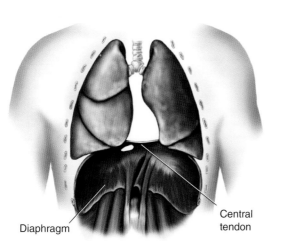

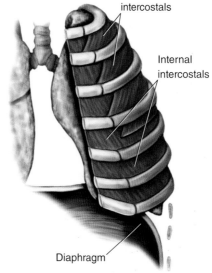

Figure 9.7 Main respiratory muscles.

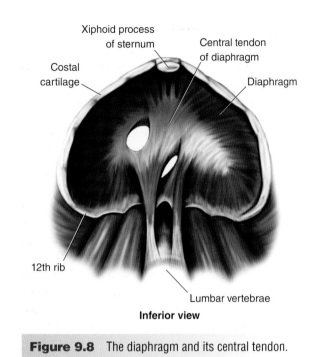

Xiphoid process
of sternum

Central tendon
of diaphragm

Costal
cartilage

Diaphragm

12th rib

Lumbar vertebrae

Inferior view

Figure 9.8 The diaphragm and its central tendon.

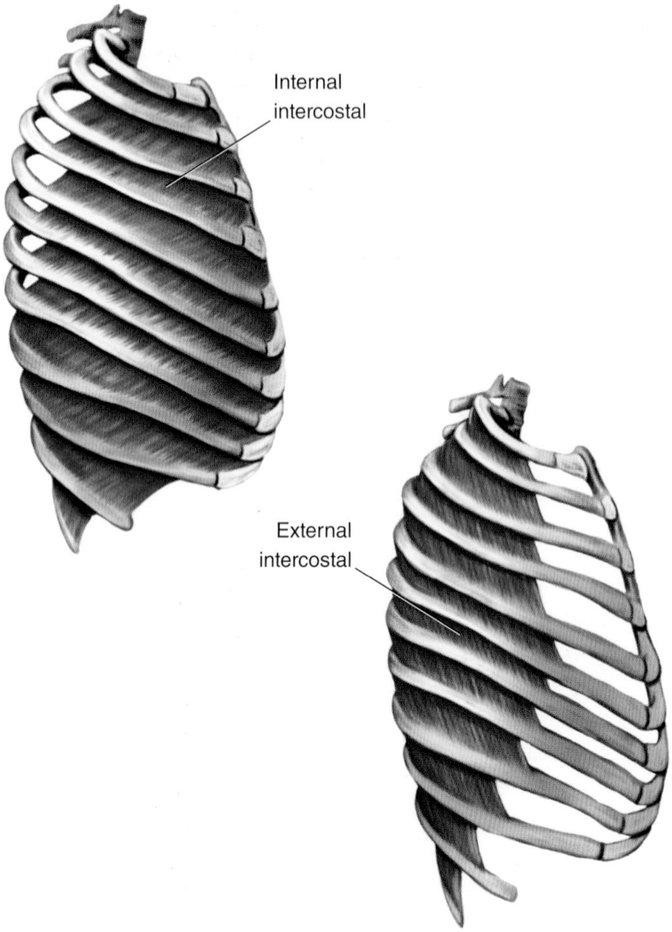

contracts, its central tendon pulls downward, increasing space in the thoracic cavity and thus changing the pressure within the cavity. As air is drawn into the lungs and they expand, both the **internal** and **external intercostal muscles** (figure 9.9) draw the ribs together, allowing the lungs to expand outward and downward.

Hands On

Raise either arm above your head; place your fingers in the space between the anterior lateral aspect of the thorax to place pressure on the intercostal muscles (figure 9.10). The external intercostals (11 pairs), on the anterior aspect of the ribs, run between the lower border of a rib to the upper border of the rib below. Note the angle made by the direction of the external intercostal fibers.

Hands On

Place your open hands in the pockets of a jacket. Note the direction your fingers are pointing. Is there any similarity between the angle your fingers point and the angle the external intercostal fibers run?

Internal
intercostal

External
intercostal

Figure 9.9 External and internal intercostal muscles, lateral view.

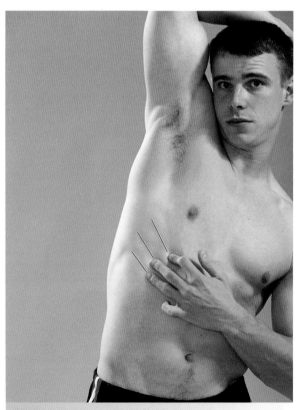

Figure 9.10 Locating the intercostals.

The internal intercostals (11 pairs), on the posterior aspect of each rib, run between the upper border of a rib to the upper border of the rib below. The movement created by contraction of both the external and internal intercostal muscles is a drawing together of the ribs. If the scalene muscles of the cervical spine contract to fix (stabilize) the first rib, the external intercostal muscles elevate the ribs to allow more volume in the thoracic cavity for the lungs to fill. If the last rib is fixed by the quadratus lumborum muscle, the lower ribs are drawn closer together to decrease the volume of the thoracic cavity.

Other thoracic muscles include the **levatores costarum** (12 pairs), which run from the transverse processes of the vertebrae C7 through T11 to the angle of the rib below (figure 9.11). Contraction assists with elevation of the rib. The **subcostal muscles** (10 pairs) run between the inner surface of a rib near its angle to the inner surface of the second or third rib below (figure 9.12). Contraction assists with elevation of the rib. The **serratus posterior** muscle has inferior and superior portions (figure 9.13). The inferior portion runs from the spinous processes of the last two thoracic and first three lumbar vertebrae to the inferior border of the last four ribs, just lateral to their angles, and depresses these ribs. The superior portion of the serratus posterior runs from the spinous processes of the seventh cervical and first three thoracic vertebrae to just lateral to the angle of the second, third, fourth, and fifth ribs and elevates these ribs. The last muscle considered a muscle of the thorax is the **transversus thoracis** (figure 9.14). It runs from the lower third of the posterior aspect of the sternum, the xiphoid process, and the costal cartilage of the fourth through seventh ribs to the inner aspect of the costal cartilage of the second to sixth ribs and elevates the ribs.

Numerous other muscles have one of their attachments to one of the bones of the thorax and usually are not involved in quiet respiration

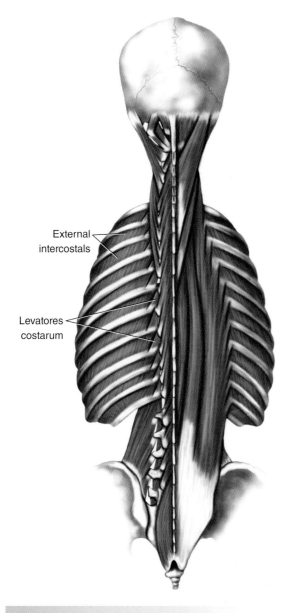

External intercostals

Levatores costarum

Figure 9.11 The levatores costarum.

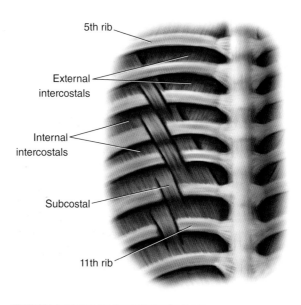

5th rib

External intercostals

Internal intercostals

Subcostal

11th rib

Figure 9.12 The subcostal muscles.

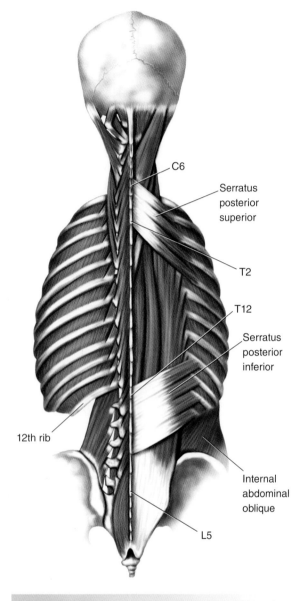

Figure 9.13   The serratus posterior, superior and inferior portions.

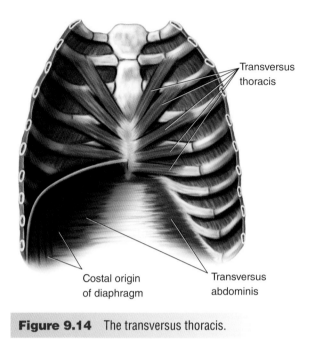

Figure 9.14   The transversus thoracis.

expanded lungs. When the diaphragm and intercostal muscles relax and return to their starting positions, **quiet expiration** occurs. In **forced inspiration**, to make space for greater filling of the lungs, additional muscles that have an attachment to the thorax and are capable of expanding the thorax are called into action. When as much air as possible needs to be expelled from the lungs, any muscle capable of squeezing the thoracic cavity as small as possible is called into action during **forced expiration**.

Structures Within the Thorax

Any discussion of the thorax needs to include the anatomy of the two major structures housed within it: the heart and the lungs.

Heart Structures

The heart is the key organ in the circulatory system. It receives deoxygenated blood from the body and pumps it to the lungs to pick up oxygen. The oxygenated blood returns from the lungs to the heart, and the heart then pumps this blood to the other organs and tissues. Various anatomical structures accomplish this blood flow to, through, and from the heart and lungs.

but can be involved in forced respiration, when lung capacity needs to increase to meet demands and therefore needs additional space within the thoracic cavity to expand. Table 9.1 describes the muscles involved in quiet respiration and those involved in forced respiration.

The information in table 9.1 implies that **quiet inspiration** requires the diaphragm to contract, and as the lungs fill with air, the external and internal intercostal muscles contract to elevate the ribs to create more space for the

TABLE 9.1

Muscles of Respiration

Muscle	QUIET RESPIRATION		FORCED	
	Inspiration	Expiration	Inspiration	Expiration
Diaphragm	●	Passive	●	
External intercostals	●	Passive	●	●
Internal intercostals	●	Passive	●	●
Latissimus dorsi			●	
Levatores costarum			●	
Pectoralis major and minor			●	
Rhomboid major and minor			●	
Scaleni anterior, middle, and posterior			●	
Serratus anterior			●	
Serratus posterior (superior aspect)			●	
Sternocleidomastoid			●	
Subcostals			●	
Trapezius			●	
Obliques, external and internal			●	
Rectus abdominis				●
Sacrospinalis group				●
Serratus posterior (inferior aspect)				●
Transversus abdominis				●
Transversus thoracis				●

The heart (figures 9.15 and 9.16) lies just above the diaphragm and between the lungs in the area known as the **mediastinum**. The heart is a hollow organ with thick muscular walls and is wider at the upper end (base) and narrower and more pointed at the lower end (apex). The heart is wrapped in a membrane known as the **pericardium**. A little larger than the size of a fist, the heart contains four specific compartments, or chambers (figure 9.17).

The two upper chambers are the **right atrium** and **left atrium**. These two chambers are separated by a wall of tissue known as the **interatrial septum**. The atrial cavity of these chambers is enlarged by an extension of the atrial wall known as the **auricle** (earlike projection). The two lower chambers, larger in size compared to the atria, are known as the **left ventricle** and the **right ventricle**. These chambers are separated by a wall known as the **interventricular septum**. Additionally, there is a wall (septum) separating the atria from the ventricles: the **atrioventricular septum**. Within the atrioventricular septum is an opening between the right atrium and right ventricle and another opening between the left atrium and left

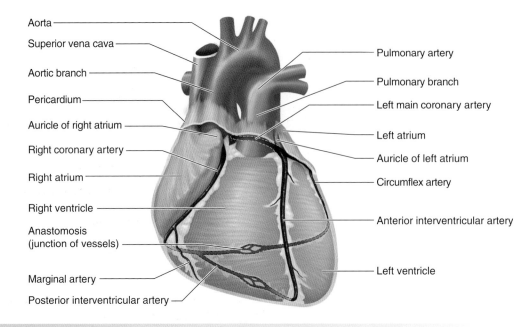

Figure 9.15 Anterior heart.

Adapted by permission from Kenney, Wilmore, and Costill 2012.

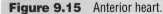

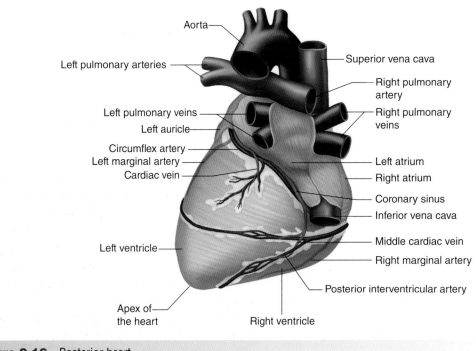

Figure 9.16 Posterior heart.

ventricle. The opening between the right atrium and right ventricle (atrioventricular orifice) has a valve with three **cusps** (flaps) covering the orifice, known as the **tricuspid valve**. The opening between the left atrium and left ven- tricle has a valve with two cusps (flaps), known as the bicuspid valve (also known as the **mitral valve**). Both valves are designed to prevent blood flow backward from the ventricles into the atria. The **chordae tendineae** are tendonlike cords

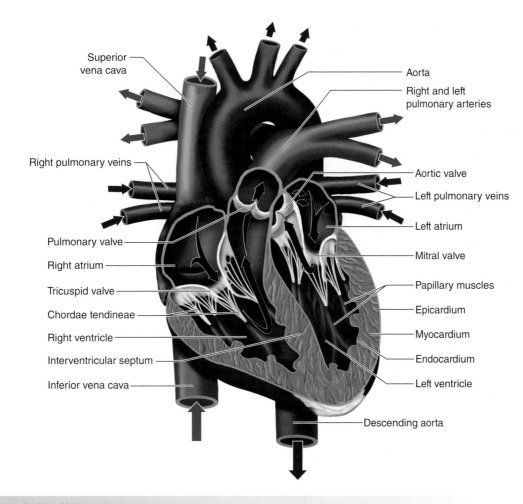

Superior vena cava

Right pulmonary veins

Pulmonary valve

Right atrium

Tricuspid valve

Chordae tendineae

Right ventricle

Interventricular septum

Inferior vena cava

Aorta

Right and left pulmonary arteries

Aortic valve

Left pulmonary veins

Left atrium

Mitral valve

Papillary muscles

Epicardium

Myocardium

Endocardium

Left ventricle

Descending aorta

Figure 9.17 Heart anatomy.

Adapted by permission from Kenney, Wilmore, and Costill 2012.

that attach to the cusps (flaps) of each valve. These cords are also attached to the ventricular walls by small muscles known as the **papillary muscles**. The papillary muscles are responsible for maintaining appropriate tension on the valves to prevent blood from flowing back into the atria from the ventricles.

Under normal conditions, the opening and closing of the heart valves produce a "lubb-dup" sound. When the valves do not function normally, sounds such as a whirring or hissing may occur. Two common defects are known as **regurgitation** and **stenosis**. Failure of a heart valve to open fully produces a sound of rushing blood during and just after the "lubb" sound. This is known as stenosis. The rushing of blood backward, regurgitation, is the result of a valve failing to completely close and can be heard

during and after the "dup" sound. The **puncta maxima** refers (collectively) to the six points on the thoracic wall where the sounds of the opening and closing of the various heart valves are best heard (figure 9.18). Although the points may not seem to be directly over the heart, they are the best place to put the stethoscope to hear a particular valve.

The ventricular walls are three layers thick. A thin outer layer of cells is known as the **epicardium**. An equally thin inner layer of cells on the ventricular walls is known as the **endocardium**. Between these two walls is a thicker muscular wall of tissue known as the **myocardium**. Because of the amount of force needed to push the blood throughout the body, the left ventricular myocardium is thicker than the right ventricular myocardium. The left ventricle forces blood from the

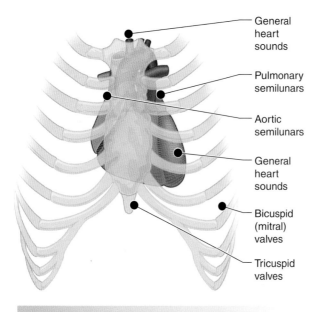

- General heart sounds
- Pulmonary semilunars
- Aortic semilunars
- General heart sounds
- Bicuspid (mitral) valves
- Tricuspid valves

Figure 9.18 Location of different heart sounds via the puncta maxima.

heart throughout the entire circulatory system, while the right ventricle must force blood only from this chamber to the nearby lungs.

Airway and Lung Structures

As air enters the mouth and nose, it passes through structures known as the **respiratory tree** (figure 9.19).

After being inhaled through the oral cavity (mouth) and the nasal cavity (nose), air passes through the **pharynx**, **larnyx**, and **trachea** (figure 9.19). The trachea divides into two **bronchi**; each bronchus divides into smaller passageways known as **bronchioles**. The bronchioles divide into smaller tubelike structures, **respiratory bronchioles**, that end as small **alveolar ducts**. These ducts lead into expansions known as **atria**. From the atria, air then flows into structures known as **air sacs**.

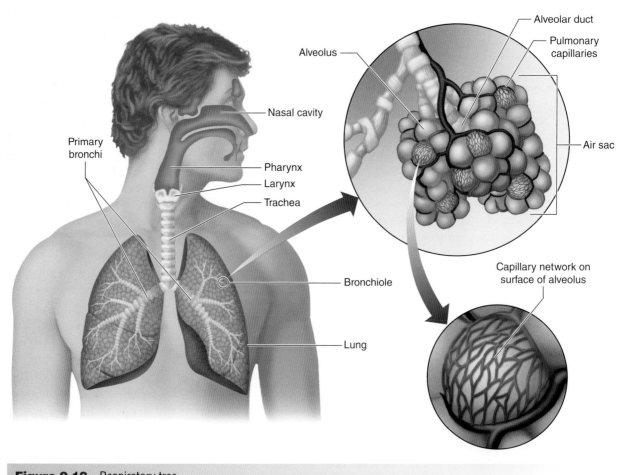

- Alveolus
- Nasal cavity
- Primary bronchi
- Pharynx
- Larynx
- Trachea
- Alveolar duct
- Pulmonary capillaries
- Air sac
- Bronchiole
- Lung
- Capillary network on surface of alveolus

Figure 9.19 Respiratory tree.

Adapted by permission from Kenney, Wilmore, and Costill 2012.

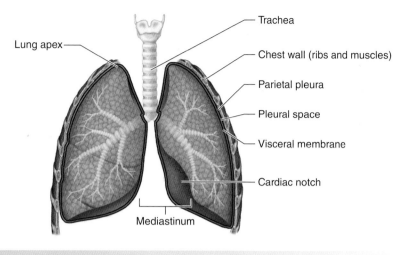

Figure 9.20 Airway and lung anatomy.

Adapted by permission from Kenney, Wilmore, and Costill 2012.

Within the thorax is the **thoracic cavity**, which contains the **pleural cavities** (**left** and **right**). These are separated by the space known as the **mediastinal septum** (figure 9.20). Once the trachea divides into the bronchioles, the bronchioles enter the pleural cavities and ultimately terminate at the air sacs in either the right or left lung. Both lungs are covered by a membrane known as the **pleural membrane**. This membrane has two layers: the **parietal pleura** lining the walls of the pleural cavity and the **visceral membrane** lining each lung. A fluid is present between the membrane layers for lubricating purposes to facilitate the movement between the lungs and the pleural walls during the process of respiration.

Although alike in most ways, the right and left lungs do differ slightly anatomically. The lungs have the following anatomical similarities:

an apex at the narrow superior portion; a base at the wide inferior portion; **costal surfaces** lying next to the ribs; **costal grooves** caused by the ribs; and **mediastinal surfaces** where the lungs touch the mediastinal septum. Located on the mediastinal surfaces is the **hilum**, which is a depression midway between the apex and base where the **bronchial** and **pulmonary arteries** and **veins** enter and leave the lungs. The major anatomical difference between the right and left lungs is the number of **lobes** in each lung. The right lung has three lobes (lower, middle, and upper), and the left lung has only two lobes (upper and lower). The left lung also has an area known as the **cardiac notch** that creates space for the heart. The right lung is shorter than the left lung because most of the liver lies under the right side of the diaphragm, and this pushes the right lung higher in the thoracic cavity than the left lung.

LEARNING AIDS

REVIEW OF TERMINOLOGY

The following terms are discussed in this chapter. Define or describe each term, and where appropriate, identify the location of the named structure either on your body or in an appropriate illustration.

air sac	auricle	bronchiole
alveolar duct	body of the sternum	bronchus
atrioventricular septum	bronchial artery	cardiac notch
atrium (in the lungs)	bronchial vein	central tendon

(continued)

REVIEW OF TERMINOLOGY *(continued)*

chondrosternal joint
chordae tendineae
clavicular notch
corpocapitate capsular articulation
costal groove
costal surface
costochondral (capsular) ligament
costochondral joint
costosternal ligament
costotransverse capsular articulation
costotransverse capsular ligament
costovertebral articulation
costovertebral radiate ligament
cusp
diaphragm
endocardium
epicardium
first costal notch
forced expiration
forced inspiration
head of a rib
hilum
interatrial septum

interchondral (interarticular) ligament
interchondral joint
intercostal muscle (internal and external)
interventricular septum
larnyx
left atrium
left ventricle
levatores costarum
lobe
manubrium
mediastinal septum
mediastinal surface
mediastinum
mitral valve
myocardium
neck of a rib
papillary muscle
parietal pleura
pericardium
pharynx
pleural cavity (left and right)

pleural membrane
pulmonary artery
pulmonary vein
puncta maxima
quiet expiration
quiet inspiration
regurgitation
respiratory bronchiole
respiratory tree
rib
right atrium
right ventricle
serratus posterior
stenosis
sternum
subcostal muscle
suprasternal notch
thoracic cavity
trachea
transversus thoracis
tricuspid valve
visceral membrane
xiphoid process

SUGGESTED LEARNING ACTIVITIES

1. On your body or your partner's body, trace the clavicle with your finger from its most lateral end to its medial end.

   a. At the medial end of both clavicles, what do you feel?

   b. What is the name of the area you are palpating?

2. Raise your or your partner's hand and arm laterally overhead. Find the clavicle, and palpate all 12 ribs.

   a. Which, if any, ribs were difficult or impossible to palpate?

   b. Why?

3. With a cloth tape measure, determine the circumference of your partner's thorax.

   a. After determining the resting (quiet) circumference, have your partner inhale to the greatest degree possible, and then remeasure the circumference. Then have your partner forcefully exhale all the air possible from the lungs, and again remeasure the thoracic circumference. (Note: Be sure to measure at the same level of the thorax each time.)

   b. What muscles caused the thorax to expand?

   c. What muscles caused the thorax to contract?

4. Assume the position for applying chest compressions as in cardiopulmonary resuscitation (CPR). CAUTION: ***Do not actually apply this technique!***

   a. Place the heel of your hand on your partner's sternum. Identify the manubrium, the body, and the xiphoid process.

   b. Identify the thorax joints that allow you to compress the thorax without fracturing ribs.

5. With the aid of a stethoscope and using figure 9.18 for reference points, listen to the following heart sounds:

   a. General heart sounds (heart rate and rhythm) just superior to the jugular notch of the sternum

   b. General heart sounds (heart rate and rhythm) between the fourth and fifth ribs, a couple of finger widths left of the sternum

c. Was there any differences in the sounds? If yes, listen to the valves opening and closing, as indicated on the illustration, in the pulmonary and semilunar valves and the tricuspid and bicuspid (mitral) valves. Hopefully no abnormal sounds (murmurs) are present.

MULTIPLE-CHOICE QUESTIONS

1. How many pairs of ribs arc typically referred to as false ribs?
   a. 2
   b. 5
   c. 7
   d. 12

2. The ligament connecting the bony portion of a rib to the costal cartilage is known as the
   a. costosternal
   b. costochondral
   c. costoclavicular
   d. costothoracic

3. Quiet expiration requires action by which of the following muscles?
   a. none
   b. diaphragm
   c. intercostals
   d. scaleni

4. The membrane lining the walls of the pleural cavity is the
   a. parietal pleural membrane
   b. visceral membrane
   c. alveolar membrane
   d. bronchial membrane

5. The tricuspid valve is located between the
   a. right atrium and left ventricle
   b. left atrium and right ventricle
   c. left atrium and left ventricle
   d. right atrium and right ventricle

6. Which of the following structures contain the valves of the heart?

   a. atrioventricular septum
   b. interatrial septum
   c. mitralatrial septum
   d. interventricular septum

7. The heart is surrounded by a membrane known as the
   a. epicardium
   b. pericardium
   c. endocardium
   d. myocardium

8. The bronchial and pulmonary arteries and veins enter and leave the lungs through an area of the heart known as the
   a. apex
   b. base
   c. cardiac notch
   d. hilum

9. Failure of a heart valve to fully open can result in a condition known as
   a. regurgitation
   b. myocardial infarction
   c. stenosis
   d. mitral valve prolapse

10. Because of the large amount of force needed to move blood throughout the entire circulatory system, the myocardium of which of the following heart chambers is thicker than the myocardium of the other chambers?
    a. left atrium
    b. left ventricle
    c. right atrium
    d. right ventricle

FILL-IN-THE-BLANK QUESTIONS

1. The thorax consists of thoracic vertebrae, ribs, costal cartilage, and the _____.

2. The superior portion of the sternum is known as the _____.

3. The chief muscle of respiration is known as the _____.

4. The right lung has _____ lobes.

5. The right lung is shorter than the left lung because the _____ lies under the diaphragm.

6. The trachea divides into two large structures known as the _____.

7. The heart lies between the lungs in an area known as the _____.

8. The tissue forming a wall between the left and right atria is known as the _____.

9. The mitral valve has _____ flaps also known as _____.

CHAPTER 10

Nerves and Blood Vessels of the Head, Spinal Column, Thorax, Heart, and Lungs

This chapter starts with an overview of the anatomical structures of the brain, including the cerebrum, brain stem, and cerebellum. The chapter then presents the nerves and blood vessels of the head and brain before shifting focus to the thorax and the two major organs contained there: the heart and lungs. For movement to occur, muscles need oxygen and other nutrients and a means of disposing the by-products that result. The heart (circulatory system) and the lungs (respiratory system) are the organs mainly responsible for providing the necessary materials essential for muscle activity. The nerves, arteries, and veins of the thorax and its major organs are presented in this chapter.

The Brain

Before discussing the three main parts of the brain, the anatomical structures covering the brain need to be examined (figure 10.1). Immediately beneath the scalp is the periosteum of the skull, and the **epidural space** is found just

beneath this. The next layer is a brain covering called the **dura mater**. Beneath the dura mater is another space known as the **subdural space**. Under the subdural space is another brain covering known as the **arachnoid**. The space below the arachnoid, the **subarachnoid space**,

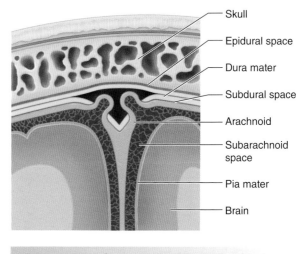

Skull

Epidural space

Dura mater

Subdural space

Arachnoid

Subarachnoid space

Pia mater

Brain

Figure 10.1 Cross section of the skull and brain.

Head Trauma

Most of the major blood vessels supplying the brain are found in the spaces between the brain's coverings. The epidural space primarily contains the arteries, and the subdural space primarily contains the veins. Frequently the terms *epidural hematoma*, *subdural hematoma*, and *subarachnoid hemorrhage* are used to describe head trauma (figure 10.2). (A hematoma is defined as a well-formed blood tumor, while a hemorrhage is defined as a more diffuse discharge of arterial, venous, or capillary blood.) An epidural hematoma usually develops rapidly (it is arterial in nature) and can have the most serious consequences. Since the skull is above the epidural space, a hematoma in this area cannot expand anywhere but downward, exerting pressure on the brain. Within both the subdural and subarachnoid spaces, expansion of a hemorrhage (which is venous in nature) is slower and has more space (above and below) in which to expand before exerting pressure on the brain. These factors are major considerations when someone suffers head trauma and part of the immediate treatment includes observation. Someone experiencing head trauma may appear symptom free after an initial period of confusion, but a subdural hematoma or subarachnoid hemorrhage may take time to develop.

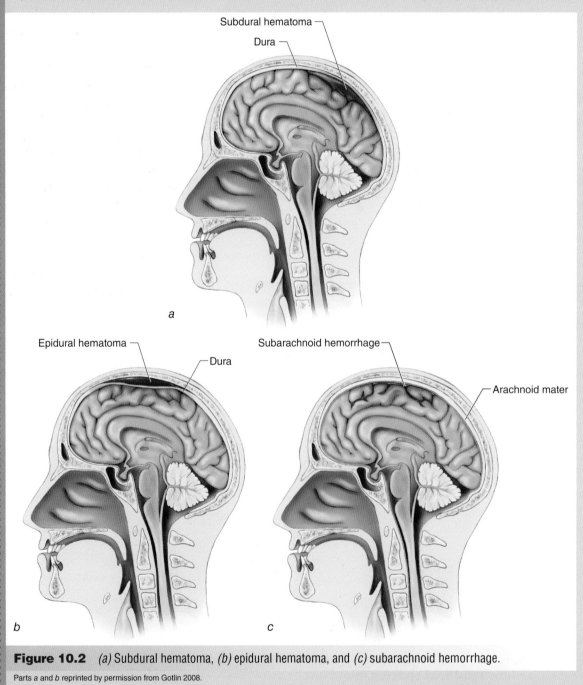

Figure 10.2 *(a)* Subdural hematoma, *(b)* epidural hematoma, and *(c)* subarachnoid hemorrhage.

contains **cerebrospinal fluid**. The subarachnoid space is between the arachnoid and **pia mater** coverings. Beneath the pia mater is the brain. These coverings are also referred to as **meninges**.

The brain itself consists of three major structures (figure 10.3): the cerebrum, the brain stem, and the cerebellum.

Cerebrum

The **cerebrum** (upper brain; figure 10.4) is the most developed portion of the central nervous system and can be considered the spread-out end of the spinal cord because it is contained in the more spacious skull area compared to the more confined space within the vertebra of the spinal column. The right and left sides are bilaterally symmetrical (interconnected by a large fiber bundle known as the **corpus callosum**. The outer layer of brain tissue contains the **gray matter** known as the **cortex**. The gray matter mainly consists of nerve cell bodies and dendrites. Beneath the gray matter is the **white matter**, which is composed of nerve axons. The cerebrum has many folds of tissue, with the raised portions known as **gyri** and the depressions between the gyri known as **sulci**. The four most

prominent sulci divide the cerebrum into four distinct lobes. The median sulcus (**longitudinal fissure**) divides the cerebrum into its bilaterally symmetrical sides. The central sulcus (**fissure of Rolando**) divides the cerebrum into the **frontal** and **parietal lobes**. The lateral sulcus (**fissure of Sylvius**) separates the parietal and **temporal lobes**. The occipitoparietal sulcus separates the parietal and **occipital lobes**. Although there are also numerous gyri, the most prominent are the precentral gyrus (frontal lobe); postcentral gyrus (parietal lobe); superior, middle, and inferior temporal gyri; and superior, middle, and inferior frontal gyri.

The cerebrum houses the area of consciousness that makes a person aware of his surroundings. There is **cross-control**, meaning the left side of the cerebrum controls the right side of the body, and the right side of the cerebrum controls the left side of the body. Although trauma to either side of the motor area of the cerebrum causes major problems to the musculature on the opposite side of the body, there can also be lesser problems to the same side of the body. Voluntary activity of the musculature is initiated in the motor area of the cerebral cortex.

The cerebrum contains specific areas responsible for various functions (figure 10.5). The

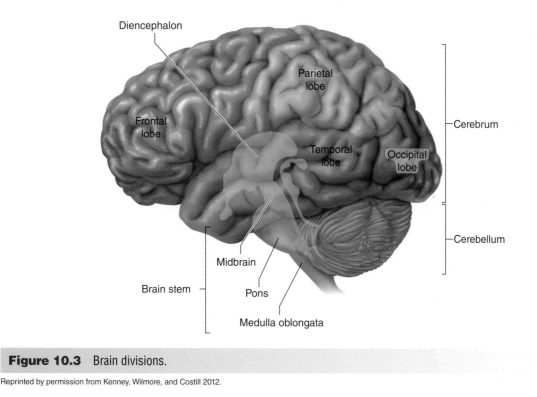

Figure 10.3   Brain divisions.

motor area, located in the precentral gyrus, contains the majority of the motor nerves of the cerebrum. The **body sense area**, located in the postcentral gyrus, also has cross-control, with the cutaneous and muscular sensations on one side of the body being recognized by the opposite side of the cerebrum. The **visual center**, located in the parietal lobe, contains nerve fibers via the optic nerve. The **olfactory center**, located in the temporal lobe, controls the ability to distinguish

different odors. The **taste center** is thought to be located posterior to the olfactory area of the cerebrum. With the taste center so close to the olfactory center, it makes sense that taste and smell are often closely associated. The **auditory center**, located in the temporal lobe, contains nerve fibers that detect sound from both ears. The **speech** and **writing centers**, in the frontal lobe, control the ability to speak and write words. The **word hearing center**, located in the pos-

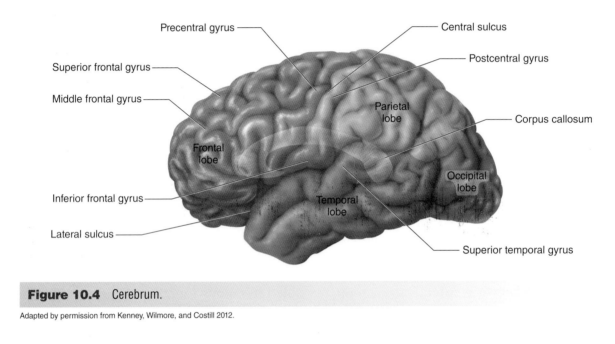

Figure 10.4 Cerebrum.

Adapted by permission from Kenney, Wilmore, and Costill 2012.

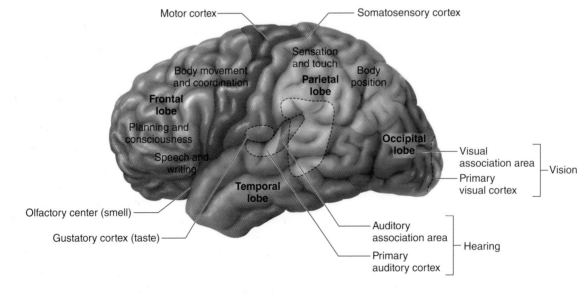

Figure 10.5 Cross section of the motor and sensory areas of the cerebrum.

Adapted by permission from Kenney, Wilmore, and Costill 2012.

terior aspect of the temporal lobe, controls the ability to recognize and understand the spoken word. The **word visual center**, located in the posterior aspect of the parietal lobe, controls the ability to recognize written words.

The remaining large portion of the surface of the cerebrum contains the **association areas**, which control the ability to respond to various stimuli. A particular sound or odor will be associated with a certain response, but depending on previous experiences, this might not be predictable from person to person. This brings about another ability of the cerebrum: the ability to discriminate, or **discriminatory power**. A high degree of discriminatory power allows people to vary in their abilities to judge tastes and smells. This leads to the retention of various impulses (e.g., sight, sound, taste, smell), known as **memory**. Although good memory is important, it should not be confused with **intelligence**. The **power of reasoning** involves the process of bringing together, via the associative powers of memories, old knowledge with new knowledge to identify solutions to various situations. Before one can reason, one must have memory. The combination of one's memory of facts and the power to reason results in one's degree of intelligence. The frontal lobe of the cerebrum is the area believed to be the center for various **emotions**, and there is no extensive evidence that this area has a major role to play regarding intelligence.

Brain Stem

The **brain stem** contains the **higher reflex centers** and is actually the expanded superior end of the spinal cord. Nerves travel to and from the cerebrum through the brain stem, and cranial nerves originate here. The brain stem consists of several parts, the most important being the **medulla oblongata** (or **myelencephalon**). This structure contains the major portion of the **respiratory center**; the **cardioinhibitory center**, which decreases heart rate through the vagus nerve; the **vasomotor center**, which controls the size of small blood vessels; the **sweat center**, which stimulates the sweat glands to secrete sweat as a result of a rise in body temperature; and the origins of the 9th, 10th, 11th, and 12th cranial nerves.

As the **motor tract** of nerves passes from the motor aspect of the brain's cortex through the medulla oblongata, a major portion of the nerve fibers cross over to the opposite side (i.e., left cortex to right medulla oblongata) as the fibers enter the spinal cord. This process is known as **pyramidal decussation**.

The medulla oblongata contains the **reflex center** where emotional activities such as fear and excitement either stimulate or inhibit visceral movements, and it also contains the center that reflexively controls activities of the organs of the alimentary tract. The medulla oblongata is where, through skeletal activity, the smooth muscle of the alimentary tract, blood vessels, and various hollow organs is innervated allowing the smooth muscle of these structures to contract or relax.

The other parts of the brain stem include the **afterbrain** (**metencephalon** or **pons**), the **middle brain** (**mesencephalon** or **midbrain**), and the **second brain** (**diencephalon**, which includes the **thalamus** and **hypothalamus**). These areas contain the origins of the first through eighth cranial nerves as well as higher reflex centers for sight, sound, smell, and muscle tonus.

Cerebellum

The third portion of the brain is the **cerebellum**. It is actually part of the afterbrain (metencephalon). As opposed to the crossed control of the cerebrum, the cerebellum (consisting of two equal sides) is **homolateral**, meaning the right half controls the right side of the body, and the left half controls the left side of the body. Muscular coordination and body **equilibrium** are under the control of the cerebellum. The cerebellum has an amplifying effect on the motor impulses coming from the cerebral cortex.

The body's voluntary muscles are connected to the motor area of the brain (precentral gyrus) via nerves (cells and axons) of either the **direct** or **crossed pyramidal tracts**. The vast majority of nerves cross from the brain to the opposite side of the body (crossed pyramidal tract) as they pass through the medulla oblongata. The remaining nerves that do not cross over pass down the same side of the body (direct pyramidal tract).

Peripheral Nervous System

In addition to the central nervous system components (brain and spinal cord), the nervous system has a peripheral component, made up of the cranial nerves, spinal nerves, and autonomic nervous system. The **cranial nerves** (12 pairs) (figure 10.6) and the **autonomic nervous system** (figure 10.7) have their origins in the cranium and are therefore closely associated with the central nervous system. The **spinal nerves** (cervical, thoracic, lumbar, sacral, coccygeal) are discussed in chapter 1 and other chapters relevant to the anatomical areas they innervate.

The autonomic nervous system has two specific systems within it: the **sympathetic** (also known as the **thoracolumbar system**) and **parasympathetic** (also known as the **craniosacral system**). These systems, when stimulated, reflexively control the activity of all structures of the body that are not voluntarily controlled. Nerves conducting impulses to voluntary muscles are referred to as **motor nerves**; nerves conducting impulses to the heart, smooth muscle tissue, and various glands of the body

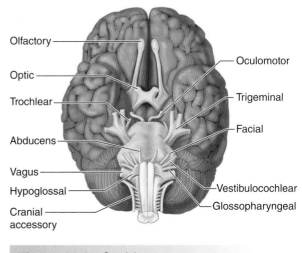

Figure 10.6 Cranial nerves.

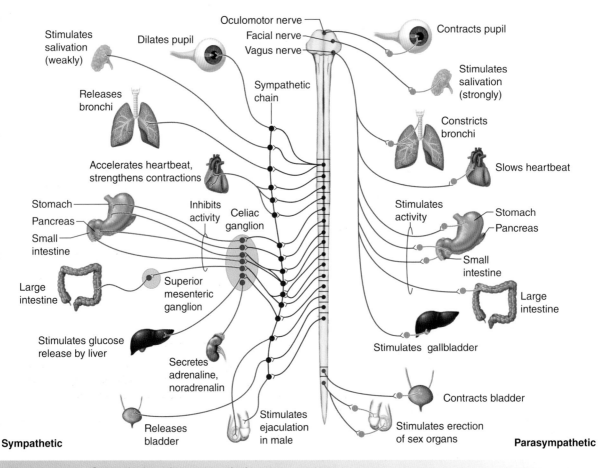

Figure 10.7 Sympathetic and parasympathetic nervous systems.

are considered part of the autonomic nervous system. The nerves consist of efferent, afferent, and mixed fibers that conduct nerve impulses both to (efferent) and from (afferent) the central nervous system.

As can be observed from their alternative names, the cells of these nerve fibers originate from various levels of the central nervous system. The nerves of the sympathetic system originate from the gray matter of the spinal cord from the first thoracic to the third lumbar level.

The 12 pairs of cranial nerves (table 10.1) contain both efferent (motor) and afferent (sensory) nerve fibers and also function in some aspects of the parasympathetic system.

The first pair of cranial nerves is the **olfactory**, running from the upper aspect of the nose to the temporal lobe where the center of smell is located. The second pair of cranial nerves is the **optic**. Impulses are conducted between the retina and the center for vision located in the occipital lobe. Both the olfactory and optic cranial nerves

TABLE 10.1

Cranial Nerves

| Name | FUNCTION | |
	Motor	Sensory
1. Olfactory		Smell
2. Optic		Sight
3. Oculomotor	Superior, medial, and ciliary sphincter pupillae; inferior recti and inferior eyeball muscles; oblique eye muscles	
4. Trochlear	Superior oblique, eyeball muscles	
5. Trigeminal	Muscles of mastication, teeth, gums, tensor tympani, face, skin, palate, veli palatini, mylohyoid, anterior 2/3 of tongue, anterior digastric	
6. Abducens	Lateral rectus muscle of the eyeball	
7. Facial	Muscles of face, scalp, anterior 2/3 of lacrimal gland, submaxillary platymus, stylohyoid, sublingual salivary gland, stapedius, posterior digastric	Tongue's taste buds
8. Auditory		Hearing
Cochlear		Hearing
Vestibular	Equilibrium via semicircular canals	
9. Glossopharyngeal	Stylopharyngeus muscle, pharynx, parotid salivary gland, lower 1/3 of tongue	Taste buds
10. Vagus	Muscles of the soft palate, pharynx, larynx, trachea, esophagus, stomach, intestines, other abdominal viscera, and the heart	Heart, lungs, trachea, pharynx, larynx, esophagus, and the gastrointestinal tract
11. Accessory	Trapezius and sternocleidomastoid muscles	
12. Hypoglossal	Tongue muscles	

are considered sensory nerves; the **oculomotor** (3rd cranial), **trochlear** (4th cranial), and **abducens** (6th cranial) are motor nerves that innervate the voluntary muscles of the eye. The **trigeminal** (5th cranial) nerve has sensory fibers that sense warmth, cold, pressure, and pain in the teeth and skin of the face and motor fibers that innervate the muscles of mastication. The **facial** (7th cranial) and the **glossopharyngeal** (9th cranial) nerves have sensory fibers connecting the brain with the taste buds of the tongue and the pharynx. The motor fibers of the glossopharyngeal innervate the muscles of the tongue and pharynx, and the facial innervates the muscles of the face. The **auditory** (8th cranial) nerve is sensory in nature and has two parts: one for hearing (located in the **cochlea**) connecting to the temporal lobe, and one for equilibrium (located in the **semicircular canals**) connecting with the cerebellum. The **vagus** or **pneumogastric** (10th cranial) nerve contains both sensory and motor fibers. The sensory fibers involve the heart, lungs, trachea, pharynx, larynx, esophagus, and gastrointestinal tract. The motor fibers of the vagus nerve innervate the muscles of the soft palate, pharynx, larynx, trachea, esophagus,

stomach, intestines, other abdominal viscera, and the heart. The **spinal accessory** (11th cranial) and the **hypoglossal** (12th cranial) are motor nerves originating in the brain stem and innervating the tongue, face, and neck.

Arteries of the Head and Brain

The epidural space contains the major arteries of the brain, and the subdural space contains the major veins of the brain. The left and right **common carotid arteries** ascend toward the head. Figure 10.8 illustrates the right common carotid artery. Arising from the **subclavian arteries** are the **vertebral arteries** that, in addition to supplying blood to the spinal cord and the vertebrae and musculature of the cervical spine, supply blood to the brain. The **basilar artery** arises between the two vertebral arteries (and branches known as **posterior cerebral arteries**) to supply blood to structures within the skull.

The right carotid artery, located at the angle of the jaw (mandible), arises from the brachioce-

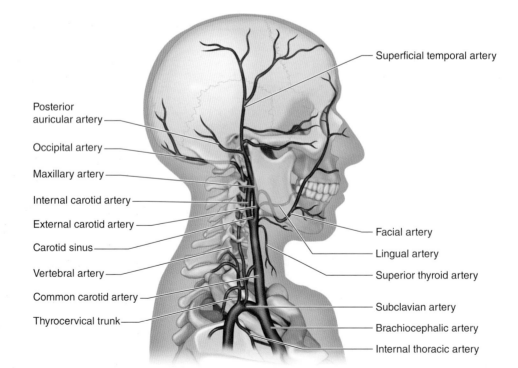

Figure 10.8 Arteries of the head, neck, and face.

phalic artery and separates into the right external and internal carotid arteries. The left carotid artery arises independently from the aortic arch and then separates into the left external and internal carotid arteries. The **external carotid artery** and its branches supply blood to the skin, muscles, tongue, glands, and other tissues of the upper aspect of the neck and the head. The branches of the external carotid artery include the following:

• Ascending pharyngeal artery: supplies the dura mater, tympanic area, pharyngeal wall, soft palate, cervical glands, and prevertebral muscles and has branches that include the inferior tympanic, posterior meningeal, prevertebral, palatine, and pharyngeal.

• External maxillary artery: supplies the mylohyoid, masseter, and buccinator muscles; the muscles of the upper and lower lip; the pharynx, auditory tube, palatine tonsils, submaxillary and sublingual glands, and the skin of the lips and nose; and the muscles and alar cartilages of the nose. Branches include buccal, angular, ascending palatine, glandular, masseteric, submental, superior and inferior labial, lateral nasal, and tonsilar.

• Lingual artery: supplies the posterior aspect of the tongue, the palatine tonsils, the sublingual gland, and the following muscles: geniohyoid, mylohyoid, and genioglossus. Branches include the deep lingual, sublingual, dorsal lingual, and hyoid.

• Occipital artery: supplies the dura mater, transverse sinus, internal jugular, medial auricular surface, and sagittal sinus and the postvertebral and sternocleidomastoid muscles. Branches include auricular, meningeal, and mastoid.

• Posterior auricular artery: supplies the auricle, posterior temporal scalp, parotid gland, external acoustic meatus, vestibule, semicircular canals, mastoid cells, and the following muscles: sternocleidomastoid, occipitalis, digastric, stylohyoid, styloglossus, stylopharyngeal, and stapedius. Branches include the occipital, parotid, auricular, and stylomastoid.

• Superior thyroid artery: supplies the thyroid gland, larynx membrane, and thyrohyoid and laryngeal muscles. Branches include hyoid, cricothyroid, and superior laryngeal.

At the terminal end of the external carotid artery are two important arteries: the internal maxillary and the superficial temporal arteries and their branches.

• Internal maxillary artery: supplies most structures of the head, excluding the brain and eyes, including the auditory tube, tympanic membrane, buccal membrane, dura mater, external acoustic meatus, maxilla, gums, teeth, lacrimal gland, eyelids, mandibular joint, maxillary sinus, sphenoidal sinus, skin of the face, hard and soft palate, palatine bone, vomer bone, parotid gland, nasal cavity, fifth cranial nerve root, and pharynx, along with the following muscles: buccinator, masseter, internal and external pterygoids, tensor veli palatini, levator veli palatini, inferior oblique, and inferior rectus. Branches include the pterygoid, sphenopalatine, accessory, meningeal, anterior tympanic, buccinator, deep auricular, deep temporal, infraoribtal, alveolar, pharyngeal, greater palatine, and middle meningeal.

• Superficial temporal artery: the posterior branch of the external carotid artery, supplying the upper head structures including the skin, fascia, external acoustic meatus, temporomandibular joint, lateral auricular surface, and parotid gland, along with the following muscles: frontalis, anterior and superior auricular, masseter, temporalis, and orbicularis oculi. Branches include the transverse facial, frontal, zygomaticoorbital, articular, anterior articular, middle temporal, parietal, and parotid.

The **internal carotid artery** is the posterior branch of the common carotid artery, supplying most of the brain, the meninges, the eyes, the frontal sinus, and the skin of the nose. The branches of the internal carotid artery are extensive and considered in more advanced coursework in human anatomy. To illustrate how numerous these deep arterial branches are, these branches and the structures they supply are outlined here, but they are not illustrated in this text:

Caroticotympanic: provides connections (anastomoses) between other cranial arteries

Cavernous: supplies blood to the cavernous sinus; the abducens; and the trochlear, trigeminal, and oculomotor nerves

Choroidal: supplies the optic tract

Hypophyseal: supplies the hypophysis

Meningeal: supplies the dura mater

Ophthalmic: supplies the eye, lacrimal gland, forehead surface, meninges, frontal sinus, skin of the upper nose, anterior and lateral scalp, and upper and lower eyelids, along with the superior and lateral recti and orbital muscles; branches are the anterior ethmoidal, anterior meningeal, central artery of retina, dorsal nasal, frontal, lacrimal, palpebral, posterior ethmoidal, and ciliary

Posterior communicating artery: supplies the thalamus and optic chiasma (forms the lateral side of the circle of Willis)

The two terminal branches of the internal carotid artery are the anterior and middle cerebral arteries.

• Anterior cerebral artery: supplies numerous structures of the brain, including the lobes of the anterior aspect of the brain, through the following branches: the anteriomedial basal artery; anterior, middle, and posterior medial frontal; and medial orbital

• Middle cerebral artery: supplies structures of the brain, including the lateral orbital surface of the frontal lobe, lateral surface of the occipital lobe, parietal lobe, temporal lobe, inferior and middle frontal gyri, and anterior and posterior central gyri, through the following branches:

temporal, ascending frontal and parietal, parietotemporal, inferior lateral frontal, and lateral orbital

The vertebral arteries also have numerous deep branches that supply structures of the brain, the meninges, the muscles of the neck, the spinal cord, and the cervical vertebrae. Again, an in-depth discussion of these structures is beyond the scope of an entry-level study of human anatomy. The deep branches are listed here along with the structures they supply, but they are not illustrated.

Anastomotic: connects occipital and deep cervical arteries

Anterior spinal: supplies the pia mater

Meningeal: connects arteries within the meninges

Muscular: supplies suboccipital muscles and deep cervical spine muscles

Posterior inferior cerebellar: supplies the medulla oblongata; the brain stem; structures within the hemisphere of the cerebellum; and the 9th, 10th, and 11th cranial nerves

Posterior spinal: supplies the pia mater

Spinal: supplies the bodies of the vertebrae and the intervertebral discs

The basilar artery is formed between the two vertebral arteries and supplies the pons; internal ear; anterior medullary velum; superior surface

FOCUS ON

Epidural Hematoma

As noted earlier, the arteries supplying blood to the brain are found primarily in the epidural space. A portion of the artery is attached to the dura mater, a portion lies in the epidural space, and the upper portion of the artery lies within actual grooves on the underside of the skull. When a body is in motion and the head is forced to suddenly stop, there can be a brief period where the skull is stopped while the brain is still moving, which results in the brain striking the skull (a concussion). The motion of the brain may cause a tearing of the artery as it moves away from its position in a groove of the skull. This could cause a laceration of the artery. A common form of this trauma involves the middle cerebral artery and its branches, primarily because of their anatomical position, resulting in an epidural hematoma (see figure 10.2). Because this form of hemorrhage causes rapid expansion between the brain's dura mater and skull (via arterial bleeding), vision, orientation, reasoning, and other brain functions can rapidly deteriorate and even endanger life. The epidural hematoma can be one of the most dangerous forms of head trauma, and its signs and symptoms should be known to anyone involved with activities that put participants at high risk of head trauma.

of the cerebellar hemispheres; visual area of the cerebral cortex; brain stem; hippocampal, fusiform, and lingual gyri; and fourth nerve root. Branches include the anterior inferior cerebellar, superior cerebellar, pontine, internal auditory, and posterior cerebral. The terminal branches of the basilar artery are known as the posterior cerebral arteries.

There is a circular formation of arteries on the base of the brain known as the **circle of Willis** (figure 10.9). This circle is formed anteriorly by the anterior cerebral arteries and the internal carotid artery branches, connected by the **anterior communicating artery**, and posteriorly by two posterior cerebral arteries and the basilar branches, connected on either side to the internal carotid by the **posterior communicating artery**.

Veins of the Head and Brain

The blood supplied to the head and brain by arteries is returned to the heart by veins (figure 10.10). There are deep veins, superficial veins, and sinuses (which are expanded spaces in the head that receive venous blood). The deep veins typically take the names of the arteries they drain, while the superficial veins usually have their own distinct names.

The **deep veins** of the head are the **internal jugular vein** and the **vertebral vein**. The internal jugular vein drains blood from the face, pharynx, thyroid gland, cochlea, and tongue into the **subclavian vein**.

Like the arteries, the veins have branches (known as tributaries). The tributaries of the

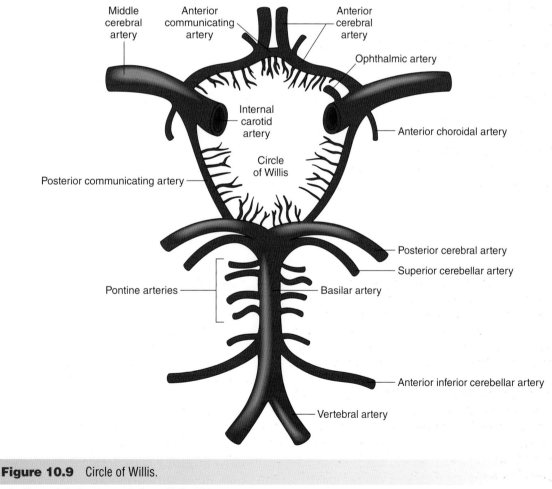

Figure 10.9   Circle of Willis.

Reprinted by permission from Denegar, Saliba, and Saliba 2010.

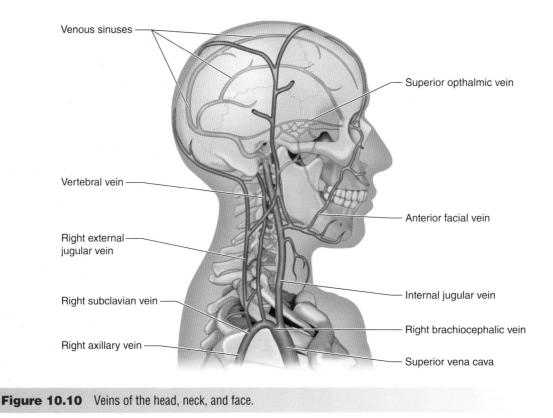

Figure 10.10 Veins of the head, neck, and face.

internal jugular vein include the cochlea vein, common facial vein, lingual vein, superior and middle thyroid vein, occipital vein, and pharyngeal vein and the inferior petrosal sinus.

The vertebral vein and its tributaries primarily drain the arteries of the cervical spine area into the **innominate vein** (which is formed by the joining of the subclavian and internal jugular veins). The primary tributaries include the anterior vertebral vein, which drains areas of the anterior cervical vertebrae; the deep cervical vein, which drains the structures of the posterior cervical area; and the posterior intercostal vein, which drains the first intercostal space.

Although there are only two deep veins in the head and neck, the **superficial veins** are numerous. The **external jugular vein** and its tributaries drain the muscles of the scalp, head, and neck and empty into the subclavian vein. The tributaries include the anterior jugular vein, which drains the anterior neck and lower lip surfaces; the posterior auricular vein, which drains the posterior and lateral surfaces of the head and neck; the posterior external jugular vein, which drains the posterior and lateral aspects of the neck and the occipital

area of the head; and the transverse cervical vein and transverse scapular vein, which drain the area supplied by the arteries of the same name.

The superficial veins that drain the scalp include the **frontal vein**, which drains the medial and anterior surfaces of the scalp and joins with the **supraorbital vein** to form the angular vein. The supraorbital vein drains the posterior and lateral surfaces of the scalp. The **occipital vein** drains the occipital and parietal surfaces of the scalp. The **posterior auricular vein** drains the posterior and lateral surfaces of the head and neck. The **superficial temporal vein** drains the frontal and lateral surfaces of the head and includes the following tributaries: the transverse facial vein draining the lateral side of the face, the orbital vein draining the lateral areas of the eyelids, the parotid vein draining the parotid gland, the middle temporal vein draining the temporalis, the articular vein draining the temporomandibular joint, and the anterior auricular vein draining the area anterior to the ear.

The superficial veins that drain the face include the **angular vein** (formed by the frontal and supraorbital veins), which drains the surface of the nose and then becomes the **anterior facial vein**. The anterior facial vein combines with the **posterior facial vein** to become the **common facial vein**, which drains the muscles and skin of the face; its tributaries include the buccinator vein, which drains the buccinator muscle; the **deep facial vein**, which drains the facial muscles and bones; the superior and inferior labial veins, which drain the upper and lower lips; the superior and inferior palpebral veins, which drain the upper and lower eyelids; the masseteric vein, which drains the masseter muscle; the submaxillary vein, which drains the submaxillary glands; the palatine vein, which drains the palate; and the submental vein, which drains the area supplied by the artery of the same name. The common facial vein (anterior and posterior facial veins) and a communicating vein from the anterior jugular vein drain into the internal jugular vein. The posterior facial vein, formed by the junction of the superficial temporal and the **internal maxillary veins** and its tributaries, has an anterior branch that joins with the anterior facial vein to form the common facial vein and a posterior branch joining the posterior auricular vein to form the external jugular vein.

Like the arteries of the head and brain, the veins of the head and brain also have an extensive number of deep vein branches that are beyond an entry-level study of human anatomy. The following is a brief description of the veins considered in this advanced category.

The veins of the cranium are contained in the spongy material (diploe) of the cranial bones and are interrelated with the veins of the pericranium (periosteum), meninges, and sinuses of the dura mater. There are four veins responsible for draining blood from the cranium. The anterior temporal diploic vein drains the frontal bone and empties into one of the deep temporal veins and the sphenoparietal sinus. The posterior temporal diploic vein drains the posterior area of the parietal bone and empties into the transverse sinus. The frontal diploic vein drains the anterior area of the frontal bone. The occipital diploic vein (the largest diploic vein) drains the occipital bone and empties into the transverse sinus.

The veins of the meninges are both deep and superficial, draining the arteries of the dura mater and emptying into the cranial sinuses. As with the arteries, these veins are embedded in the dura mater and also lie in the grooves on the interior surface of the skull. This position makes the potential for an intercranial hemorrhage very high, particularly when associated with a fracture of the skull.

The veins of the brain drain into the cranial sinuses and are described as either cerebral or cerebellar. The cerebral veins are both superficial (external) and deep (internal). The superficial veins include the inferior cerebral veins, which drain the inferior surface of the brain into the transverse, superior, cavernous, and petrosal sinuses. The inferior cerebral veins have two tributaries: the middle cerebral vein, which drains the cerebrum into the cavernous sinus, and the ophthalmic vein, which connects the cerebral veins with the orbit and empties into the superior ophthalmic vein. The superior cerebral veins (12 plus) are located in the pia mater and the subarachnoid space; they drain the superior and lateral areas of the cerebrum and empty into the superior sagittal sinus. The basal vein is a combination of the anterior cerebral, deep middle cerebral, and inferior striate veins. It drains the midbrain, the hippocampal gyrus, the inferior portion of the lateral ventricle, and the interpeduncular fossa. The anterior cerebral vein drains the corpus callosum and cingulate gyrus and empties into the basal vein. The deep middle cerebral vein drains the insula and opercula and empties into the basal vein. The inferior striate vein drains the corpus striatum and empties into the basal vein.

Two deep cerebral veins drain the internal aspects of the cerebrum and empty into the great cerebral vein. Receiving blood from the basal vein, the choroid vein and the vena terminalis combine to form the great cerebral vein.

The veins of the cerebellum are grouped into superior and inferior veins. The inferior cerebellar veins drain the inferior area of the cerebellum and empty into the dura mater sinuses. The superior cerebellar veins drain the superior area of the cerebellum and empty into the transverse and superior petrosal sinuses. The deep cerebellar veins drain the cerebellum and empty into the superficial veins.

The veins of the medulla oblongata drain into a superficial plexus of veins that drain into the anterior and posterior median and radicular veins. The veins of the pons empty into the superficial plexus.

The veins of the ear include those of the external acoustic meatus and external ear that drain into the posterior facial and posterior auricular veins. The veins of the tympanic membrane empty into the posterior facial vein and the superior petrosal sinus. The labyrinth of the ear is drained by the internal auditory veins into the inferior petrosal sinus and vestibular vein.

The veins of the nose drain the walls of the nasal cavity and form the sphenopalatine vein, which empties into the pterygoid plexus. The superior labial and the lateral nasal veins are tributaries of the anterior facial vein.

The veins of the orbit are branches of the ophthalmic artery and the frontal and supraorbital veins. These veins combine to form the superior and inferior ophthalmic veins and drain into the cavernous sinus.

The **sinuses** of the dura mater are widened venous areas lying between the periosteal and meningeal layers of the dura mater. The sinuses receive blood from the veins of the brain and the meningeal and diploic veins. The cranial sinuses are divided into paired lateral sinuses and unpaired median sinuses.

The **paired sinuses** include the superficial plexus on either side of the sphenoid bone; as the superficial plexus extends toward the temporal bone, it divides into the superior and inferior petrosal sinuses. The inferior petrosal sinus is located at the petrous area of the temporal bone and empties into the internal jugular vein. Tributaries of the inferior petrosal sinus include the veins of the internal ear and the inferior cerebellar veins. The superior petrosal sinus extends from the apex of the petrous area of the temporal bone at the posterior end of the cavernous sinus to the transverse sinus at the superior edge of the petrous area of the temporal bone. The occipital sinus is attached to the dura mater at the posterolateral borders of the foramen magnum. The occipital sinus has tributaries from the inferior cerebellar veins and empties into the transverse sinus. The cavernous sinus lies on each side of the sphenoid bone and has tributary veins including the inferior cerebral and ophthalmic veins and veins from the sphenoparietal sinuses. The sphenoparietal sinuses are found in the dura mater under the sphenoid bones. This is where the veins from the dura mater empty into the cavernous sinus. Tributaries of the sphenoparietal sinuses include the diploic, tympanic, and superior and inferior cerebellar veins. The transverse sinuses extend from the external occipital protuberance to the internal jugular vein. The transverse sinuses have tributary veins including the inferior cerebral and cerebellar veins, the diploic veins, and the superior petrosal sinuses.

The **unpaired sinuses** include the basilar plexus in the dura mater over the basilar area of the occipital bone and extending to the cavernous sinus along the edge of the foramen magnum. It connects the two inferior petrosal sinuses. The circular sinus encompasses the two cavernous sinuses and the connecting anterior and posterior intercavernous sinuses. The inferior sagittal sinus is found in the posterior margins of the cerebri and connects to the great cerebral vein. Tributaries of the inferior sagittal sinus include the veins from the cerebri and the medial aspects of the hemispheres. The occipital sinus is formed by the inferior sagittal sinus and the great cerebral vein and empties into the transverse sinus. Tributaries of the occipital sinus include veins from the cerebri, the occipital lobes, and the superior cerebellum. The superior sagittal sinus is in the median groove of the inner area of the cranium along the cerebri and extends from the foramen cecum to the internal occipital protuberance. Its tributaries are the superior cerebral veins; it empties into the transverse sinus.

Nerves of the Thorax and Trunk

As presented in the chapters on the upper and lower extremities (chapters 6 and 14, respectively), the nerves innervating the thorax and trunk originate from plexuses (networks) of nerves from the spinal cord. In particular, one plexus that innervates the lungs is vital for respiration and the maintenance of life. Additionally, the major nerves that serve the entire lower extremity originate at the lower end of the spinal column (lumbosacral plexus). This section of the chapter concentrates on the nerves that supply the muscles of the thorax, trunk, heart, and lungs. These structures are identified in the summary table found at the end of this section for the head, spinal column, and thorax.

The nerves that innervate the head and cervical spine arise from the **cervical plexus** (figure 10.11). The sensory nerves of the plexus are superficial, whereas the motor nerves are the deep branches. The deep branches innervate the muscles of the scalp and face and also the cervical spine muscles, such as the sternocleidomastoid, the trapezius, and the levator scapulae. Branches from a cranial nerve, the spinal accessory, also communicate with C2, C3, and C4 and innervate the sternocleidomastoid and trapezius muscles. Additional deep branches of the cervical plexus

innervate the rectus capitis anterior and lateralis muscles, the longus capitis and colli muscles, the prevertebral muscles, the levator scapulae, the scalenus medius muscle, the sternocleidomastoid, and the trapezius. The most important nerve of the cervical plexus for respiration is the **phrenic nerve** (C3, C4, C5), which innervates the **diaphragm**. Other nerves of the brachial plexus that innervate the muscles of the cervical spine and the lateral aspect of the thorax are described in chapter 6.

There are 12 **thoracic nerves**, often described as 11 **intercostal nerves** and one **subcostal nerve**, that arise from the thoracic section (T1–T12) of the spinal column (figure 10.12). The 1st thoracic intercostal nerve innervates the levatores costarum muscles, the intercostal muscles, and the superior portion of the serratus posterior muscles. The 2nd thoracic intercostal nerve innervates the intercostal muscles, the levatores

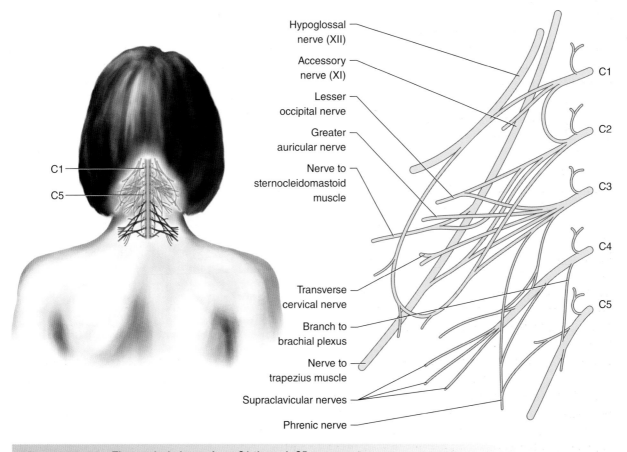

Figure 10.11   The cervical plexus, from C1 through C5 nerve roots.

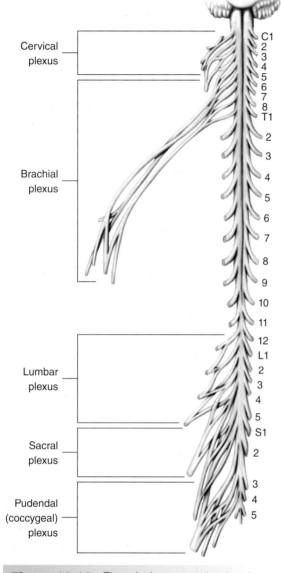

Cervical plexus

Brachial plexus

Lumbar plexus

Sacral plexus

Pudendal (coccygeal) plexus

C1
2
3
4
5
6
7
8
T1
2
3
4
5
6
7
8
9
10
11
12
L1
2
3
4
5
S1
2
3
4
5

Figure 10.12 The spinal nerves and major plexuses.

Cervical Spine Injury

When a person suffers a cervical spine injury, it is extremely important to assess the person's vital signs, including respiration. Trauma to the phrenic nerve can disrupt breathing, necessitating the first-aid technique known as rescue breathing. Disruption of breathing combined with cessation of the pulse calls for the initiation of cardio-pulmonary resuscitation (CPR). Obviously, when someone uses a CPR technique (chest compressions only) that does not include rescue breathing, any consideration of phrenic nerve trauma that may disrupt breathing is not a factor to consider.

serratus posterior, and the abdominal muscles. The 12th thoracic subcostal nerve innervates the abdominal muscles.

The **lumbosacral plexus** is formed by the anterior rami of the nerves of the **lumbar plexus**, the **sacral plexus**, and the **pudendal** (coccygeal) **plexus** (figure 10.12). Most of these nerves innervate the structures of the lower extremity—including the hips, buttocks, groin, and organs of the pelvic region—and are therefore discussed in chapter 14, the nerves and blood vessels of the lower extremity.

Arteries of the Thorax and Trunk

The **ascending aorta** arises from the heart and has right and left **coronary arteries**, which have branches to the heart itself. The area between the ascending and **descending aorta** is known as the **aortic arch** (figure 10.13), which first gives rise to the **brachiocephalic artery** that supplies blood to the right arm, the thorax, and the right side of the head and neck. The first branches from the brachiocephalic artery are the right common carotid and right subclavian arteries. The right common carotid artery, just below the jaw (mandible), divides into the external and internal carotid arteries. The left carotid artery arises independently from

costarum muscles, and the subcostal muscles. The 3rd thoracic intercostal nerve innervates the levatores costarum, intercostal, and subcostal muscles. The 4th thoracic intercostal nerve innervates the levatores costarum, intercostal, subcostal, and transversus thoracis muscles. The 5th and 6th thoracic intercostal nerves innervate the intercostal, subcostal, and transversus thoracis muscles. The 7th and 8th thoracic intercostal nerves innervate the intercostal muscles. The 9th, 10th, and 11th thoracic intercostal nerves innervate the intercostals, the inferior portion of the

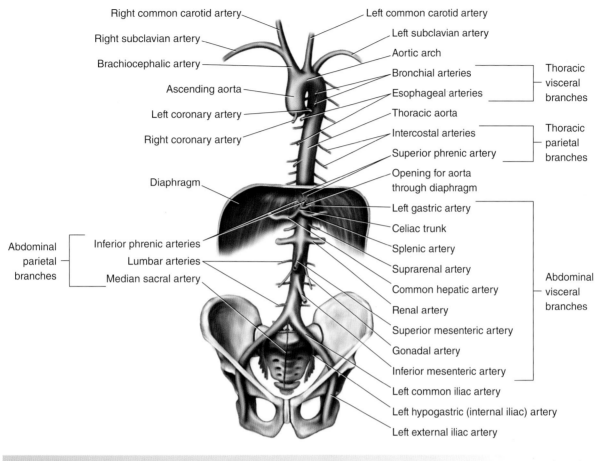

Figure 10.13   Major arteries of the trunk, thorax, and abdomen.

the aortic arch and then separates into the left external and internal carotid arteries.

✋ Hands On

To obtain a carotid artery pulse, place your index and middle fingers on your Adam's apple and move them laterally between this structure and the anterior muscles of your neck (figure 10.14). Apply gentle pressure with your fingertips. Do you feel a pulse? What vessel are you palpating to get that pulse? What muscle are you palpating to find this pulse?

The external carotid artery and its branches supply the muscles of the head and upper portion of the neck. The internal carotid artery and its branches supply the brain and other structures of the head. Additionally, the vertebral arteries, arising from the subclavian arteries, supply the brain, upper portions of the spinal cord, cervical vertebrae, and deep muscles of the neck. An

artery between the two vertebral arteries (known as the basilar artery) and its branches (the posterior cerebral arteries) supply structures within the skull.

The descending aorta, initially designated as the **thoracic aorta**, has both **thoracic** and **abdominal parietal** and **thoracic** and **abdominal visceral branches** (figure 10.13). The thoracic parietal branches (intercostal, subcostal, superior phrenic) supply muscles such as the intercostals, the vertebral column muscles, and the diaphragm. The thoracic visceral branches (bronchial, esophageal) supply the structures within the thoracic cavity. Below the diaphragm, the descending aorta becomes known as the **abdominal aorta** (figure 10.13). The abdominal parietal branches (inferior phrenic, lumbar, median sacral) supply the diaphragm and the musculature of the abdominal, lumbar, and sacroiliac areas. The abdominal visceral

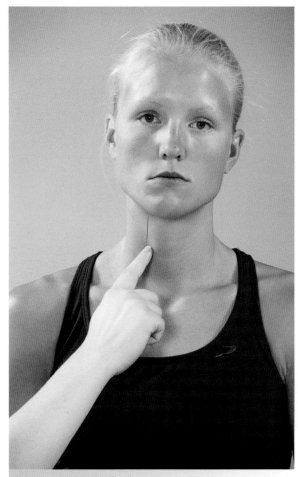

Figure 10.14 Finding the carotid artery.

nerves and blood vessels of the lower extremity. The hypogastric (internal) iliac artery has branches that supply muscles of the pelvic girdle and the hip joint.

Veins of the Thorax and Trunk

With a few exceptions, the veins of the thorax and trunk (figure 10.15) parallel the arteries they drain and also have the same or similar names. The **superior vena cava** is the major vein that drains the head, neck, shoulders, upper extremity, and parts of the thorax and abdomen into the heart. The **brachiocephalic vein** receives blood from the subclavian veins and drains into the superior vena cava. The veins draining the head and neck include the internal jugular, the vertebral, and the external jugular veins and their branches. The internal jugular drains the face and structures of the throat into the subclavian vein. The vertebral vein drains the structures of the skull, the posterior neck muscles, and the first intercostal area into the brachiocephalic vein. The external jugular and its more superficial branches drain the superficial muscles of the head and neck into the subclavian vein.

The **inferior vena cava**, which empties into the heart, is formed by the juncture of the left and right common iliac veins. It drains the lower extremities, the pelvis, and the abdominal region. The common iliac vein is formed by the junction of the external and internal iliac veins. The external iliac vein, which is a continuation of the femoral vein (discussed in chapter 14) drains the lower extremity. The hypogastric (internal) iliac vein and its branches drain structures of the lower abdominal and pelvic regions.

Nerves, Arteries, and Veins of the Heart and Lungs

The two major organs housed within the thorax are the heart and the lungs. As with all other anatomical structures, the heart and lungs need nerves and blood vessels to accomplish their functions.

branches (gastric, celiac, splenic, suprarenal, renal, common hepatic, superior and inferior mesenteric, gonadal) supply the organs of the abdominal cavity, including the liver, spleen, stomach, suprarenals, kidneys, pancreas, intestines, and reproductive structures.

At approximately the L4 level of the vertebral column, the abdominal aorta divides into the right and left **common iliac arteries** (figure 10.13). The arteries in the lower abdominal cavity and pelvic region are divisions of the common iliac artery: the **external iliac** and the **hypogastric (internal iliac) arteries**. The external iliac artery supplies the abdominal muscles, the iliacus, the psoas muscles, the sartorius, and the tensor fasciae latae of the lower extremity. At the level of the inguinal ligament, the external iliac artery becomes known as the **femoral artery**, which is discussed further in chapter 14 on the

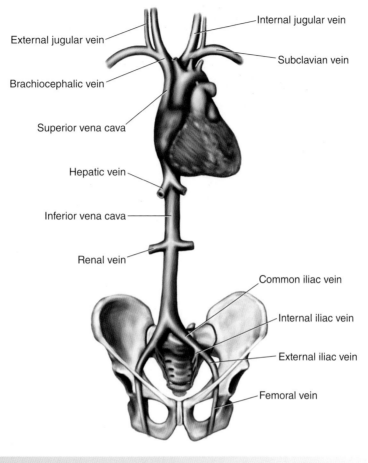

Figure 10.15 Major veins of the trunk, thorax, and abdomen.

Heart Nerves

The nerves of the heart are cardiac branches of the vagus nerve fibers arising from trunks of the sympathetic nervous system. When stimulated, the **sinoatrial node** (SA node) of the heart, found in the area of the right atrium near the superior vena cava, sends the impulse to the right and left **atria myocardium**. Specialized tissue (myocardial cells) of the **atrioventricular bundle** (AV node) (located in the lower aspect of the interatrial septum) receives the impulse after it has passed through the atrium (figure 10.16). The impulse continues through the **bundle of His** (atrioventricular bundle), which divides into left and right branches (becoming known as myofibers of conduction, or **Purkinje fibers**) that enter the muscular walls (myocardium) of the ventricles and the papillary muscles. The

result is atrial contraction rapidly followed by ventricular contraction.

Respiratory Nerves

The respiratory center is a group of cell bodies located on each side of the medulla oblongata. Arising from these centers are two sets of nerves (figure 10.17): (1) the phrenic nerves arising from the cervical plexus and leading to the diaphragm and (2) the intercostal nerves that innervate the intercostal muscles. When the respiratory center is stimulated by carbon dioxide, it sends an impulse over the phrenic nerves, causing the diaphragm to contract (pulling downward) and increasing space in the thoracic cavity. At the same time, the intercostal nerves cause the intercostal muscles to contract, lifting the ribs and also increasing the space within the thoracic cavity. This change in

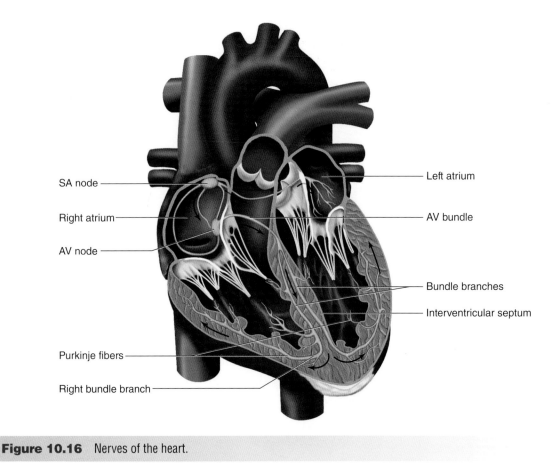

SA node

Right atrium

AV node

Purkinje fibers

Right bundle branch

Left atrium

AV bundle

Bundle branches

Interventricular septum

Figure 10.16 Nerves of the heart.

Adapted by permission from Kenney, Wilmore, and Costill 2012.

capacity of the thoracic cavity creates a change in atmospheric pressure, causing air to rush in and distend the lungs (inspiration). When the lungs have been distended to a certain point, sensory nerves running from the air sacs, via the vagus nerve, send an impulse to the respiratory center to inhibit it. This stops the center's impulses to the phrenic and intercostal nerves, causing the diaphragm and intercostal muscles to relax and resulting in a reduction in the size of the thoracic cavity, forcing air out of the lungs (expiration).

Heart Arteries

Two of the largest arteries of the heart are the **pulmonary artery** coming from the right ventricle and the **aorta** coming from the left ventricle (figure 10.18). These arteries have valves at their ventricular ends to prevent any backflow of blood. The pulmonary artery has a three-flap valve known as the **pulmonary semilunar valve**, and the aorta has a similar valve known as the **aortic semilunar valve**.

The right and left coronary arteries come from the aorta and are located on the outer surface of the heart, supplying blood flow to the muscular walls of the heart (myocardium). The branches of the right coronary artery include the posterior (dorsal) interventricular and marginal arteries supplying the anterior surface of the right ventricle, the aortic and pulmonary branches supplying the aorta and pulmonary arteries, the interventricular supplying both ventricles, the right atrial supplying the right atrium surface, and the right marginal supplying the inferior surfaces of both ventricles. The branches of the left coronary artery include the aortic and pulmonary branches supplying the aorta and pulmonary arteries, the **circumflex** supplying the left atrium and ventricles, the anterior (ventral) interventricular supplying both ventricles, and the left atrial supplying the left atrium.

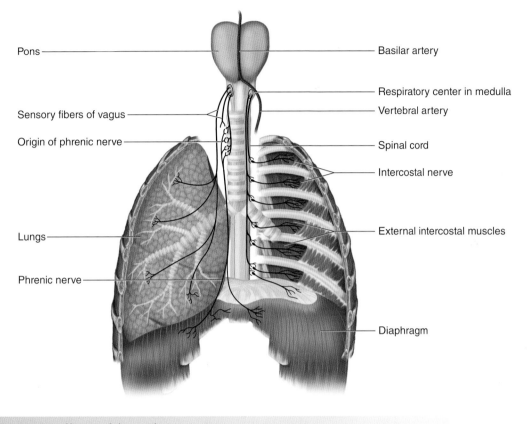

Pons — Basilar artery

Respiratory center in medulla

Sensory fibers of vagus — Vertebral artery

Origin of phrenic nerve — Spinal cord

Intercostal nerve

External intercostal muscles

Lungs —

Phrenic nerve —

Diaphragm

Figure 10.17 Nerves of the respiratory system.

Heart Veins

Blood from the body drains to the heart via two major veins: The superior vena cava (and its tributaries) drains the upper extremities, head, neck, shoulders, thorax, and a portion of the abdominal wall into the heart's right atrium. The inferior vena cava (and its tributaries) drains the lower extremities, pelvis, abdominal viscera, and a portion of the abdominal wall into the right atrium. The left atrium contains the opening for the **pulmonary veins** that bring the blood from the lungs to the heart.

The right and left coronary veins drain into the **coronary sinus**, which empties into the right atrium. Tributaries of the coronary sinus include the **great cardiac vein** draining the left atrium and both ventricles into the coronary sinus. The great cardiac vein also had a tributary: the left **margin vein** draining the left margin of the heart. Other coronary sinus tributaries include the **inferior cardiac vein** of the left ventricle

draining the inferior surface of the left ventricle, the **middle cardiac vein** draining both ventricles and emptying into the coronary sinus, the **oblique vein** of the left atrium draining the left atrium into the coronary sinus, and the **small cardiac vein** draining the right atrium and right ventricle into the coronary sinus (figure 10.18).

Respiratory Arteries and Veins

Blood flow to and from the lung tissues is accomplished through branches of the **bronchial arteries** and **bronchial veins**. More in-depth discussions of oxygen and carbon dioxide levels and changes in atmospheric pressures are found in coursework and texts in human physiology.

Arterial blood supplies the body cells with oxygen and is therefore well oxygenated. Venous blood is less oxygenated as it returns to the lungs. The blood in the pulmonary veins returning from the lungs to the heart is highly oxygenated, however (figure 10.19). In other words, the pulmonary veins are the only veins in the body that carry oxygen-rich blood.

A few veins of the heart do not drain into the coronary sinus. These include the **anterior cardiac veins**, arising from the wall of the right ventricle and emptying into the right atrium, and the **venae cordis minimae**, which are small veins in the heart walls that drain into the atria.

It is in the capillaries of the pulmonary vessels, in the walls of the **air sacs** in the lungs, where the **respiratory exchange** of oxygen and carbon dioxide takes place (figure 10.20).

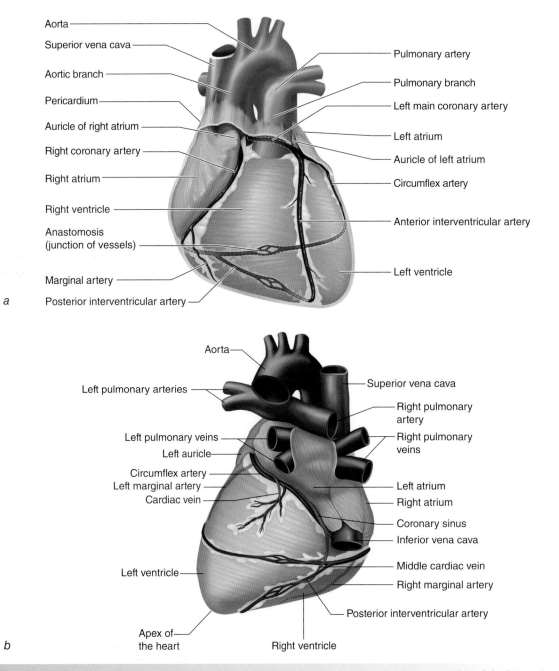

Figure 10.18   Heart blood vessels: *(a)* anterior heart, *(b)* posterior heart, and *(c)* cross section of the heart. *(continued)*

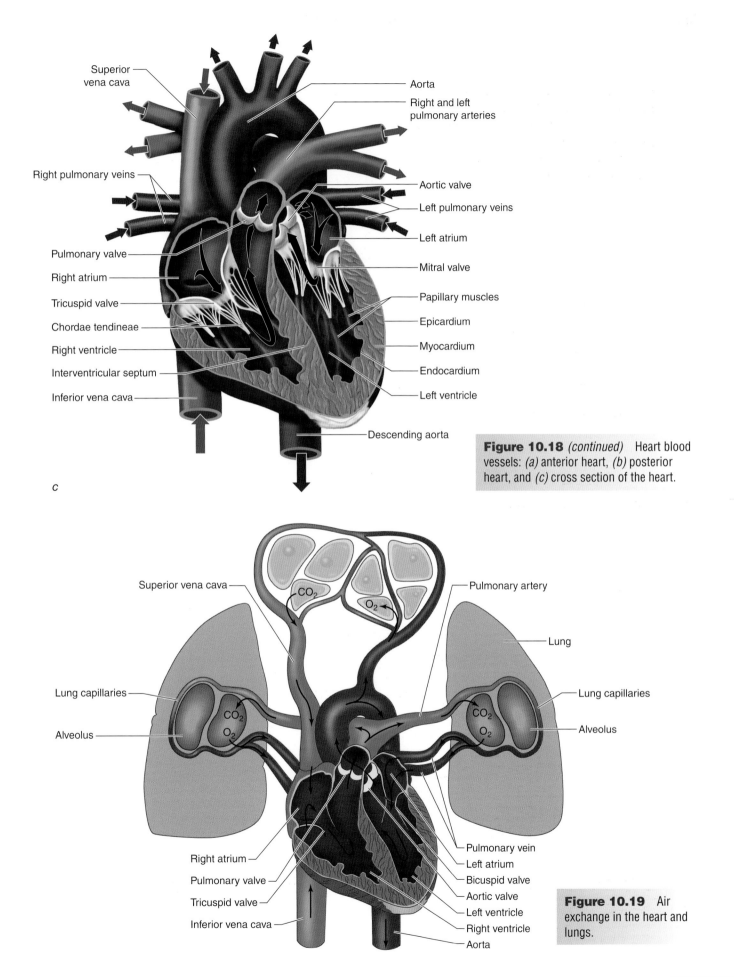

Superior vena cava

Aorta

Right and left pulmonary arteries

Right pulmonary veins

Aortic valve

Left pulmonary veins

Left atrium

Pulmonary valve

Mitral valve

Right atrium

Papillary muscles

Tricuspid valve

Epicardium

Chordae tendineae

Myocardium

Right ventricle

Endocardium

Interventricular septum

Left ventricle

Inferior vena cava

Descending aorta

c

Figure 10.18 *(continued)* Heart blood vessels: *(a)* anterior heart, *(b)* posterior heart, and *(c)* cross section of the heart.

Superior vena cava

Pulmonary artery

CO_2

O_2

Lung

Lung capillaries

Lung capillaries

Alveolus

CO_2

CO_2

O_2

O_2

Alveolus

Pulmonary vein

Right atrium

Left atrium

Pulmonary valve

Bicuspid valve

Tricuspid valve

Aortic valve

Left ventricle

Inferior vena cava

Right ventricle

Aorta

Figure 10.19 Air exchange in the heart and lungs.

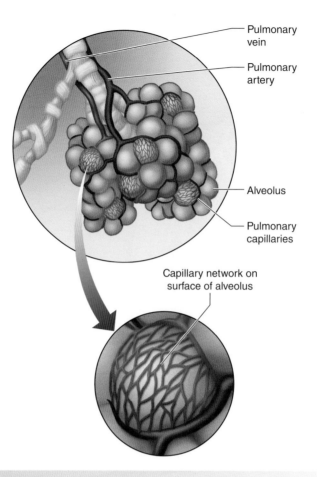

Pulmonary vein

Pulmonary artery

Alveolus

Pulmonary capillaries

Capillary network on surface of alveolus

Figure 10.20 Air sac artery and vein arrangement.

Adapted by permission from Kenney, Wilmore, and Costill 2012.

LEARNING AIDS

REVIEW OF TERMINOLOGY

The following terms are discussed in this chapter. Define or describe each term, and where appropriate, identify the location of the named structure either on your body or in an appropriate illustration.

abdominal aorta
abdominal parietal branch
abdominal visceral branch
abducens nerve
afterbrain
air sac
angular vein
anterior cardiac vein
anterior communicating artery
anterior facial vein
aorta
aortic arch
aortic semilunar valve
arachnoid
ascending aorta
association area

atria myocardium
atrioventricular bundle
auditory center
auditory nerve
autonomic nervous system
basilar artery
body sense area
brachiocephalic artery
brachiocephalic vein
bradycardia
brain stem
bronchial artery
bronchial vein
bundle of His
cardioinhibitory center
cerebellum

cerebrospinal fluid
cerebrum
cervical plexus
circle of Willis
circumflex
cochlea
common carotid artery (left and right)
common facial vein
common iliac artery
coronary artery
coronary sinus
corpus callosum
cortex
cranial nerve
craniosacral system

cross-control
crossed pyramidal tract
deep facial vein
deep vein
descending aorta
diaphragm
diencephalon
direct pyramidal tract
discriminatory power
dura mater
emotions
epidural space
equilibrium
external carotid artery
external iliac artery
external jugular vein
facial nerve
femoral artery
fissure of Rolando
fissure of Sylvius
frontal lobe
frontal vein
glossopharyngeal nerve
gray matter
great cardiac vein
gyrus
higher reflex center
homolateral
hypogastric (internal iliac) artery
hypoglossal nerve
hypothalamus
inferior cardiac vein
inferior vena cava
innominate vein
intelligence
intercostal nerve
internal carotid artery
internal jugular vein
internal maxillary vein
longitudinal fissure
lumbar plexus
lumbosacral plexus

margin vein
medulla oblongata
memory
meninges
mesencephalon
metencephalon
midbrain
middle brain
middle cardiac vein
motor area
motor nerve
motor tract
myelencephalon
oblique vein
occipital lobe
occipital vein
oculomotor nerve
olfactory center
olfactory nerve
optic nerve
paired sinuses
parasympathetic system
parietal lobe
phrenic nerve
pia mater
pneumogastric nerve
pons
posterior auricular vein
posterior cerebral artery
posterior communicating artery
posterior facial vein
power of reasoning
pudendal plexus
pulmonary artery
pulmonary semilunar valve
pulmonary vein
Purkinje fiber
pyramidal decussation
reflex center
respiratory center
respiratory exchange
sacral plexus

second brain
semicircular canals
sinoatrial node
sinus
small cardiac vein
speech center
spinal accessory nerve
spinal nerve
subarachnoid space
subclavian artery
subclavian vein
subcostal nerve
subdural space
sulcus
superficial temporal vein
superficial vein
superior vena cava
supraorbital vein
sweat center
sympathetic system
tachycardia
taste center
temporal lobe
thalamus
thoracic aorta
thoracic nerve
thoracic parietal branch
thoracic visceral branch
thoracolumbar system
trigeminal nerve
trochlear nerve
unpaired sinuses
vagus nerve
vasomotor center
venae cordis minimae
vertebral artery
vertebral vein
visual center
white matter
word hearing center
word visual center
writing center

SUGGESTED LEARNING ACTIVITIES

1. Heart rate is typically determined using one of two methods: palpation or auscultation. Using the palpation method, apply pressure to the radial artery with your fingers. Once you locate the pulse, count the number of beats for 20 seconds, and then multiply this number by 3 to determine your heart rate (beats per minute). At rest (both mental and physical), people's heart rates vary greatly (possibly as low as 40 beats per minute and up to 100 beats per minute). A very low heart rate (below 50 beats per minute) is often called **bradycardia**, while heart rates above 90 beats per minute are referred to as **tachycardia**.

2. The auscultation method used to determine heart rate utilizes a stethoscope placed over the area of the fourth or fifth intercostal space. If possible, establish heart rates using both palpation and auscultation techniques. If there is any difference between heart rates using both techniques, what anatomical factors might be involved to cause the differences?

MULTIPLE-CHOICE QUESTIONS

1. Which of the following structures does *not* belong in the same group as the other three structures?

    a. ascending aorta
    b. descending aorta
    c. abdominal aorta
    d. thoracic aorta

2. Which of the following structures does *not* drain into the inferior vena cava?

    a. internal iliac vein
    b. jugular vein
    c. common iliac vein
    d. femoral vein

3. The brain lies directly below which of the following meninges?

    a. arachnoid
    b. pia mater
    c. dura mater
    d. periosteum

4. The part of the brain stem known as the pons is also called the

    a. diencephalon
    b. metencephalon
    c. mesencephalon
    d. myelencephalon

5. The artery most frequently involved in an epidural hematoma, because of its anatomical position, is the

    a. anterior cerebral artery
    b. posterior cerebral artery
    c. middle cerebral artery
    d. meningeal artery

6. The bundle of His is also known as the

    a. AV node
    b. Purkinje fibers
    c. SA node
    d. myocardial fibers

7. The portion of the brain that makes a person consciously aware of her surroundings is the

    a. cerebrum
    b. cerebellum
    c. brain stem
    d. pons

FILL-IN-THE-BLANK QUESTIONS

1. The diaphragm is innervated by the _____ nerve.

2. The vertebral arteries arise from the _____ arteries to supply blood to the brain.

3. The brain consists of three major structures: the cerebrum, the cerebellum, and the _____.

4. One's emotions are believed to originate in the _____ lobe of the cerebrum.

5. The _____ are small veins in the walls of the heart that drain blood into the atria.

6. The majority of arteries supplying blood to the brain are located in the _____ space.

7. The motor area of the brain is located in the _____ gyrus.

Articulations of the Head, Spinal Column, Thorax, and Pelvis

Joint	Type	Bones	Ligaments	Movement
Head				
Temporomandibular	Diarthrodial	Mandible and temporal bone	• Capsule • Temporomandibular • Articular disc • Support from the stylomandibular and sphenomandibular	Elevation, depression, protrusion, medial and lateral mandibular motion
Spinal column				
Atlantooccipital	Condyloid	Occipital and atlas	• Capsules • Anterior atlantooccipital • Posterior atlantooccipital • Lateral atlantooccipital	Flexion, extension, abduction, adduction
Occipitoaxial	None	Axis and occipital	• Tectorial • Apical • Alar	None
Atlantoaxial	Pivot	Atlas and axis	• Capsules • Anterior atlantoaxial • Posterior atlantoaxial • Transverse	Rotation
Intervertebral	Amphiarthrodial (connected by cartilage)	Vertebral bodies	• Anterior longitudinal (occipital to sacrum) • Posterior longitudinal (occipital to coccyx) • Intervertebral discs	Slight gliding (limited flexion, extension, rotation)
Intervertebral	Arthrodial (plane/gliding)	Vertebral articular processes and arches	• Capsules • Ligamentum flavum (axis to sacrum) • Ligamentum nuchae (occipital to C7) • Supraspinous (thoracic to lumbar) • Interspinous (lumbar) • Intertransverse (lumbar)	Gliding

(continued)

Joint	Type	Bones	Ligaments	Movement
Thorax				
Costovertebral	Arthrodial (plane/gliding)	Vertebral bodies and heads of ribs	• Capsule • Costovertebral radiate • Interarticular	Gliding
Costotransverse	Arthrodial (plane/gliding)	Vertebral transverse processes and the necks and tubercles of ribs	• Capsule • Anterior costotransverse • Posterior costotransverse • Ligament of the neck • Ligament of the tubercle	Gliding
Sternocostal	Synarthrodial (fixed, joined by cartilage)	Manubrium of sternum and 1st rib	• Articular disc	None
Sternocostal	Arthrodial (plane/gliding)	Sternum and 2nd through 7th ribs	• Costosternal capsule • Costosternal radiate • Interarticular • Costoxiphoid	Gliding
Costochondral	Synarthrodial (fixed)	Ribs and costocartilage	• Capsule • Periosteum	None
Interchondral	Synarthrodial (fixed)	6th, 7th, 8th, 9th ribs and costocartilages	• Capsule • Interchondral	None
Pelvis				
Sacroiliac	Amphiarthrodial (connected by cartilage)	Sacrum and ilium	• Anterior sacroiliac • Posterior sacroiliac (3 bands: long, short, and interosseous) • Iliolumbar	Very limited (slight in pregnancy)
Pubic symphysis	Amphiarthrodial (connected by cartilage)	Pubic bones	• Interpubic disc • Anterior pubic • Posterior pubic • Superior pubic • Inferior pubic (arcuate)	(Slight in pregnancy)
Nonarticular ligaments of the pelvis				
		Sacrum, coccyx, and ilium	• Sacrotuberous • Sacrospinous	• (Stabilize the sacrum and coccyx)
		Anterior superior iliac spine to the pubic tubercle	• Inguinal ligament	• A landmark that separates the anterior abdominal wall from the thigh (the groin area)

Muscles, Nerves, and Blood Supply of the Head, Thorax, and Spinal Column

Muscle	Origin	Insertion	Action	Nerve	Blood supply
Cranial (epicranius)					
Frontalis	Galea aponeurotica	Procerus and orbicularis oculi	Elevates eyebrows and pulls scalp forward	Facial	Ophthalmic and superficial branch of external carotid
Occipitalis	Lateral 2/3 of superior nuchal lines	Galea aponeurotica	Pulls scalp backward	Facial	Occipital artery
Cranial (external ear)					
Anterior auricularis	Temporal fascia	Ear cartilage	Anterior ear movement	Facial	Superficial temporal branch
Posterior auricularis	Mastoid process	Cranial ear surface	Backward ear movement	Facial	Posterior auricular
Superior auricularis	Galea aponeurotica	Medial ear surface	Upward ear	Facial	Superficial temporal
Cranial (middle ear)					
Stapedius	Pyramidal eminence of temporal bone	Posterior aspect of neck of stapes	Backward movement of head of stapes	Facial (stapedius branch)	Posterior auricular artery
Tensor tympani	Roof of auditory tube and great wing of sphenoid	Anterior aspect of handle of malleus	Tenses tympanic membrane	Otic ganglion	Internal maxillary branch
Facial muscles (eyelid)					
Corrugator supercilii	Medial aspect of superciliary arch	Orbicularis oculi	Pulls eyebrow down and wrinkles forehead	Facial	Superficial external branch of external carotid and ophthalmic
Levator palpebrae superioris	Small wing of sphenoid	Superior tarsal plate of eyelid	Elevates upper eyelid	Oculomotor	Superficial temporal branch of external carotid and ophthalmic branch of internal carotid
Orbicularis oculi	Sphincter muscle in and around eyelid	Sphincter muscle in and around eyelid	Closes eyelid, compresses lacrimal sac, and wrinkles forehead	Facial	Labial branch of external axillary

(continued)

PART III Summary Tables (continued)

Muscle	Origin	Insertion	Action	Nerve	Blood supply
Facial muscles (nose)					
Caput angulare	Frontal process of maxilla	Ala of nose and upper lip	Elevates lateral half of upper lip and ala of nose	Zygomatic branch of facial	External and internal branches of maxillaries and ophthalmic
Depressor alaeque nasalis	Incisive fossa of maxilla	Septum	Pulls nostrils down, constricts nostrils, and depresses tip of nose	Facial	Lateral nasal branch of external maxillary branch of external carotid
Dilator naris anterior	Greater alar cartilage	Margin of nostril skin	Dilates external nostrils	Facial	Lateral nasal branch of external maxillary
Dilator naris posterior	Nasal notch of maxilla	Margin of nostril skin	Dilates external nostrils	Facial	Lateral nasal branch of external maxillary
Nasalis alar	Greater alar cartilage	Tissue of tip of nose	Pulls nostrils down and constricts nostrils	Facial	Lateral nasal branch of external maxillary branch of external carotid
Nasalis transverse	Maxilla	Membrane over nose	Pulls nostrils up and laterally	Facial	Lateral nasal branch of external maxillary branch of external carotid
Procerus	Tissue stretched over nose and nasal bone	Skin over root of nose and fibers attached to frontalis	Pulls skin over central portion of forehead downward and wrinkles skin at root of nose	Infraorbital branch of facial	Angular branch of external maxillary branch and nasal branch of ophthalmic
Facial muscles (mouth)					
Buccinator	Buccinator crest of mandible and molar aspect of outer alveolar process of maxilla	Fibers to the orbicularis oris in both upper and lower lip	Pulls corners of mouth laterally, pulls lips against teeth, flattens cheeks and assists mastication	Buccal branch of facial	Buccinator branch of internal maxillary
Caninus	Canine fossa of maxilla	Orbicularis oris	Elevates and medially pulls corners of mouth	Zygomatic branch of facial	Branches of external and internal maxillaries
Mentalis	Incisor fossa of mandible	Skin of chin	Draws up skin of chin, protrudes lower lip	Mandibular branch of facial	External maxillary

Orbicularis oris	Muscles of mouth area	Lip skin and mucosa	Sphincter closing mouth	Facial	Labial branch of external maxillary
Quadratus labii inferior	Lateral surface of mandible and medial to triangularis	Orbicularis oris and lower lip	Protrudes and pulls up lower lip	Mandibular branch of facial	Mental branch of external maxillary
Quadratus labii superior	Three parts: caput muscles				
Caput angulare	Frontal process of maxilla	Ala of nose and upper lip	Raises lateral 1/2 upper lip and ala of nose	Zygomatic branch of facial	Branches of external and internal maxillaries and ophthalmic
Caput infraorbitale	Maxilla below infraorbital foramen	Skin of upper lip	Elevates upper lip	Zygomatic branch of facial	Branches of external and internal maxillaries
Caput zygomaticum	Lower external surface of zygomatic bone	Skin of upper lip	Elevates upper lip	Zygomatic branch of facial	Branches of external and internal maxillaries
Risorius	Parotid masseteric fascia	Skin of angle of mouth	Pulls angle of mouth laterally	Buccal branch of facial	Branches of external and internal maxillaries
Triangularis	Oblique line of mandible	Orbicularis oris and skin of lower lip	Pulls corners of mouth both downward and laterally	Buccal branch of facial	External maxillary
Zygomaticus	Temporal process of zygomatic bone	Orbicularis oris and skin of upper lip	Elevates and pulls corners of mouth laterally	Zygomatic branch of facial	Branches of external and internal maxillaries and ophthalmic
Facial muscles (mastication)					
Internal pterygoids	Palatine bone and maxillary tuberosity	Medial aspect of ramus and angle of mandible	Protraction, elevation, and lateral movement of mandible	Mandibular branch of 5th nerve	Pterygoid branch of internal maxillary
External pterygoids	Great wing and lateral pterygoid plate of the sphenoid	Neck of mandible	Protraction, closing, and lateral movement of mandible	Mandibular branch of 5th nerve	Pterygoid branch of internal maxillary
Masseter	Inferior border of zygomatic arch	Lateral aspect of ramus and angle of mandible	Elevation and protraction of mandible	Mandibular branch of 5th nerve	Internal maxillary branch of external carotid
Temporalis	Temporal fossa	Coronoid process and ramus of mandible	Elevation and retraction of mandible	Mandibular branch of 5th nerve	Temporal branch of external carotid

(continued)

Muscle	Origin	Insertion	Action	Nerve	Blood supply
Eyeball muscles (extrinsic)					
Superior obliquus	Superior aspect of optic foramen	Sclera of lateral surface of eyeball	Moves eyeball laterally and inferiorly	Trochlea	Ophthalmic
Inferior obliquus	Orbital aspect of maxilla	Sclera of lateral surface of eyeball	Moves eyeball laterally and superiorly	Oculomotor	Ophthalmic and infraorbital branch of internal maxillary
Orbitalis	Spans inferior orbital fissure	Spans inferior orbital fissure	Protrudes eye	Sympathetic	Ophthalmic branch of internal carotid
Rectus superior	Fibrous ring around optic foramen	Sclera of eyeball	Medial and superior movement of eyeball	Oculomotor	Ophthalmic
Rectus medialis	Fibrous ring around optic foramen	Sclera of eyeball	Medial movement of eyeball	Oculomotor	Ophthalmic and infraorbital branch of maxillary
Rectus lateralis	Fibrous ring around optic foramen	Sclera of eyeball	Lateral movement of eyeball	Abducens	Lacrimal branch of ophthalmic
Rectus inferior	Fibrous ring around optic foramen	Sclera of eyeball	Medial and inferior movement of eyeball	Oculomotor	Ophthalmic and infraorbital branch of internal maxillary
Eyeball muscles (intrinsic)					
Ciliary	Posterior aspect of sclera	Ciliary process	Accommodation	Short ciliary nerve of oculomotor	Ophthalmic branch of internal carotid
Dilator pupillae	Involuntary and radiating fibers converging toward center of eyeball and blending with circular fibers	Involuntary and radiating fibers converging toward center of eyeball and blending with circular fibers	Dilates pupil	Sympathetic fibers from superior cervical ganglion	Ophthalmic branch of internal carotid
Sphincter pupillae	Involuntary and circular fibers surrounding pupil margins	Involuntary and circular fibers surrounding pupil margins	Constricts pupil	Oculomotor (3rd cranial) through ciliary ganglion	Ophthalmic branch of internal maxillary
Tongue muscles (extrinsic)					
Chondroglossus	Lesser cornu of hyoid bone	Side of tongue	Tongue depression	Hypoglossal (12th cranial)	Lingual artery of external carotid

Muscle	Origin	Insertion	Action	Innervation	Blood supply
Genioglossus	Upper genial tubercle of the mandible	Tip of tongue and hyoid bone	Depression, protrusion (posterior fibers), and retraction (anterior fibers) of tongue	Hypoglossal (12th cranial)	Lingual artery of external carotid
Hyoglossus	Body and greater cornu of hyoid bone	Side of tongue	Tongue depression	Hypoglossal (12th cranial)	Lingual artery of external carotid
	Anterior border of styloid process and stylohyoid ligament	Side and inferior surface of tongue	Elevation and retraction of tongue	Hypoglossal (12th cranial)	Deep lingual branch of lingual and posterior auricular artery
Styloglossus					
Tongue muscles (intrinsic)					
Superior longitudinalis	Base of tongue	Tip of tongue	Controls shape of tongue	Hypoglossal (12th cranial)	Deep lingual branch of lingual
Inferior longitudinalis	Inferior portion of tongue	Tip of tongue	Controls shape of tongue	Hypoglossal (12th cranial)	Deep lingual branch of lingual
Transversus linguae	Median septum of tongue	Dorsum of tongue	Controls shape of tongue	Hypoglossal (12th cranial)	Deep lingual branch of lingual
Verticalis linguae	Dorsum of tongue	Extrinsic tongue muscles	Controls shape of tongue	Hypoglossal (12th cranial)	Deep lingual branch of lingual
Soft palate muscles					
Glossopalatinus	Dorsum and side of tongue	Soft palate	Pulls tongue up and back, depresses sides of soft palate	Glossopharyngeus	Lingual branch of external carotid
Uvulae	Posterior nasal spine and soft palate aponeurosis	Uvula	Elevates uvula	Pharyngeal plexus	Palatine branch of internal maxillary
Levator veli palatini	Cartilaginous aspect of auditory tube and petrous part of temporal bone	Soft palate aponeurosis	Elevates soft palate, narrows pharyngeal opening, and expands isthmus of auditory tube	Vagus	Branch of internal maxillary

(continued)

Muscle	Origin	Insertion	Action	Nerve	Blood supply
Larynx muscles (intrinsic)					
Aryepiglotticus	Apex of arytenoid cartilage	Thyroepiglottic ligament and lateral edge of epiglottic cartilage	Pulls down epiglottis	Superior laryngeal	Superior laryngeal branch of superior thyroid
Arytenoid oblique	Muscular process of arytenoid cartilage	Apex of opposite arytenoid cartilage and angle of thyroid	Constricts laryngeal aperture and vestibule of larynx	Anterior branch of inferior laryngeal nerve	Superior laryngeal branch of superior thyroid branch of external carotid
Arytenoid transverse	Posterior concave area of one arytenoid cartilage	Posterior concave area of other arytenoid cartilage	Narrows glottis	Posterior branch of inferior laryngeal nerve	Superior laryngeal branch of superior thyroid branch of external carotid
Cricoarytenoid posterior	Posterior surface of cricoid lamina	Muscular process of arytenoid cartilage	Tenses vocal ligaments and widens glottis	Posterior branch of inferior laryngeal nerve	Inferior laryngeal branch of inferior thyroid
Cricoarytenoid lateralis	Upper margin and outer surface of cricoid arch	Anterior surface of muscular process of arytenoid cartilage	Stretches and approximates vocal ligaments	Anterior branch of inferior laryngeal nerve	Superior thyroid branch of external carotid
Cricothyroid	Arch of cricoid cartilage	Caudal margin of thyroid cartilage	Elevates cricoid cartilage and depresses lamina and arytenoid cartilage	External branch of inferior laryngeal nerve	Branch of superior thyroid
Thyroarytenoid	Lamina of thyroid cartilage	Margin of arytenoid cartilage	Relaxes vocal ligaments	Anterior branch of inferior laryngeal nerve	Superior laryngeal branch of superior thyroid branch of external carotid
Thyroepiglotticus	Inner surface of thyroid lamina	Quad membrane and lateral border of epiglottis	Closes epiglottis	Anterior branch of inferior laryngeal nerve	Superior laryngeal branch of superior thyroid branch of external carotid
Ventricularis	Lateral edge of arytenoid cartilage	Lateral margin of epiglottic cartilage	Pulls epiglottis downward	Internal branch of superior laryngeal nerve	Superior laryngeal branch of superior thyroid branch of external carotid
Vocal muscle	Angle of thyroid lamina	Vocal process and oblong fovea of arytenoid cartilage	Relaxes vocal ligaments	Anterior branch of inferior laryngeal nerve	Superior laryngeal branch of superior thyroid branch of external carotid
Thorax					
Diaphragm	Sternum (xiphoid), last 7 ribs, first 4 lumbar vertebrae	Central tendon	Pulls central tendon down	Phrenic	Superior and inferior phrenic

Muscle	Origin	Insertion	Action	Nerve level	Nerve
Internal intercostals	Upper portion of costal groove	Upper edge of rib cartilage below	Elevation of rib	T2–T6	Intercostals
External intercostals	Lower edge of rib cartilage	Upper edge of rib below	Elevation of rib	T2–T6	Intercostals
Levatores costarum	Transverse process of C7 and T1–T11	Angle of rib below	Elevation of rib	T2–T6	Intercostals
Subcostals	At angle on inner aspect of rib	Inner aspect of 2nd and 3rd ribs below	Elevation of rib	T2–T6	Intercostals
Serratus posterior superior	Spines of C7 and T1–T3	Angles of 2nd, 3rd, 4th, and 5th ribs	Elevation of rib	1st–3rd intercostal	Intercostal and transverse cervical
Serratus posterior inferior	Spines of T11–T12 and L1–L3	Inferior edge of last 4 ribs	Depression of rib	10th and 11th intercostal	Intercostal and transverse cervical
Transversus thoracis	Distal 1/3 of posterior aspect of sternum and cartilage of 4th, 5th, 6th, and 7th ribs	Posterior aspect of 2nd, 3rd, 4th, 5th, and 6th rib cartilage	Elevation of rib	2nd–6th thoracic	Intercostals
Spinal column, cervical (anterior)					
Sternocleidomastoid	Middle 1/3 of clavicle and sternum	Mastoid process	Flexion, lateral flexion, and rotation of cervical spine	Spinal accessory, 2nd and 3rd cervical	Transverse cervical
Prevertebrals					
Rectus capitis anterior	Lateral aspect of atlas	Occipital bone	Flexion and lateral flexion of cervical spine	C1 and C2	Pharyngeal and inferior thyroid
Rectus capitis lateralis	Transverse process of atlas	Inferior aspect of occipital bone	Flexion and lateral flexion of cervical spine	C1 and C2	Pharyngeal and occipital
Longus colli					
Vertical	Body of C5–C7 and T1–T3	Bodies of C2–C5	Flexion of cervical spine	C2–C8	Inferior and superior thyroid
Superior oblique	Anterior aspect of C3–C5 transverse processes	Anterior atlas	Flexion of cervical spine	C2–C8	Inferior and superior thyroid
Inferior oblique	T1–T3 bodies	C5–C6 transverse processes	Flexion of cervical spine	C2–C8	Inferior and superior thyroid

(continued)

PART III Summary Tables *(continued)*

Muscle	Origin	Insertion	Action	Nerve	Blood supply
Spinal column, cervical (anterior) *(continued)*					
Scaleni					
Scalenus anterior	C3–C7 transverse processes	First rib	Lateral flexion of cervical spine and elevation of 1st rib	C4–C8	Transverse cervical and inferior thyroid
Scalenus medius	C2–C7 transverse processes	First rib	Lateral flexion of cervical spine and elevation of 1st rib	C4–C8	Cervical and inferior thyroid
Scalenus posterior	C4–C6 transverse processes	Second rib	Elevation of 2nd rib	C4–C8	Cervical and inferior thyroid
Spinal column, cervical (posterior)					
Rectus capitis posterior major	Spine of axis (C2)	Occipital bone	Extension, lateral flexion, and rotation of cervical spine	C1	Vertebral
Rectus capitis posterior minor	Posterior atlas (C1)	Occipital bone	Extension, lateral flexion, and rotation of cervical spine	C1	Vertebral
Obliquus capitis superior	Transverse process of atlas (C1)	Occipital bone	Extension of cervical spine	C1	Vertebral and occipital
Obliquus capitis inferior	Spine of axis (C2)	Transverse process of the (C1)	Rotation of cervical spine	C1	Vertebral
Splenius capitis	C7–T6 spinous processes	Mastoid process	Extension, lateral flexion, and rotation of cervical spine	C2–C4	Occipital and transverse cervical
Splenius cervicis	T4–T6 spinous processes	C1–C4 transverse processes	Extension, lateral flexion, and rotation of cervical spine	C2–C4	Occipital and transverse cervical
Spinal column, thoracic/lumbar (anterior)					
Rectus abdominis	Pubic symphysis	5th–7th rib cartilage and sternum	Trunk flexion and pelvic girdle flexion		
Obliquus internus	Lateral inguinal ligament, middle of iliac crest			Last 6 intercostals	Lower intercostals and epigastric

Muscle	Origin	Insertion	Action	Nerve	Blood Supply
Obliquus externus	Inferior edge of last 8 ribs	Pubic bone to 7th, 8th, and 9th ribs	Trunk flexion and rotation	Last 3 intercostals, iliohypogastric, and ilioinguinal	Inferior epigastric branch of external iliac
Transversus abdominis	Inner aspect of last 6 ribs, internal aspect of middle 1/2 of iliac crest, lateral 1/3 of inguinal ligament	Outer edge of middle 1/2 of iliac crest	Trunk flexion, rotation, and pelvic girdle flexion	Last 7 intercostals, iliohypogastric	Lower intercostals and inferior epigastric branch of external iliac
		Pubic bone and linea alba	Flexion and rotation of trunk	Last 6 intercostals, iliohypogastric, ilioinguinal, and genitofemoral	Inferior epigastric branch of external iliac
Spinals					
Spinalis capitis	C7 and T1–T7 transverse processes	Occipital bone	Extension of cervical spine	T6–T9	Occipital
Spinalis cervicis	T4–T6 spinous processes	C2, C3, and axis spinous processes	Extension of cervical spine	T6–T9	Vertebral and transverse cervical
Spinalis dorsi (thoracis)	T11, T12, L1, and L2 spinous processes	T4–T8 spinous processes	Extension of vertebral column	T6–T9	Intercostals
Semispinals					
Semispinalis capitis	C4–C7 articular processes and T1–T6 transverse processes	Occipital bone	Extension, lateral flexion, and rotation of cervical spine	C1–C5	Occipital, transverse cervical, vertebral, and deep cervical
Semispinalis cervicis	T1–T6 transverse processes	C2–C5 spinous processes	Extension and lateral flexion of cervical spine	T3–T6	Deep cervical, occipital, and vertebral
Semispinalis dorsi (thoracis)	T6–T10 transverse processes	C6–C7 and T1–T4 spinous processes	Extension of cervical spine	T3–T6	Intercostals
Iliocostals					
Iliocostalis cervicis	Angle of 3rd, 4th, 5th, and 6th ribs	C4–C6 transverse processes	Lateral flexion and rotation of cervical spine	Cervical	Thyrocervical
Iliocostalis dorsi (thoracis)	Angle of last 6 ribs	Upper edge of first 6 ribs and C7 transverse process	Lateral flexion and rotation of spine, depression of ribs	Thoracic	Intercostals

(continued)

PART III Summary Tables *(continued)*

Spinal column, thoracic/lumbar (anterior) *(continued)*

	Origin	Insertion	Action		
Iliocostalis lumborum	Posterior aspect of iliac crest	Angle of last 7 ribs	Extension and rotation of spine	Lumbar	Lumbar
Longissimi					
Longissimus capitis	C4–C7 articular processes and T1–T5 transverse processes	Mastoid process	Extension, lateral flexion, and rotation of cervical spine	C1–L5	Transverse cervical, external carotid, and vertebral
Longissimus cervicis	T1–T5 transverse processes	C2–C6 transverse processes	Extension, lateral flexion, and rotation of cervical spine	C1–L5	Transverse cervical, external carotid, and vertebral
Longissimus dorsi (thoracis)	L1–L5 transverse processes and posterior iliac crest	3rd–12th ribs and T1–T12 transverse processes	Extension, lateral flexion, and rotation of cervical spine; rib depression	C1–L5	Intercostals and lumbar
Sacrospinalis (erector spinae)	Lumbar and sacral spinous processes	Angle of last 7 ribs, 3rd–12th ribs and T1–T12 transverse processes, T4–T8 spinous processes	Extension, lateral flexion, and rotation of spine	Thoracic, lumbar, sacral	Lumbar and intercostal
Quadratus lumborum	Lower lumbar transverse processes, posterior aspect of iliac crest	Lowest edge of 12th rib and L1–L4 transverse processes	Extension and lateral flexion of spine, depression of 12th rib	L1–L4	Lumbar and lower intercostal
Multifidus	C4–C7 articular processes, T1–T12 and L1–L5 transverse processes, posterior superior iliac spine, and posterior sacrum	S5–C4 spinous processes	Extension and rotation of spine	C4–S5	Lumbar

Lower Extremity

The Hip and Thigh

Any discussion of the hip must include the bones of the pelvis and thigh and the muscles of the thigh as well as the ligaments and muscles of the hip. Several muscles that cross the hip joint also cross the knee joint, and discussion of these muscles in this chapter is restricted to their function at the hip joint. Their function at the knee joint is discussed in the next chapter.

Bones of the Hip Joint and Thigh

The hip joint is the articulation between the pelvis and the **femur**, or thigh bone. The pelvis is discussed in chapter 8. The important struc-ture of the pelvis regarding the hip joint is the anatomical area labeled the **acetabulum** (figure 11.1; also see figure 8.32). Remember that the acetabulum is formed by the articulation of the three bones making up the **innominate bone** (the **ilium**, the **ischium**, and the **pubic bone**). The hip joint is classified as a ball-and-socket or triaxial joint. The acetabulum is considered the socket of the joint, and the ball of the hip joint is the structure known as the **head** of the femur (figure 11.2). Although the hip joint and the shoulder joint are often compared because of their similarities as ball-and-socket or triaxial joints, the hip joint is a much more stable joint than the shoulder because of the depth of the acetabulum compared with the very shallow glenoid fossa of the scapula. It is through the hip joint that the entire weight of the trunk and upper extremities is transferred to the lower extremities.

The femur, the longest and largest bone in the body, is discussed in this chapter only as it pertains to the hip joint (at the proximal end). Further discussion of the distal end of the femur is presented in chapter 12 on the knee.

The ball, or head, of the femur is separated from the shaft of the bone by the **neck** of the femur. At the proximal end of the shaft of the femur, just distal to the neck, are two large projections of bone known as the **greater** and **lesser trochanters**. These structures serve as points of insertion for the major muscles responsible for hip joint actions.

Hands On

Place your hand over your hip joint and rotate one leg toward the other leg (internally rotate your hip joint) and feel the greater trochanter as it moves against your skin (figure 11.3).

The lesser trochanter is distal, medial, and somewhat posterior to the greater trochanter and more readily observed in a posterior view of the femur. The ridge of bone running between the lesser and greater trochanters is known as the **intertrochanteric ridge** (see figure 11.2). A large ridge or crest of bone running down the posterior aspect of the shaft of the femur is known as the **linea aspera**, Latin for "rough line" (see figure 11.2). The linea aspera serves as a source

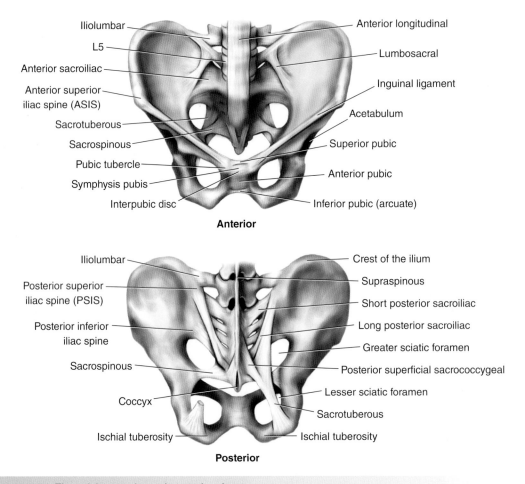

Anterior

Iliolumbar — L5 — Anterior sacroiliac — Anterior superior iliac spine (ASIS) — Sacrotuberous — Sacrospinous — Pubic tubercle — Symphysis pubis — Interpubic disc

Anterior longitudinal — Lumbosacral — Inguinal ligament — Acetabulum — Superior pubic — Anterior pubic — Inferior pubic (arcuate)

Posterior

Iliolumbar — Posterior superior iliac spine (PSIS) — Posterior inferior iliac spine — Sacrospinous — Coccyx — Ischial tuberosity

Crest of the ilium — Supraspinous — Short posterior sacroiliac — Long posterior sacroiliac — Greater sciatic foramen — Posterior superficial sacrococcygeal — Lesser sciatic foramen — Sacrotuberous — Ischial tuberosity

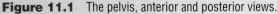

Figure 11.1   The pelvis, anterior and posterior views.

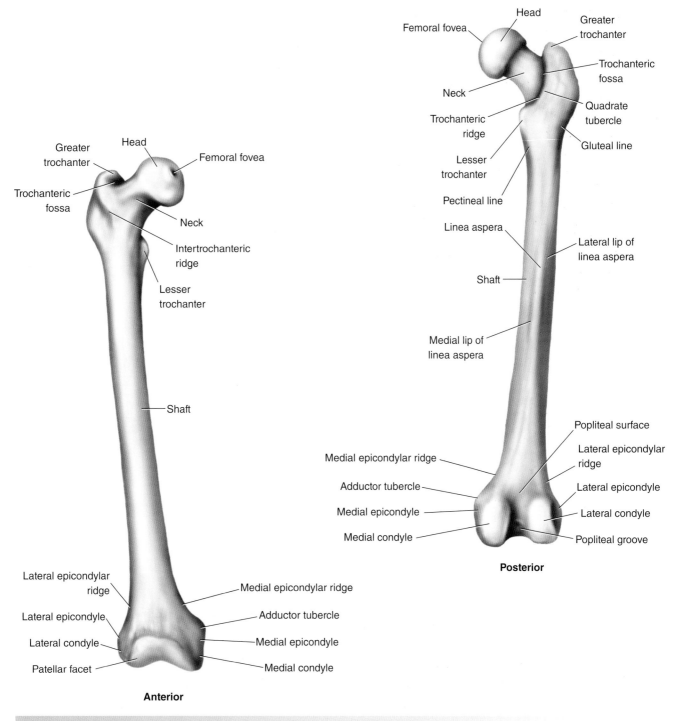

Anterior labels:
Greater trochanter
Head
Femoral fovea
Trochanteric fossa
Neck
Intertrochanteric ridge
Lesser trochanter
Shaft
Lateral epicondylar ridge
Medial epicondylar ridge
Lateral epicondyle
Adductor tubercle
Lateral condyle
Medial epicondyle
Patellar facet
Medial condyle

Anterior

Posterior labels:
Femoral fovea
Head
Greater trochanter
Trochanteric fossa
Neck
Quadrate tubercle
Trochanteric ridge
Gluteal line
Lesser trochanter
Pectineal line
Linea aspera
Lateral lip of linea aspera
Shaft
Medial lip of linea aspera
Popliteal surface
Lateral epicondylar ridge
Medial epicondylar ridge
Adductor tubercle
Lateral epicondyle
Medial epicondyle
Lateral condyle
Medial condyle
Popliteal groove

Posterior

Figure 11.2 The femur, anterior and posterior views.

the **pubofemoral ligaments** (figures 11.5 and 11.6). The iliofemoral ligament looks like an upside-down Y and therefore is frequently referred to as the Y ligament. Because of the anatomical relationships between the ilium and the femur and between the pubic bone and the femur, the iliofemoral and pubofemoral ligaments cross the anterior aspect of the hip joint and attach to the neck of the femur. Likewise,

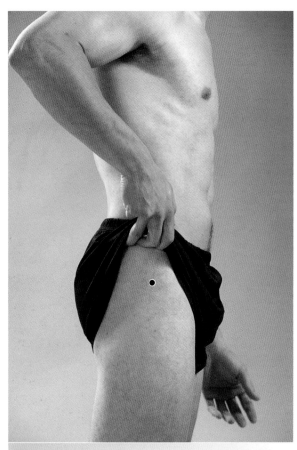

Figure 11.3 Locating the greater trochanter of the femur.

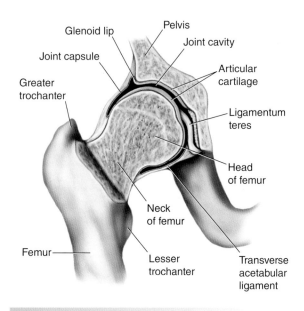

Figure 11.4 A longitudinal section of the hip joint.

of muscular attachment for numerous muscles of both the hip and knee joints.

Ligaments of the Hip Joint

Although the hip joint (figure 11.4) derives most of its stability from the bony formation between the head of the femur and the acetabulum of the pelvis, it is also reinforced by several strong ligaments. There are seven ligaments of the hip joint. As with all synovial joints, there is a **capsular ligament** running from the edge of the acetabulum of the pelvis to the neck of the femur. The **glenoid lip** (acetabular labrum), as in the shoulder joint, is a fibrocartilaginous rim on the outer edge of the acetabulum that helps deepen its socket. Three ligaments that strengthen the capsular ligament are easily identified because their names represent the bones they tie together: the **iliofemoral**, the **ischiofemoral**, and

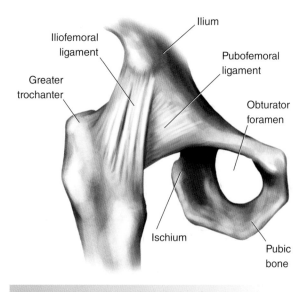

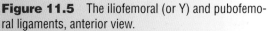

Figure 11.5 The iliofemoral (or Y) and pubofemoral ligaments, anterior view.

because the ischium is posterior to the femur, the ischiofemoral ligament crosses the posterior aspect of the hip joint and attaches to the neck of the femur. The **transverse acetabular ligament** spans a small notch in the lower portion of the glenoid lip and creates a foramen that blood vessels to and from the hip joint pass through (figure 11.7). The last ligament, the **liga-mentum capitis femoris** (figure 11.7), runs between the transverse acetabular ligament and a hole in the superior edge of the femoral head known as the **femoral fovea** (see figure 11.2). This ligament attaches the head of the femur to the acetabulum and plays the least significant role of any of the ligaments of the hip joint.

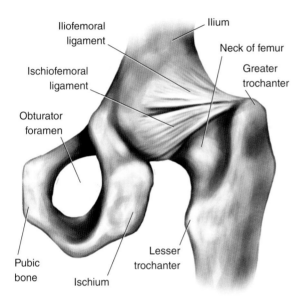

Figure 11.6 The ischiofemoral ligament, posterior view.

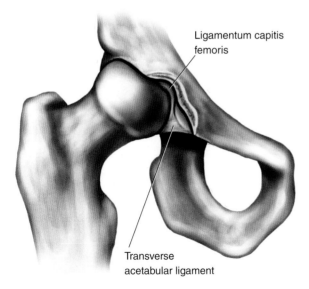

Figure 11.7 The transverse acetabular ligament and ligamentum capitis femoris, anterior view.

Fundamental Movements of the Hip Joint

The hip joint is classified as a **triaxial joint**, having movement in all three planes about all three axes: flexion and extension in the sagittal plane about a frontal horizontal axis, abduction and adduction in the frontal plane about a sagittal horizontal axis, and internal and external rotation in the horizontal plane about a vertical axis. The musculature that creates those six fundamental movements includes flexors, extensors, abductors, adductors, and rotators. They are presented in groups based on their position: anterior, posterior, lateral, or medial to the hip joint.

🖑 Hands On

As you stand in the anatomical position, flex your hip. What group of muscles primarily flexed your hip (anterior, posterior, lateral, medial)? Extend your hip joint to return to the anatomical position. Again, what muscle group (anterior, posterior, lateral, medial) was primarily involved in extending your hip? Now abduct and then adduct your hip joint. Determine which muscle groups are primarily responsible for these actions of the hip joint. To determine which muscle groups are primarily responsible for internal and external rotation of the hip joint, lie on your back, and internally and externally rotate your hip joint while palpating the muscles surrounding your hip joint. Now that you have completed the hip exercises, we can look at the muscles that cross the hip joint.

Muscles of the Hip Joint and Upper Leg

Muscles that cross the hip joint are grouped by position as anterior, posterior, medial, or lateral to the joint.

Anterior Muscles

From the previous exercise, you know that the muscles that performed flexion of your hip are located anterior to the hip joint. Interestingly enough, some of these muscles also cross the knee joint. In this chapter, however, only their function at the hip joint is considered. The anterior muscles are listed here (figure 11.8).

- **Iliopsoas**: The iliopsoas muscle is a combination of the iliacus and psoas major muscles. These muscles, along with the psoas minor, are commonly referred to as the true groin muscles (or hip flexor muscle group): The term *groin* refers to the anterior aspect of the hip joint. Their main function is to flex the hip joint, but if the femur is stabilized, these muscles cause flexion of the lumbar spine when they contract. The medial muscles of the hip joint (the adductors) are often referred to as the common groin muscles (or hip adductor muscle group) and are discussed later in this chapter.

- **Psoas major**: This muscle originates on the transverse processes of all five **lumbar vertebrae** and the bodies and **intervertebral discs** of the 12th **thoracic vertebra** and all five lumbar vertebrae. The psoas major inserts on the lesser trochanter of the femur. Along with the iliacus muscle, this muscle flexes the hip joint, assists with adduction and external rotation of the hip joint, and assists with flexion and rotation of the lumbar spine (figure 11.9).

- **Iliacus**: This muscle originates from the iliac fossa of the pelvis to the tendon of the psoas major and inserts just distal to the lesser trochanter of the femur. Along with the psoas major muscle, this muscle flexes the hip joint, assists with adduction and external rotation of the hip joint, and, when the femur is stabilized, assists with flexion of the lumbar spine (figure 11.9).

- **Psoas minor**: This muscle is not present on one or both sides in more than 50% of the population. The muscle originates from the 12th thoracic and first lumbar vertebrae and the intervertebral disc between these two vertebrae and inserts on the pubic bone. Therefore, the psoas minor is *not* considered a muscle of the hip joint.

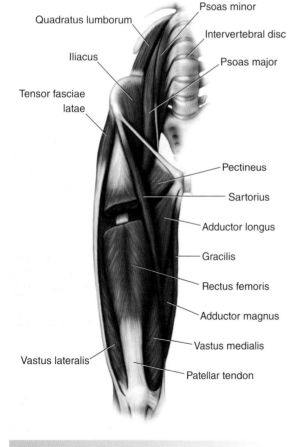

Figure 11.8 Anterior muscles of the hip.

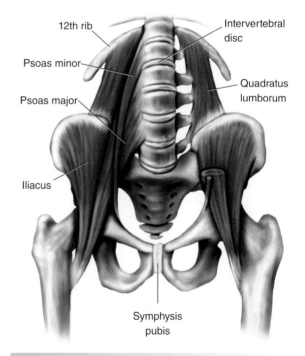

Figure 11.9 The psoas major, psoas minor, and iliacus.

Its function is to assist the psoas major with flexion of the lumbar spine (figure 11.9).

- **Sartorius**: This muscle, the longest muscle of the body, originates from the anterior superior spine of the ilium and inserts on the medial aspect of the proximal end of the tibia, a bone of the lower leg (see chapter 12, figure 12.5), just inferior to the medial condyle (figure 11.8). Note that this muscle crosses both the hip and knee joints. The functions of the sartorius at the hip joint are flexion, abduction, and external rotation.

✋ Hands On

Sitting on the floor with your ankles crossed and your knees apart (the tailor's position), you should be able to palpate your sartorius just beneath the anterior superior iliac spine (figure 11.10). The terms *sartorius* (from a Latin word for *tailor*) and *tailor's position* both refer to tailors' practice in bygone days of sitting for hours to do their work in a cross-legged position, which often caused soreness in this muscle after a long day of sewing.

- **Rectus femoris**: This muscle is part of a powerful group of four muscles known as the **quadriceps femoris**. Three of these muscles, the vastus lateralis, the vastus intermedius, and the vastus medialis, cross only the knee joint, but the rectus femoris crosses both the knee and hip joints. The muscle originates on the anterior inferior iliac spine and inserts on the tibial tuberosity. The rectus femoris flexes the hip joint and also acts as a very weak abductor.

✋ Hands On

Lie on a table with your hips in extension and your knees flexed over the end of the table, and extend your knee against a partner's manual resistance. Note the prominence of muscular tissue running the length of the anterior thigh area (figure 11.11). This prominence is the rectus femoris contracting against your resistance.

Figure 11.10 Finding the sartorius.

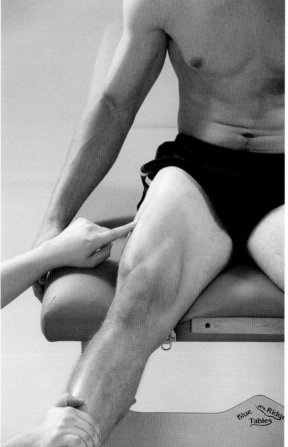
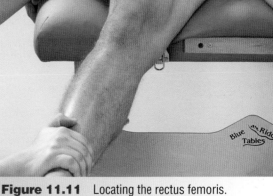

Figure 11.11 Locating the rectus femoris.

- **Tensor fasciae latae**: This muscle originates on the external rim of the crest of the ilium and combines with the gluteus maximus muscle to form another structure, the **iliotibial band** or **tract**, which inserts in the area of the lateral condyle of the tibia (figures 11.8, 11.12, and 12.5). The tensor fasciae latae flexes and abducts the hip joint. Even though the term *latae* might lead one to believe this is a lateral muscle of the hip joint, it is actually considered an anterior hip joint muscle, although a very lateral anterior muscle. More about this muscle and its involvement with the iliotibial band is presented later in this chapter and in chapter 12 on the knee joint.

- **Pectineus**: This muscle originates between the anterior superior aspect of the pubic bone and inserts on the femur just distal to the lesser trochanter (figure 11.8). Besides crossing the hip anteriorly, making it a hip flexor, it is a strong adductor and external rotator of the hip because of its angle of pull.

Posterior Muscles

Ten posterior muscles cross the hip joint (figure 11.13), including a group of three (the **hamstrings**) and a group commonly referred to as the six **deep external rotators**. The hamstrings are discussed only as they function at the hip joint. They also cross the knee joint and are discussed again in chapter 12.

- **Biceps femoris**: Although all three hamstring muscles (biceps femoris, semitendinosus, and semimembranosus) originate on the **ischial tuberosity** and have the same functions—extension of the hip joint and flexion of the knee joint—the biceps femoris is unique in that it has two heads: The **long head** originates on the ischial tuberosity with the semitendinosus and semimembranosus, and the **short head** originates from the linea aspera of the femur. Both heads combine into the belly of the muscle, which then inserts on the head of the **fibula**, a bone of the lower leg, and the lateral condyle of the tibia

Figure 11.12 Demonstrating the tensor fasciae latae and iliotibial band.

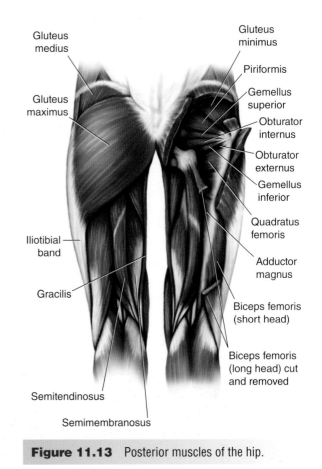

Figure 11.13 Posterior muscles of the hip.

(figure 12.5). The biceps femoris functions specifically at the hip joint to extend the joint and assist in adduction and external rotation of the joint.

👋 Hands On

The tendon of insertion of the biceps femoris can be observed on the posterior lateral aspect of the knee joint when you lie in the position illustrated in figure 11.14.

• **Semitendinosus**: Another muscle of the hamstring group, the semitendinosus, also originates on the ischial tuberosity and inserts on the medial aspect of the tibia just below the medial condyle (figure 12.5).

👋 Hands On

The tendon of insertion of the semitendinosus has a very prominent, cordlike structure (figure 11.15). Have your partner lie in a prone (facedown) position and flex his knee joint, and then locate this tendon on the posterior medial aspect of the knee joint. The muscle extends and assists with internal rotation and adduction of the hip joint.

• **Semimembranosus**: The final muscle of the hamstring group is the semimembranosus, which, like the semitendinosus, originates on the ischial tuberosity and inserts on the medial

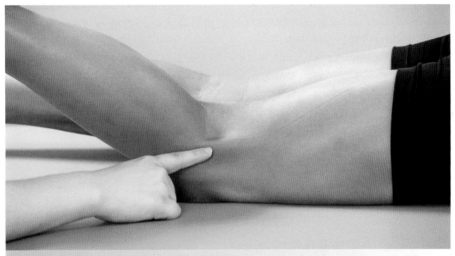

Figure 11.14   Locating the biceps femoris.

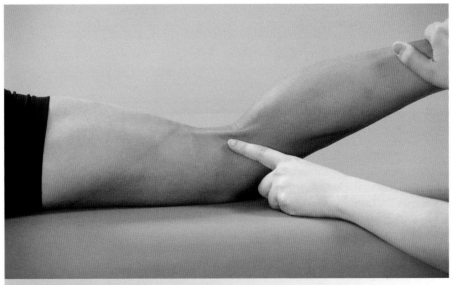

Figure 11.15   Finding the semitendinosus.

condyle of the tibia (figure 12.5). This muscle extends the hip joint and assists with internal rotation and adduction of the hip joint.

✋ Hands On ▬▬▬

The semimembranosus tendon of insertion is a very broad structure just medial to the semitendinosus tendon of insertion (figure 11.16). Ask your partner to assume the same position as when you located the semitendinosus, and then locate the semimembranosus.

• **Gluteus maximus**: This rather large muscle is often identified as the buttocks muscle. It originates from the posterior aspects of the ilium, **sacrum**, and **coccyx** bones and inserts by blending with the tensor fasciae latae muscle to form the iliotibial band (figure 11.13). The muscle extends and externally rotates the hip joint.

Because one portion of the belly of the muscle crosses above the hip joint and another portion below the hip joint, the muscle is involved in both abduction and adduction of the hip. Look carefully at the hip joint in figure 11.13, and determine which portion of the gluteus maximus assists with abduction and which portion assists with adduction of the hip joint.

✋ Hands On ▬▬▬

Deep below the gluteal muscles is a group of six muscles that perform external rotation of the hip joint. Lying on your back, legs straight, have your partner stabilize one of your feet in the anatomically neutral position. Attempt to move your foot away from the other foot by externally rotating the entire lower extremity. Apply hand pressure into the belly of the gluteus maximus muscle of the rotating leg. Did you feel contraction of muscles

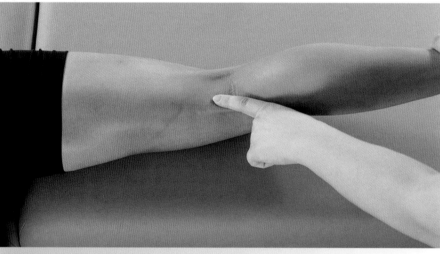

Figure 11.16   Locating the semimembranosus.

Hamstring Strain

FOCUS ON

The term *hamstring strain* is often used and often misunderstood. There are three muscles (biceps femoris, semitendinosus, and semimembranosus) that could be strained as a result of hip hyperflexion or knee hyperextension or, more likely, a combination of both hip flexion and knee extension. Although determining which of the three muscles (or combination of muscles) is strained is of little importance to the general public, someone with a good knowledge of human anatomy should be able to understand that *hamstring strain* is a somewhat nonspecific term. Flexibility is a major factor in preventing hamstring strains because many athletic activities require the hip joint to flex at the same time the knee joint is extending, which puts this muscle group under tension. (Picture a hurdler's lead leg as it goes over a hurdle.)

deep within the joint? What muscles were contracting? Let's look for these answers in the discussion of the six deep external rotator muscles of the hip joint (figure 11.13), which are presented anatomically from superior to inferior.

- **Piriformis**: This muscle originates on the upper portion of the sacrum, crosses the **greater sciatic notch**, and inserts on the upper surface of the greater trochanter. If contracted in spasm (sudden, involuntary contraction) as the result of strain or physical trauma, this muscle can impinge the sciatic nerve as it passes through the **sciatic notch** of the pelvis, causing pain to radiate down the lower extremity along the course of the sciatic nerve and its branches.
- **Gemellus superior**: Originating from the **spine of the ischium** and inserting on the medial aspect of the greater trochanter, the gemellus superior externally rotates the hip joint.
- **Internal obturator (obturator internus)**: Originating from the inner surface of the **obturator foramen** of the pelvis and inserting on the medial aspect of the greater trochanter, the internal obturator externally rotates the hip joint.
- **Gemellus inferior**: Originating from the ischial tuberosity and inserting on the medial aspect of the greater trochanter, the gemellus inferior externally rotates the hip joint.
- **External obturator (obturator externus)**: Originating from the outer surface of the pubic bone and ischium around the obturator foramen and inserting on the **trochanteric fossa** of the femur, the external obturator externally rotates the hip joint.
- **Quadratus femoris**: The quadratus femoris originates on the ischial tuberosity, inserts on the intertrochanteric ridge of the femur, and externally rotates the hip joint.

Medial Muscles

Three of the four adductor muscles of the hip joint give their function away by their names. The fourth adductor, the gracilis, is the only one of the four that also crosses the knee joint. These four muscles (figure 11.17), as mentioned earlier, are known as the common groin muscles (or hip adductor muscle group). Compared with the true groin muscles (the iliopsoas muscle group), these groin muscles are more commonly affected when one experiences a groin strain.

- **Adductor longus**: This muscle originates from the body of the pubic bone and inserts on the middle third of the linea aspera of the femur. Its major function is adduction of the hip, and it assists with flexion and external rotation.
- **Adductor brevis**: This muscle originates from the body and inferior surface of the pubic bone and inserts on the proximal third of the linea aspera of the femur. Its major function is adduction of the hip, and it assists with external rotation and flexion.
- **Adductor magnus**: This muscle (figures 11.17 and 11.18) originates on the inferior surface of the pubic bone, the ischium, and the ischial tuberosity and inserts on the entire length of the linea aspera and the **adductor tubercle** of the femur. The adductor magnus has two distinct portions: the anterior portion, which adducts, flexes, and externally rotates the hip joint, and the posterior portion, which adducts, extends, and internally rotates the hip joint.
- **Gracilis**: This muscle originates on the lower portions of the anterior aspect of the **pubic symphysis** and the pubic bone and inserts on the tibia just distal to the medial condyle. At the hip, the gracilis adducts and flexes the joint.

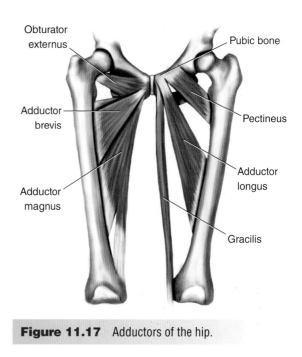

Figure 11.17 Adductors of the hip.

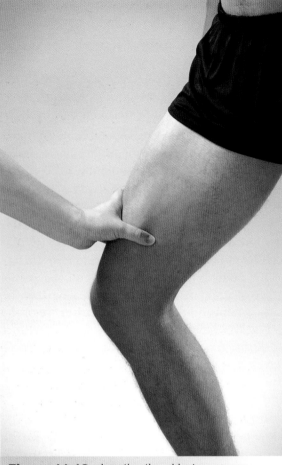

Figure 11.18 Locating the adductor magnus.

Hands On

Sitting in the lotus, or butterfly, position (on the floor, with the soles of your feet together, your knees flexed, and your hips externally rotated), you can easily palpate your adductor longus and gracilis muscles in the groin area. The gracilis is the more lateral of the two tendons. Additionally, you can palpate the gracilis by pressing with your finger on the area just medial to the semimembranosus tendon of insertion on the medial aspect of the knee joint (figure 11.19).

Lateral Muscles

There are only two lateral muscles of the hip joint, not including the tensor fasciae latae, which was identified as an anterior, not lateral, muscle of the hip joint. Even though the gluteus maximus shares a name with the lateral muscles of the hip joint, it is a posterior muscle. The lateral muscles of the hip joint are the gluteus medius and the gluteus minimus (figure 11.13).

- **Gluteus medius**: The gluteus medius originates from the middle of the external aspect of the ilium to the iliac crest and inserts on the posterior lateral aspect of the greater trochanter (figures 11.13 and 11.20). The major function of this muscle is hip abduction. The anterior portion

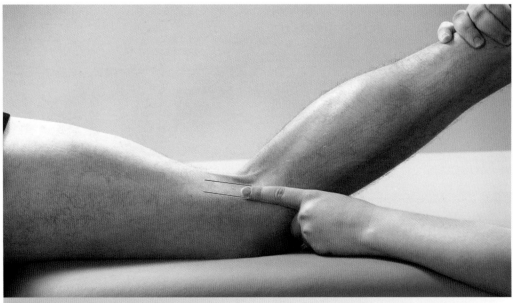

Figure 11.19   Identifying the gracilis.

tion. The anterior portion assists with flexion, and the posterior portion assists with extension.

Iliotibial Band

The iliotibial band is formed by two muscles that cross the hip joint: the tensor fasciae latae and the gluteus maximus (figure 11.13). It passes back and forth across the greater trochanter of the femur as the hip joint flexes and extends. To facilitate this movement of the iliotibial band over the greater trochanter, a large trochanteric bursa is located between the iliotibial band and the greater trochanter. Excessive flexion or extension or direct trauma to this area can cause inflammation of the bursa (trochanteric bursitis).

Figure 11.20 Locating the gluteus medius.

assists with flexion and internal rotation, and the posterior portion assists with extension and external rotation.

- **Gluteus minimus**: Originating in the same area on the ilium as the gluteus medius, the gluteus minimus inserts on the anterior surface of the greater trochanter. The major functions of this muscle are hip abduction and internal rota-

FOCUS ON

Groin Strain

Often we hear about athletes suffering from a **groin strain**. What muscle group is affected in this injury? This chapter presents two groups of groin muscles: the iliopsoas muscle group (iliacus and psoas major muscles), often referred to as the true groin muscles, and the adductor muscle group (adductor longus, adductor brevis, adductor magnus, and gracilis), referred to as the common groin muscles. The true groin muscles (the iliopsoas group) are typically strained as the result of hyperextension of the hip joint. Injury to the common groin muscles (the adductor group) is usually the result of overstretching these muscles; the adductor group is more commonly strained in athletic endeavors, hence the term *common groin strain*.

REVIEW OF TERMINOLOGY

The following terms are discussed in this chapter. Define or describe each term, and where appropriate, identify the location of the named structure either on your body or in an appropriate illustration.

acetabulum	adductor magnus (anterior and	biceps femoris
adductor brevis	posterior portions)	capsular ligament
adductor longus	adductor tubercle	coccyx

(continued)

REVIEW OF TERMINOLOGY *(continued)*

deep external rotator
external obturator (obturator externus)
femoral fovea
femur
fibula
gemellus inferior
gemellus superior
glenoid lip
gluteus maximus
gluteus medius
gluteus minimus
gracilis
greater sciatic notch
greater trochanter
groin strain
hamstrings
head of the femur
iliacus
iliofemoral ligament

iliopsoas
iliotibial band (tract)
ilium
innominate bone
internal obturator (obturator internus)
intertrochanteric ridge
intervertebral disc
ischial tuberosity
ischiofemoral ligament
ischium
lesser trochanter
ligamentum capitis femoris
linea aspera
long head of the biceps femoris
lumbar vertebra
neck of the femur
obturator foramen
pectineus
piriformis

psoas major
psoas minor
pubic bone
pubic symphysis
pubofemoral ligament
quadratus femoris
quadriceps femoris
rectus femoris
sacrum
sartorius
sciatic notch
semimembranosus
semitendinosus
short head of the biceps femoris
spine of the ischium
tensor fasciae latae
thoracic vertebra
transverse acetabular ligament
triaxial joint
trochanteric fossa

SUGGESTED LEARNING ACTIVITIES

1. Take the position of a runner in the starting blocks for a short running race. Place one foot forward and one foot back while placing your hands on the starting line.

   a. In what position is the hip joint of the forward leg (flexed, extended)?

   b. In what position is the hip joint of the back leg (flexed, extended)?

   c. Now imagine the starter has fired the starting gun. Into what position did the hip joint of the forward leg move, and what muscles caused that movement?

   d. Into what position did the hip joint of the back leg move, and what muscles caused that movement?

2. Perform the classic jumping jack exercise several times, observing the movements of the hip joint during the counts of 1 and 2.

   a. Describe the position of both hip joints on the count of 1 and the muscles that were used to put the hips in that position.

   b. Describe the position of both hip joints on the count of 2 and the muscles that were used to put the hips in that position.

3. Take one step forward with your right leg. Cross your left leg over your right leg so that your left foot is perpendicular to your right foot. Your left heel should now be near the outer edge of your right foot.

   a. Describe the position of your left hip.

   b. Describe the position of your right hip.

MULTIPLE-CHOICE QUESTIONS

1. Which of the following muscles is not considered a lateral muscle of the hip joint?

   a. tensor fasciae latae

   b. gluteus medius

   c. sartorius

   d. gluteus minimus

2. The greater trochanter of the femur is located where anatomically to the hip joint?

   a. lateral

   b. medial

   c. anterior

   d. posterior

3. Which of the following is not considered a fundamental movement of the hip joint?

  a. flexion
  b. circumduction
  c. extension
  d. adduction

4. Which of the following ligaments is not considered a ligament of the hip joint?

  a. iliofemoral
  b. pubofemoral
  c. ischiofemoral
  d. sacroiliac

5. Which of the following muscles is absent on one or both sides in approximately 50% of human beings?

  a. psoas major
  b. psoas minor
  c. iliacus
  d. iliopsoas

6. Which of the following muscles is not part of the muscle group known as the iliopsoas?

  a. iliacus
  b. sacroiliac
  c. psoas minor
  d. psoas major

7. Which of the following muscles is not considered an anterior muscle of the hip joint?

  a. rectus femoris
  b. biceps femoris
  c. sartorius
  d. tensor fasciae latae

8. Which of the following muscles is not an adductor of the hip joint?

  a. gracilis
  b. gluteus maximus
  c. pectineus
  d. gluteus medius

9. Which of the following muscles of the hip joint does not attach to the pubic bone?

  a. pectineus
  b. adductor brevis
  c. gracilis
  d. iliacus

10. Which of the following muscles is not considered one of the six deep external rotators of the hip joint?

  a. quadratus femoris
  b. rectus femoris
  c. piriformis
  d. internal obturator

FILL-IN-THE-BLANK QUESTIONS

1. A joint defined as a triaxial ball-and-socket joint of the lower extremity is the _____.

2. The ligament of the hip often referred to as the Y ligament is the _____ ligament.

3. The iliotibial band consists of the combined tendons from the tensor fasciae latae and the _____.

4. The large tuberosity on which the hamstrings originate is part of the _____.

5. The psoas major and psoas minor muscles originate on the _____.

6. The six deep external rotators of the hip joint are located on the _____ aspect of the joint.

FUNCTIONAL MOVEMENT EXERCISE

From the prone position, to perform a sit-up, the hip joint must flex and the pelvis must tilt forward. List one muscle acting as a prime mover, one as an antagonist, one as a fixator, and one as a synergist during hip flexion and forward pelvic tilt during a sit-up.

	Hip flexion	Forward pelvic tilt
Prime mover		
Antagonist		
Fixator		
Synergist		

CHAPTER 12

The Knee

The knee joint, one of the largest joints in the body, is a uniaxial synovial joint and is often referred to as a hinge joint. In reality, the knee joint is *not* a true hinge joint but rather a modified hinge joint. A true hinge joint, like a door hinge, opens and closes about a single constant axis. In the knee joint, the tibia (distal bone) glides around the distal end of the femur (proximal bone), and although the movement remains in one plane (the sagittal plane), it occurs about an ever-changing axis. With each degree of movement in the sagittal plane, the frontal horizontal axis changes. Hence, the term *modified hinge joint* is more appropriate than just *hinge*

joint for the knee joint. Although the knee joint appears to be structurally sound, it was not built to withstand many of the stresses placed on it by athletic activities. As we examine the normal anatomy of the joint, we also draw attention to the common results of abnormal stresses.

Bones of the Knee

The **femur** (figure 12.1) is presented in chapter 11 on the hip joint, but its role at the knee is discussed in this chapter. At the distal end of the shaft of the femur, the bone flares out, forming

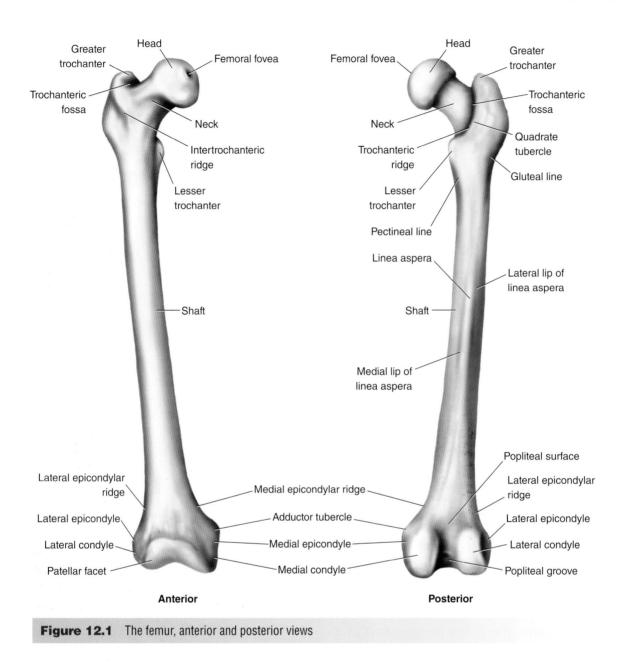

Figure 12.1   The femur, anterior and posterior views

a **medial** and **lateral epicondylar ridge**, similar to those in the humerus just proximal to the elbow joint. Just distal to these ridges are the **medial** and **lateral epicondyles** (figure 12.2). Two other large prominences of bone, easily palpated just proximal to the joint line (an imaginary line between the femur and the tibia), are the **medial** and **lateral condyles** (figure 12.3). It is these two condyles that articulate with the tibia to transfer the body weight from the femur to the lower leg. The medial condyle is slightly more distal than the lateral condyle.

Hands On

After locating the lateral and medial femoral condyles (figure 12.3), move your fingers distal to (below) the condyles. You should feel a space between the distal ends of the femur and the proximal ends of the tibia. This space is commonly known as the joint line. Under your fingers on the lateral and medial joint lines lie the anterior horns of the lateral and medial meniscus.

Just proximal to the medial epicondyle is a small prominence known as the **adductor tubercle** (figure 12.4). On the anterior surface

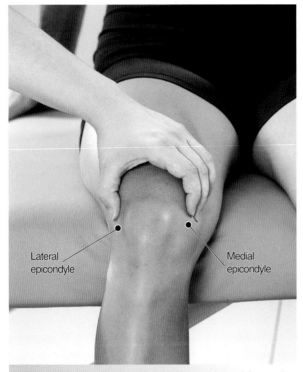

Figure 12.2 Finding the medial and lateral femoral epicondyles.

Lateral epicondyle

Medial epicondyle

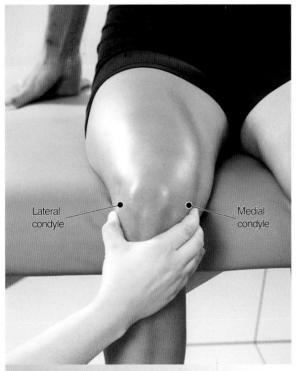

Lateral condyle

Medial condyle

Figure 12.3 Locating the medial and lateral femoral condyles.

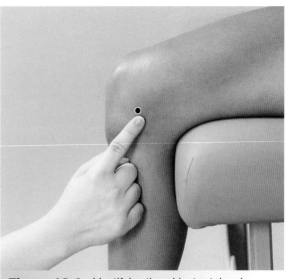

Figure 12.4 Identifying the adductor tubercle.

of the distal end of the femur, between the lateral and medial condyles, is a smooth surface covered with articular cartilage known as the **patellar facet**. This surface articulates with the posterior aspect of the patella (kneecap, discussed later), forming the **patellofemoral joint**. On the posterior side of the distal end of the femur, between the lateral and medial condyles, is the **popliteal groove**. The area just proximal to this groove is known as the **popliteal surface**. The word *popliteal* refers to the posterior knee joint and is applied to bones, ligaments, muscles, nerves, or blood vessels of the posterior knee. The space between the femoral condyles is referred to as the **intercondylar notch** or the **femoral notch**.

The two bones of the lower leg are the **fibula** and **tibia** (figure 12.5). In this chapter on the knee joint, only the proximal ends of these two bones are described. Their distal ends are considered in chapter 13.

The fibula, the smaller lateral bone, is essentially a non-weight-bearing bone that serves as a major source of soft tissue attachment at both the proximal and distal ends. Three prominent structures at the proximal end of the fibula are the **apex**, the **head**, and the **neck**. The tibia, the larger medial bone of the lower leg, articulates with the femur and bears the weight of the body from the femur to the foot. At the very proximal end of the shaft of the tibia are the **lateral** and

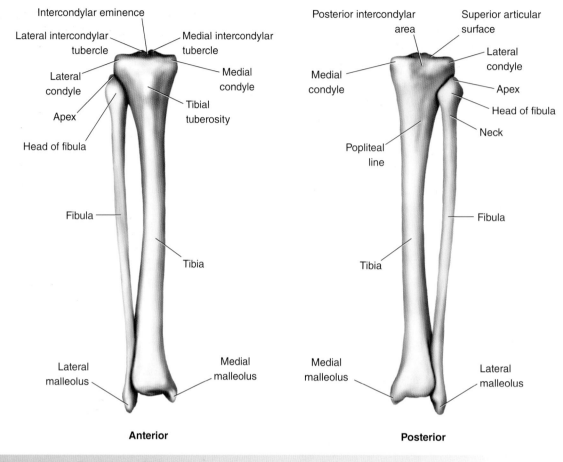

Anterior

Posterior

Figure 12.5 The tibia and fibula, anterior and posterior views.

medial condyles, and between them, on the proximal anterior surface of the tibia, lies a very large prominence known as the **tibial tuberosity**.

🖐 Hands On

You can easily palpate the tibial tuberosity by feeling for the large bump on the anterior surface of your lower leg just below your knee (figure 12.6).

A smaller prominence, located on the anterior aspect of the lateral condyle of the tibia, is **Gerdy's tubercle** (figure 12.7), where the iliotibial band inserts. On the posterior aspect of the proximal tibia, just distal to the space between the lateral and medial condyles, is the area known as the **popliteal surface**. Just beneath this surface is a small diagonal line called the **popliteal line**, also known as the linea musculi solei (see figure 12.5). The superior view of the tibia reveals several structures. The two large surfaces, the

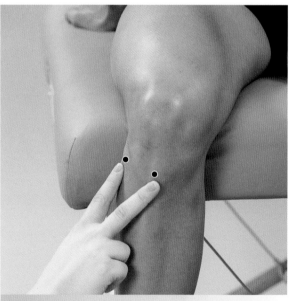

Figure 12.6 Locating the tibial tuberosity and head of the fibula.

medial and **lateral condylar surfaces**, are where the condyles of the femur articulate with the tibia to form the knee joint (figure 12.8). The large prominence of bone between the two condylar surfaces is the **intercondylar eminence**, and the smaller prominences alongside the intercondylar eminence are the **medial** and **lateral intercondylar tubercles** (see figure 12.5).

The final bone of the knee joint is vital for proper movement of the joint. The **patella** (kneecap) is the largest sesamoid (free-floating; see chapter 1) bone in the body (figure 12.9). It is not directly attached to other bones to form a joint. It is embedded in the tendon of insertion of the quadriceps femoris, the anterior knee muscle group (discussed later in the section on muscles). This muscle group inserts on the patella through a broad fibrous sheath, which in turn attaches to the tibial tuberosity through the patellar ligament and tendon. The patella not only protects the structures beneath it but also changes the angle of pull of the quadriceps femoris to create a greater rotary force (in flexion and extension) of the knee joint as compared to the stabilizing force of the quadriceps femoris that pulls the tibia into the femur. The proximal end of the patella is known as the **base**, and the distal end is known as the **apex**. The posterior surface has **lateral** and **medial articular surfaces** that are covered with articular cartilage and articulate with the medial and lateral condyles of the femur to form the patellofemoral articulation.

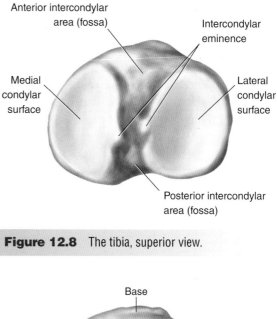

Figure 12.8 The tibia, superior view.

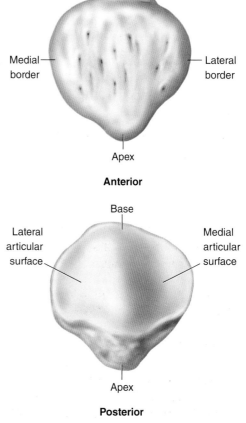

Figure 12.9 The patella, anterior and posterior views.

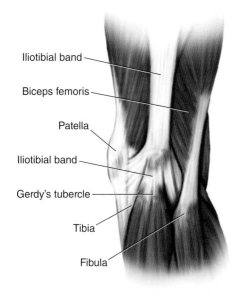

Figure 12.7 Gerdy's tubercle and the iliotibial band.

Ligaments of the Knee

Like all synovial joints, the knee joint has a **capsular ligament**. This capsular ligament is unlike others because it consists of portions of other ligaments and fibrous expansions of other structures that cross the knee joint and become part of the capsule (figures 12.10 and 12.11). The components of the capsule include portions of the **medial** and **lateral collateral ligaments** and fibrous expansions of the quadriceps femoris, the iliotibial band, the vastus muscles, the sartorius muscle, and the semimembranosus muscle. Probably the best illustration of the pure capsular ligament is seen at the posterior (**popliteal space**) of the knee joint.

The medial collateral ligament (MCL) and lateral collateral ligament (LCL) of the knee provide stability to either side of the joint, essentially preventing abduction and adduction of the joint and making it a uniaxial joint that flexes and extends in the sagittal plane. The medial collateral ligament runs from the medial condyle of the femur to the medial condyle of the tibia, with some deep fibers attaching to the medial meniscus (discussed later in this chapter). The lateral collateral ligament runs from the lateral condyle of the femur to the head of the fibula. Note that the lateral collateral ligament, unlike the medial collateral ligament, does *not* have fibers attaching to the lateral meniscus.

👋 Hands On

As you sit in a chair, place the ankle of one leg on top of the other leg's knee joint. Palpate the lateral condyles of both the femur and tibia of the leg crossed on top of the other leg. Find the joint space between the condyles. Move your finger anterior and posterior through the joint line until you feel a cordlike structure. This structure is the lateral collateral ligament. Now consider the role of the medial and lateral collateral ligaments in the mechanics of the knee. Remember that the knee

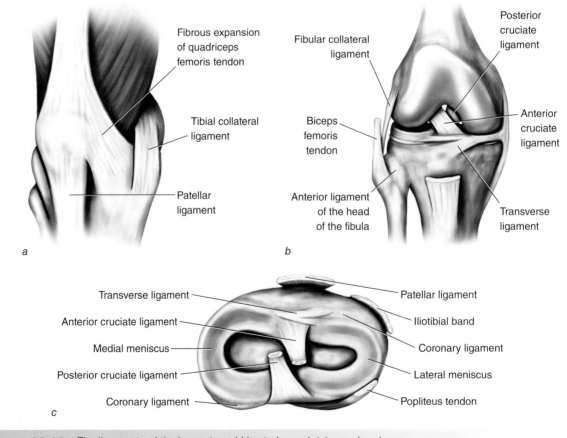

Figure 12.10 The ligaments of the knee, *(a and b)* anterior and *(c)* superior views.

normally only flexes and extends. What ligament would come under stress if a force was applied to the lateral side of the knee joint (often referred to as a valgus force), forcing the joint into abduction (i.e., increasing the space between the tibia and the femur on the medial side of the joint)? (See figure 12.12a for the answer.) What ligament would come under stress if a force was applied to the medial side of the knee joint (a varus force), forcing the joint into adduction (increasing the space between the tibia and the femur on the lateral side of the joint)? (See figure 12.12b.) These stresses placed on the MCL and LCL ligaments are commonly the mechanism of sprains of the knee joint.

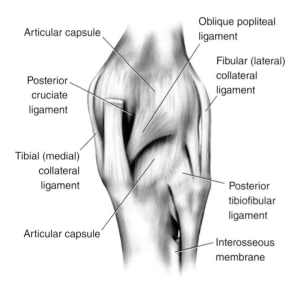

Figure 12.11 The ligaments of the knee, posterior view.

In the middle of the knee joint are two ligaments known as the **anterior cruciate ligament** (ACL) and the **posterior cruciate ligament** (PCL) (see figures 12.10 and 12.11). The term *cruciate* means "cross," and these two ligaments actually cross each other as they pass through the middle of the knee joint. The anterior cruciate ligament runs from just anterior to the intercondylar eminence of the tibia to the posterior medial surface of the lateral condyle of the femur. The primary function of the anterior cruciate ligament is to prevent anterior displacement of the tibia off the distal end of the femur. This ligament, along with the resistance of the posterior muscles crossing the knee joint, prevents the normal knee from hyperextending. The posterior cruciate ligament runs from just posterior to the intercondylar eminence of the tibia to the anterior portion of the medial surface of the medial condyle of the femur. The primary function of the posterior cruciate ligament is to prevent posterior displacement of the tibia off the distal end of the femur. Excessive squatting may place the knee joint into such a degree of flexion that this ligament comes under stress.

Three ligaments of the knee joint are found exclusively on the posterior aspect of the joint: the **oblique popliteal ligament**, the **arcuate ligament**, and the **ligament of Wrisberg** (figure 12.13). The oblique popliteal ligament runs from the posterior aspect of the lateral condyle of the femur to the posterior edge of the medial condyle of the tibia. The arcuate ligament

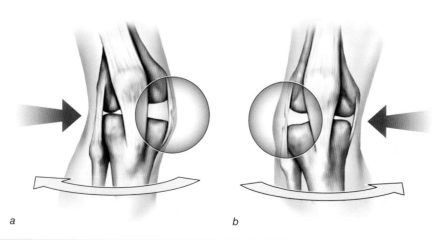

a   b

Figure 12.12 (a) A lateral force to the right knee places a valgus stress on the medial collateral ligament. (b) A medial force to the right knee places a varus stress on the lateral collateral ligament.

Anterior Cruciate Ligament

The ACL has received much attention recently in athletics. Many sport activities apply external and internal forces to the knee joint structures, including the ACL. There also has been great interest in the difference between female and male athletes in the number of ACL injuries. The theories regarding the larger number of female ACL injuries include lower-extremity malalignment (wider female pelvis, increased Q angle); ligament laxity; quadriceps–hamstring strength ratio imbalance; smaller female intercondylar (femoral) notch and femoral condyle size; the effects of estrogen on ligament tissue; and maturation rates and the effect on landing positions (straight-knee landing, one-step stop landing). Ankle braces transferring stress to the knee and the interface between footwear and playing surfaces can also contribute to ACL injuries. Other factors considered are the intensity of training, the type of activity, coaching techniques, and enough additional theories to fill an entire textbook. Simply said, physicians, biomechanists, physical therapists, athletic trainers, and other interested parties continue to search for the reasons for the differences in the number of ACL injuries between female and male athletes.

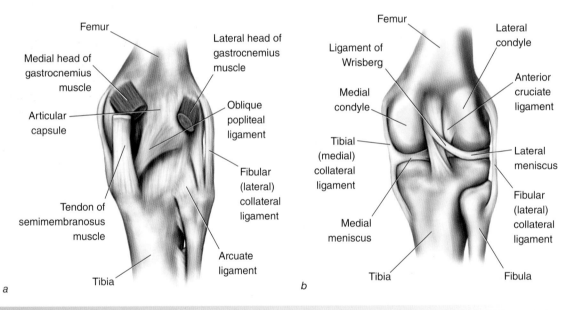

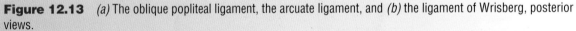

Figure 12.13   *(a)* The oblique popliteal ligament, the arcuate ligament, and *(b)* the ligament of Wrisberg, posterior views.

runs from the posterior aspect of the lateral condyle of the femur to the posterior surface of the capsular ligament. The ligament of Wrisberg runs between the posterior horn of the lateral meniscus (see the section on menisci later in this chapter) and the posterior aspect of the medial condyle of the femur.

On the anterior side of the knee, running between the apex of the patella and the tibial tuberosity, is the **patellar ligament** (figures 12.10 and 12.14). Because the patella is embedded in the patellar tendon, some anatomists consider this structure to be an extension of the quadriceps femoris tendon of insertion. Others, however, label this structure the patellar ligament because it ties bone to bone (patella to tibia). In this text, the structure is known as the patellar ligament. Even though some anatomists hold that the quadriceps tendon inserts on the tibial tuberosity, the simple approach is that the quadriceps muscles attach to the patella and the patella attaches to the tibial tuberosity via the patellar ligament.

Two additional ligaments of the knee joint are unique in that they do not tie bone to bone, the

normal function of ligaments. The **coronary ligament**, actually a portion of the capsular ligament, is responsible for connecting the outer edges of the menisci (see the next section) to the proximal end of the tibia (see figure 12.10). The **transverse ligament** runs between the anterior horns of the medial and lateral menisci (figure 12.14). This ligament prevents the anterior horn of each meniscus from moving forward when the knee joint moves into extension and the condylar surfaces of both the femur and the tibia exert pressure on the menisci.

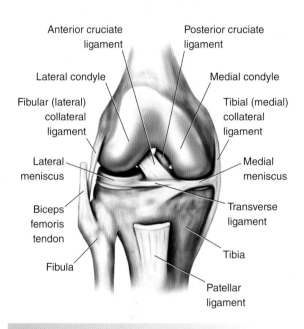

Figure 12.14   The ligaments of the knee, anterior view.

Menisci of the Knee

Two semilunar (crescent-shaped) fibrocartilaginous structures sit on the proximal end of the tibia, on the medial and lateral condylar surfaces. These structures are known as the **lateral meniscus**, which is nearly circular in shape, and the semicircular **medial meniscus**. These structures help deepen the condylar surfaces of the tibia where the condyles of the femur articulate. Their outer borders are thick and convex and are attached to the tibia by the coronary ligament. The inner edges of the menisci are paper thin and lie freely (unattached) on the floor of the

Disruption of the Meniscus

The menisci, more the medial than the lateral, often come under stress through athletic participation. Stress to the deep fibers of the MCL that attach to the medial meniscus could disrupt the meniscus. Excessive rotation of the femur on a fixed tibia (rigidly planted or pinned on the ground or playing surface) could cause stress to the menisci. Two of the most common disruptions of the menisci are called parrot-beak and bucket-handle tears (figure 12.15). The most common, the bucket-handle tear, actually causes the middle portion of the meniscus (between the anterior and posterior horns) to split, causing the outer portion of the meniscus to look like the handle of a bucket as

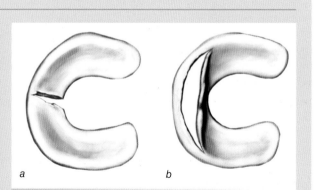

Figure 12.15   Two common tears of the meniscus of the knee: *(a)* parrot-beak tear and *(b)* bucket-handle tear.

the inner portion separates from the main body of the meniscus. Other types of tears of the menisci also occur. Equipment design, playing surfaces, footwear, and specific exercises are all areas of interest for decreasing the incidence of this serious condition.

condylar surfaces of the tibia. The inner surfaces of the menisci are concave to accommodate the condyles of the femur. The anterior and posterior aspects of each meniscus are often referred to as the **anterior** and **posterior horns** of the meniscus.

Fundamental Movements of the Knee and Lower Leg

Because the knee joint is a uniaxial joint capable of movement in the sagittal plane about an ever-changing frontal horizontal axis, the only two movements the joint is capable of are flexion and extension. However, because of the sizes and shapes of the femoral condyles and the soft tissue configurations, when the knee flexes and extends, the lower leg (tibia and fibula) rotates. When the knee extends, the leg externally rotates. When the knee flexes, the leg internally rotates. As the knee joint "locks" into extension and "unlocks" when moving into flexion, the rotation of the leg is difficult to see when weight is not being borne by the leg. The locking into full extension is often referred to as the "screw home" movement.

🖐 Hands On

Sit in a chair with your knees flexed to 90°. With your shoes off, place your feet on a smooth surface (e.g., tile or cement—not carpeting). Make a fist with one hand and place it between your knees. Internally rotate your lower legs so that your toes (first metatarsophalangeal joints; see chapter 13) on both feet push against each other. As you internally rotate your lower legs to push your feet against each other, with your free hand feel the semimembranosus (medial thigh, just proximal to the knee) muscles posterior to your knee joints. These muscles (and those of the pes anserinus—see later in this chapter) are contracting to cause the internal rotation of the lower leg. Now, from the same starting position, have your partner place her feet along the sides or the lateral aspect of both of your feet. Find the biceps femoris tendon on the posterior lateral aspect of your knee with your free hand. Now try to externally rotate your lower legs to force your feet to push your partner's feet away. What did you feel? As the muscles that cross the knee joint are reviewed, their bony attachments will help identify what, if any, movement of the leg they perform.

Muscles of the Knee and Lower Leg

The muscles crossing the knee joint can easily be divided into those crossing the joint anteriorly and those posteriorly. The anterior group is reviewed first.

Anterior Muscles

Because of their position anterior to the knee joint (figure 12.16), the logical conclusion is that these muscles extend the joint. This is primarily true, but other functions are also involved, as noted in the following discussions.

- **Sartorius**: This muscle, the longest in the body, is examined in the hip chapter because it is a flexor of the hip. The sartorius originates on the anterior superior iliac spine (ASIS), crosses the hip joint, passes posterior to the medial condyle of the femur, and inserts just inferior to the proxi-

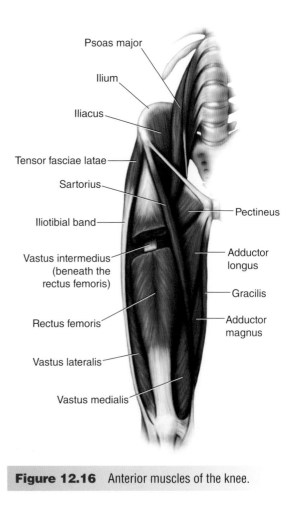

Psoas major
Ilium
Iliacus
Tensor fasciae latae
Sartorius
Iliotibial band
Vastus intermedius (beneath the rectus femoris)
Rectus femoris
Vastus lateralis
Vastus medialis
Pectineus
Adductor longus
Gracilis
Adductor magnus

Figure 12.16 Anterior muscles of the knee.

mal end of the medial surface of the tibia (figure 12.16). At the knee joint, the sartorius flexes the knee and internally rotates the lower leg.

The **quadriceps femoris**, frequently referred to as "the quads," is a group of four muscles (figures 12.16 and 12.17). One of these muscles, the rectus femoris, crosses both the knee and the hip joint and is reviewed in the hip chapter as it pertains to hip joint action. Here we look at its function as a knee extensor. The other three muscles—the vastus intermedius, vastus medius, and vastus lateralis—cross only the knee joint and have only one function: extension at the knee.

• **Rectus femoris**: This is the one quad muscle that crosses both the hip and knee joints. It is the most superficial of the anterior thigh muscles. Its straight head originates on the anterior inferior iliac spine, its reflected head originates on the acetabulum, and it inserts on the base of the patella (figure 12.16). The rectus femoris is an extensor of the knee joint.

• **Vastus lateralis**: The largest of the three vastus muscles, the vastus lateralis, originates on the proximal half of the linea aspera, the **intertrochanteric line**, and the greater trochanter of the femur and inserts on the lateral border of the patella. The vastus lateralis extends the knee joint.

• **Vastus medialis**: Originating on the medial lip of the linea aspera, the vastus medialis inserts on the medial border of the patella. The vastus medialis extends the knee joint.

Hands On

With resistance to the lower leg (from either a partner, an iron boot, or a flexion–extension machine), extend your knee (see figure 12.18). Observe and palpate the medial and lateral aspects of your thigh. Identify the musculature you are observing.

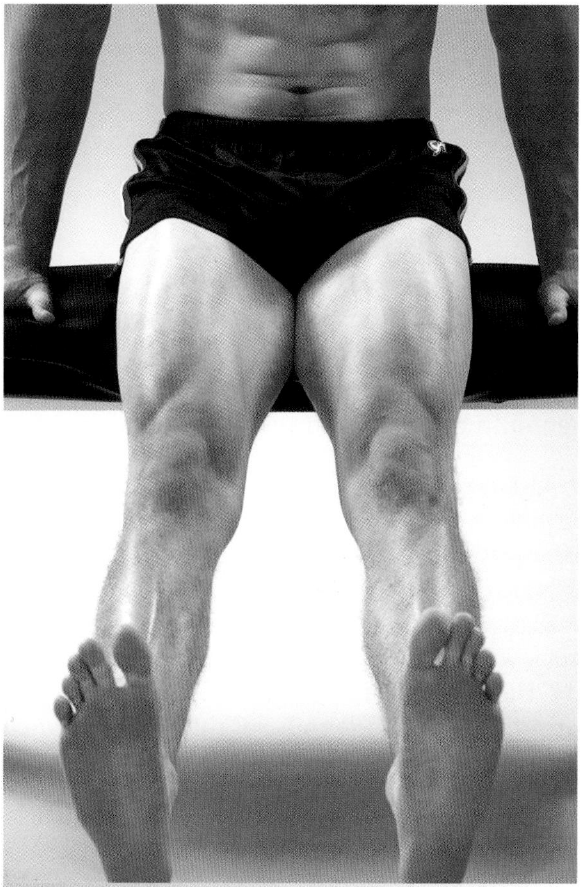

Figure 12.17 Viewing the quadriceps femoris muscle group.

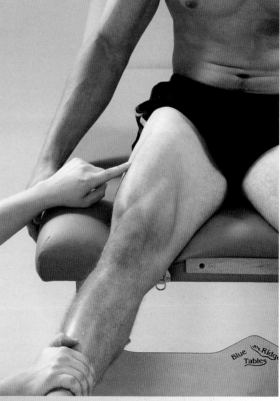

Figure 12.18 Locating the rectus femoris.

Q Angle and Patellofemoral Syndrome

Chapter 8 points out the differences in the shape of the female and male pelvis. If you draw a line through the patella and the tibial tuberosity and another line between the patella and the anterior superior iliac spine (ASIS), you will note an angle between these two lines (figure 12.19). This angle is the Q angle, defined as the angle between the line of the quadriceps muscle pull and the line of insertion of the patellar tendon. In females this angle normally is between 15° and 20°, whereas in males it ranges from 10° to 15°. A Q angle greater than 20° can result in any of a multitude of problems causing pain in and around the patellofemoral joint. This all-inclusive evaluation of pain is often identified as patellofemoral syndrome (PFS). As mentioned, there are multiple reasons for anterior knee joint pain, with an excessive Q angle as a possible contributing factor. We will not go into an extended medical diagnosis in an entry-level anatomy textbook, but suffice it to say that those experiencing PFS often complain about pain while climbing stairs. Consider the relationship of the patella and the femur when the knee is moved during stair climbing.

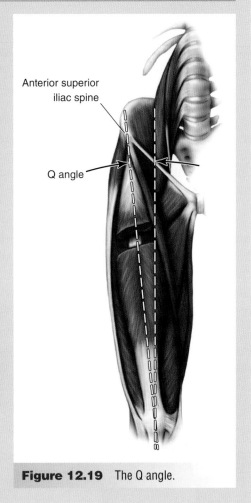

Figure 12.19 The Q angle.

• **Vastus intermedius**: Beneath the rectus femoris lies the vastus intermedius. This muscle originates on the proximal two-thirds of the anterior surface of the femur and inserts on the inferior surface of the patella. The vastus intermedius extends the knee joint.

Beneath the quadriceps femoris muscles lies another anterior muscle, the genu articularis.

• **Genu articularis**: This muscle, deep beneath the vastus intermedius, originates on the anterior surface of the femur just proximal to the condyles and inserts not on another bone but on the **synovial membrane** of the knee joint. As the knee moves into extension, this muscle contracts, pulling the articular capsule of the knee proximally to prevent the synovial

membrane from becoming impinged between the femur, the patella, and the tibia (figure 12.20).

Posterior Muscles

The posterior muscles of the knee joint (figure 12.21) include muscles, such as the hamstrings and the gracilis, whose actions at the hip joint are reviewed in the discussion of muscles in chapter 11. Here we look at their actions at the knee joint and also at the other muscles that cross the knee posteriorly.

• **Biceps femoris**: This hamstring muscle has two heads: One, the long head, originates on the ischial tuberosity, and the other, the short head, originates on the lateral aspect of the linea aspera (figure 12.21). The muscle inserts on the head of the fibula. The biceps femoris flexes the knee, and as the knee reaches active full flexion, it externally rotates the lower leg.

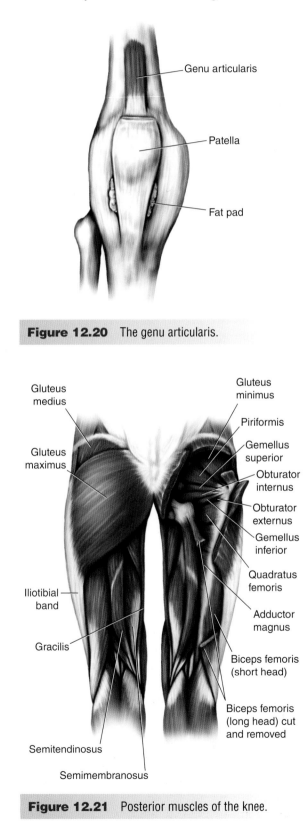

Figure 12.20 The genu articularis.

Figure 12.21 Posterior muscles of the knee.

- **Semitendinosus**: The second hamstring muscle, the semitendinosus, originates on the ischial tuberosity and inserts on the proximal aspect of the medial tibia (figure 12.21). The semitendinosus flexes the knee and internally rotates the lower leg.

- **Semimembranosus**: The third hamstring muscle, the semimembranosus, originates on the ischial tuberosity and inserts on the posterior medial aspect of the medial condyle of the tibia (figure 12.21). The semimembranosus flexes the knee and assists with internal rotation of the lower leg.

- **Gracilis**: The gracilis is the only adductor muscle of the hip joint that also crosses the knee joint. It originates on the inferior surface of the pubic symphysis, runs posterior to the medial condyle of the femur, and inserts just posterior to the medial aspect of the proximal end of the tibia (figure 12.21). The gracilis flexes the knee joint and internally rotates the lower leg.

The gracilis, semitendinosus, and sartorius all insert in the same general area, just below the proximal end of the tibia on its medial aspect. The insertion of the three closely grouped tendons is commonly identified as the **pes anserinus** (figure 12.22). All three components of the

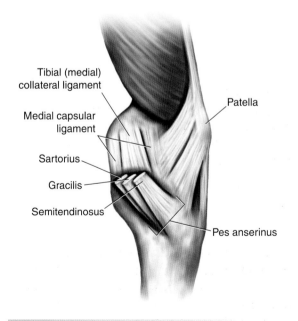

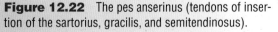
Figure 12.22 The pes anserinus (tendons of insertion of the sartorius, gracilis, and semitendinosus).

pes anserinus flex the knee joint and internally rotate the lower leg.

• **Popliteus**: Diagonally crossing the popliteal space of the knee joint, the popliteus runs between the lateral aspect of the lateral condyle of the femur and the popliteal line on the proximal third of the posterior surface of the tibia (figure 12.23). Observe the arrangement of the fibers of this muscle. Besides flexion of the knee, what obvious motion of the lower leg is likely to occur when this muscle contracts? (Answer: internal rotation)

• **Iliotibial band**: This structure (figure 12.24*a*), also discussed in the hip chapter, is a combination of the **gluteus maximus** and **tensor fasciae latae** tendons of insertion (figure 12.21). It crosses the knee in the area of the lateral condyle of the femur and inserts onto a bony prominence just inferior and anterior to the lateral condyle of the tibia known as Gerdy's tubercle (see figure 12.7). This structure both flexes and extends the knee joint, depending on the angle of the knee joint at any particular moment. When the knee joint is between full extension and 10° to 15° of flexion, the iliotibial band is anterior to the lateral femoral condyle and assists with extension of the knee joint (figure 12.24*b*). As the knee joint continues to flex beyond 10° to 15°, the iliotibial band shifts to a position posterior to the lateral femoral condyle and becomes a flexor of the knee joint (figure 12.24*c*).

Hands On

Palpate the lateral side of your knee joint. Move your hand proximally, just above the lateral condyle of the femur. Flex and extend your knee joint, and feel the iliotibial band move anterior and posterior to the femoral condyle.

Two muscles of the lower leg originate above the knee joint and play a role in knee joint function, although their primary functions involve the ankle joint.

• **Gastrocnemius**: This muscle has two heads: one originating on the posterior aspect of the lateral condyle of the femur and the other originating on the posterior aspect of the medial condyle of the femur (figure 12.23). Both heads combine into a single tendon of insertion that attaches to the **calcaneus** (heel bone).

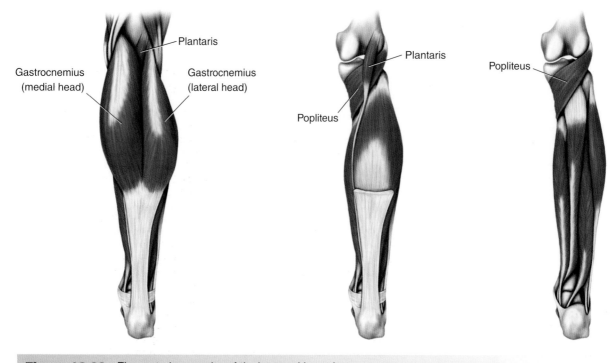

Gastrocnemius (medial head)

Plantaris

Gastrocnemius (lateral head)

Plantaris

Popliteus

Popliteus

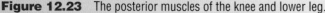

Figure 12.23 The posterior muscles of the knee and lower leg.

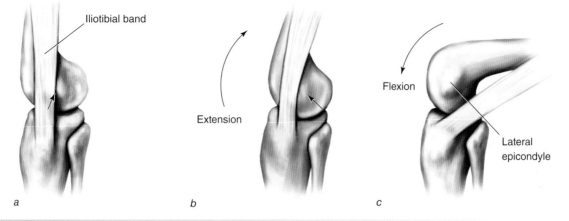

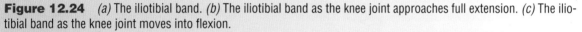

Figure 12.24 *(a)* The iliotibial band. *(b)* The iliotibial band as the knee joint approaches full extension. *(c)* The iliotibial band as the knee joint moves into flexion.

The gastrocnemius flexes the knee joint. In the rare instance where the gastrocnemius cannot perform its primary function (plantar flexion of the ankle joint) because the foot is held in a fixed position and cannot move, contraction of the gastrocnemius can cause extension of the knee joint.

• **Plantaris**: This short-bellied muscle with a long tendon of insertion originates on the lateral linea aspera and the oblique popliteal ligament and inserts on the calcaneus (figures 12.23). The plantaris muscle assists with knee flexion but is of little importance in humans compared with the other muscles that perform the same function.

LEARNING AIDS

REVIEW OF TERMINOLOGY

The following terms are discussed in this chapter. Define or describe each term, and where appropriate, identify the location of the named structure either on your body or in an appropriate illustration.

adductor tubercle
anterior cruciate ligament
anterior horn of the meniscus
apex of the fibula
apex of the patella
arcuate ligament
base of the patella
biceps femoris
calcaneus
capsular ligament
coronary ligament
femoral notch
femur
fibula
gastrocnemius
genu articularis
Gerdy's tubercle
gluteus maximus
gracilis

head of the fibula
iliotibial band
intercondylar eminence
intercondylar notch
intertrochanteric line
lateral articular surface of the patella
lateral collateral ligament
lateral condylar surface of the tibia
lateral condyle of the femur
lateral condyle of the tibia
lateral epicondylar ridge of the femur
lateral epicondyle of the femur
lateral intercondylar tubercle
lateral meniscus
ligament of Wrisberg
medial articular surface of the patella
medial collateral ligament
medial condylar surface of the tibia

medial condyle of the femur
medial condyle of the tibia
medial epicondylar ridge of the femur
medial epicondyle of the femur
medial intercondylar tubercle
medial meniscus
neck of the fibula
oblique popliteal ligament
patella
patellar facet
patellar ligament
patellofemoral joint
pes anserinus
plantaris
popliteal groove
popliteal line
popliteal space
popliteal surface of the femur

(continued)

REVIEW OF TERMINOLOGY *(continued)*

popliteal surface of the tibia
popliteus
posterior cruciate ligament
posterior horn of the meniscus
quadriceps femoris
rectus femoris

sartorius
semimembranosus
semitendinosus
synovial membrane
tensor fasciae latae
tibia

tibial tuberosity
transverse ligament
vastus intermedius
vastus lateralis
vastus medialis

SUGGESTED LEARNING ACTIVITIES

1. Lying supine on a table with your lower legs off the end of the table, extend both knee joints to full extension.
   a. What muscles performed this action?
   b. Which of these muscles played the most prominent role in the joint action?
   c. Did you have difficulty fully extending your knee joints? If so, why?

2. Sitting up on a table with your lower legs off the end of the table, extend both knee joints to full extension.
   a. What muscles performed this action?
   b. Which of these muscles played the most prominent role in the joint action?
   c. Did you have difficulty fully extending your knee joints? If so, why?

3. Sitting up on a table with your lower legs off the end of the table and your feet next to each other, extend both knee joints to full extension. As your knees moved to full extension, what did the lower legs (and the relationship of your feet to each other) do?

4. Lying prone with your entire body on a table, bring your heels to your buttocks.
   a. What muscles performed this action of the knee joints?
   b. Did your pelvis rise off the table as you performed this activity? If so, why?

MULTIPLE-CHOICE QUESTIONS

1. Which of the following ligaments has deep fibers attaching to a meniscus?
   a. medial collateral
   b. lateral collateral
   c. anterior cruciate
   d. posterior cruciate

2. Which of the following structures attaches to Gerdy's tubercle on the proximal anterior lateral aspect of the tibia?
   a. quadriceps tendon
   b. lateral collateral ligament
   c. lateral meniscus
   d. iliotibial band ligament

3. The popliteal space is found on what aspect of the knee joint?

   a. anterior
   b. posterior
   c. lateral
   d. medial

4. Which of the following ligaments is responsible for preventing posterior displacement of the proximal tibia off the distal end of the femur?
   a. anterior cruciate
   b. posterior cruciate
   c. medial collateral
   d. lateral collateral

5. When the knee joint moves into extension, which muscle becomes the external rotator of the lower leg?

a. plantaris
b. popliteus
c. biceps femoris
d. rectus femoris

6. Which of the following hamstring muscles is part of the structure known as the pes anserinus?
    a. semitendinosus
    b. semimembranosus
    c. rectus femoris
    d. biceps femoris

7. Which of the following muscles crosses only the knee joint?
    a. sartorius
    b. biceps femoris
    c. rectus femoris
    d. popliteus

8. Which of the following muscles, although considered a weak flexor of the knee, is really of little significance in human anatomy?
    a. popliteus
    b. plantaris
    c. biceps femoris
    d. sartorius

9. Which of the following hip adductor muscles also crosses the knee joint?
    a. pectineus
    b. adductor magnus
    c. gracilis
    d. adductor longus

10. Which of the following muscles is not considered a primary flexor of the knee?
    a. semitendinosus
    b. semimembranosus
    c. biceps femoris
    d. rectus femoris

11. Which of the following muscles is the longest muscle in the human body?
    a. sartorius
    b. rectus femoris
    c. semitendinosus
    d. tensor fasciae latae

12. Which of the following muscle groups contains the largest sesamoid bone in the body within its tendon of insertion?
    a. iliopsoas
    b. quadriceps
    c. hamstrings
    d. adductors

13. Which of the following muscles is considered the lateral hamstring muscle?
    a. semitendinosus
    b. semimembranosus
    c. biceps femoris
    d. rectus femoris

14. Which of the following muscles is found beneath the rectus femoris muscle?
    a. vastus lateralis
    b. vastus intermedius
    c. vastus medialis
    d. vastus femoris

FILL-IN-THE-BLANK QUESTIONS

1. The most distal aspect of the femur bone is its _____.

2. The medial and lateral menisci are attached to the _____.

3. The ligament attaching the anterior horns of the medial and lateral meniscus to prevent meniscal distortion during knee extension is the _____ ligament.

4. The gastrocnemius muscle originates on the _____.

(continued)

FILL-IN-THE-BLANK QUESTIONS *(continued)*

5. The ligament designed to prevent forward displacement of the tibia off the distal end of the femur is the _____ ligament.

6. A joint defined as a modified hinge joint of the lower extremity is the _____ joint.

7. The iliotibial band consists of the combined tendons of the tensor fasciae latae and the _____.

FUNCTIONAL MOVEMENT EXERCISE

Muscles crossing the knee joint function during both flexion and extension of the joint. List one muscle acting as a prime mover, one as an antagonist, one as a fixator, and one as a synergist in both flexion and extension of the knee joint.

	Knee flexion	Knee extension
Prime mover		
Antagonist		
Fixator		
Synergist		

CHAPTER 13

The Lower Leg, Ankle, and Foot

This chapter may seem familiar because the lower leg, ankle, and foot are very similar anatomically to the forearm, wrist, and hand. The bones, ligaments, and muscles may or may not have similar names, but their structures are similar. In contrast to the upper extremity, the lower extremity is constructed to bear the weight of the body and absorb the force applied by that body weight with each foot strike.

Bones of the Lower Leg

The two bones of the lower leg are the **tibia** (medially) and the **fibula** (laterally). Much debate has been conducted over the weight-bearing aspects of both bones, but in this text, we assume that the tibia is the major weight-bearing bone and that the fibula has little or no weight-bearing function.

The prominent bony markings of the proximal end of these bones are presented in chapter 12 on the knee. At the distal ends of the shafts of both the tibia and the fibula are large bumps known as **malleoli** (figures 13.1 and 13.2). The prominence at the distal end of the fibula is known as the **lateral malleolus**, and the prominence at the distal end of the tibia is known as the **medial malleolus**. Palpate these bumps on either side of the distal end of your lower leg, and note which one is more distal than the other. This is important to remember later, when the fundamental movements of the ankle joint are discussed.

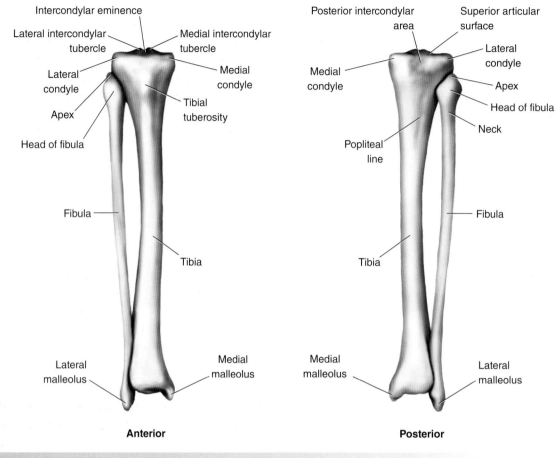

Intercondylar eminence

Lateral intercondylar tubercle

Medial intercondylar tubercle

Lateral condyle

Medial condyle

Apex

Tibial tuberosity

Head of fibula

Fibula

Tibia

Lateral malleolus

Medial malleolus

Anterior

Posterior intercondylar area

Superior articular surface

Medial condyle

Lateral condyle

Apex

Head of fibula

Neck

Popliteal line

Fibula

Tibia

Medial malleolus

Lateral malleolus

Posterior

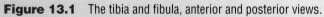

Figure 13.1   The tibia and fibula, anterior and posterior views.

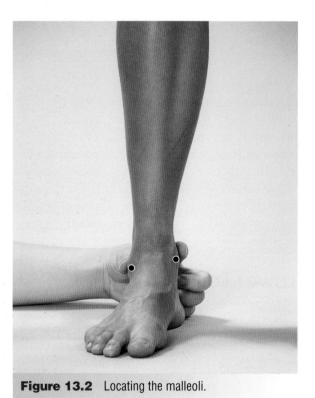

Figure 13.2   Locating the malleoli.

On the lateral surface of the tibia, just proximal to the distal end, is a notch known as the **fibular notch of the tibia**, where the tibia and fibula articulate to form the distal **tibiofibular joint**.

The distal surfaces of both the tibia and the fibula have facets (or smooth surfaces) that articulate with the talus, one of the bones of the foot (figure 13.3).

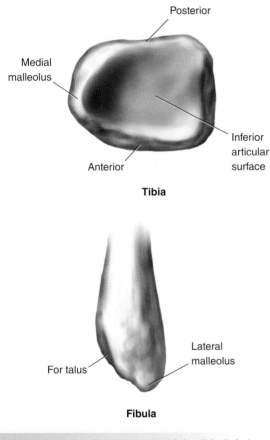

Tibia

Fibula

Figure 13.3 The distal ends of the tibia (inferior view) and the fibula (posterior view).

Bones of the Foot

The 26 bones of the foot (figure 13.4) are usually separated into three distinct segments: the **forefoot** (19), the **midfoot** (5), and the **hindfoot** (2).

The forefoot consists of 14 **phalanges**: three per toe (proximal, middle, and distal phalanges), except for the great toe, which has only a proximal and a distal phalanx. The phalanges and the metatarsal bone of the first, or great, toe are larger than those of the other four toes (second, middle,

fourth, and little) for a specific purpose. When the foot bears the weight of the body, as in walking, the great toe must bear most of the weight. This is the same arrangement we found in the hand: three phalanges per finger and only two in the thumb. The other five bones of the forefoot are the five long bones of the foot, known as the metatarsal bones. Again, these bones are similar to the five metacarpal bones of the hand. The **metatarsal bones** consist of a **head** (distal end), a **shaft**, and a **base** (proximal end). The most prominent of these structures is the base of the fifth metatarsal bone.

Hands On

Run your finger along the lateral edge of your foot, and palpate a rather large bump about half to two-thirds of the length of your foot from your fifth (little) toe. This is the base (also known as the tuberosity) of the fifth metatarsal bone (see figure 13.4).

The remaining seven bones of the foot are collectively known as the **tarsal bones**. Five of these tarsal bones (the cuboid, the navicular, and the medial, intermediate, and lateral cuneiforms) make up the midfoot, and the remaining two tarsal bones (talus and calcaneus) make up the hindfoot.

Just proximal to the medial metatarsal bones are three tarsal bones known as the **cuneiform bones** (figures 13.4 and 13.5). These are often referred to as the first, second, and third cuneiforms, but it is easier to remember them by their anatomical position: the medial, intermediate, and lateral cuneiform bones. Proximal to the cuneiform bones is the fourth midfoot bone, the **navicular**. As in the wrist, the navicular bone is also known as the **scaphoid**. The fifth tarsal bone of the midfoot is also the most lateral of the five midfoot tarsals: the **cuboid** bone. Careful examination of this bone reveals that its lateral border is concave. This groove provides a space for a tendon of the foot and ankle (the peroneus longus, discussed later in this chapter).

The final two tarsal bones, the **talus** and the **calcaneus**, make up the bones of the hindfoot (figures 13.4, 13.6, and 13.7). These two bones articulate with the tibia and fibula, similarly to how the scaphoid and lunate bones articulate with the ulna and radius of the forearm. The talus

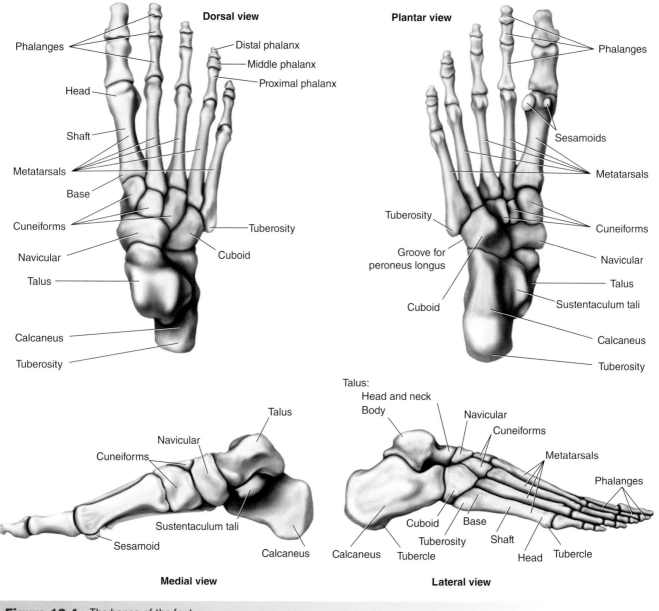

Figure 13.4 The bones of the foot.

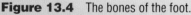

literally sits on top of (proximal to) the calcaneus, which is often referred to as the heel bone and is the largest of the tarsal bones. Note on the posterior view of the calcaneus (figure 13.8) a bony prominence extending medially from its superior surface. This prominence, the **sustentaculum tali**, serves as a platform for a portion of the talus to sit on. These two bones, articulating together and with the lower-leg bones, form the two joints (talocrural and talocalcaneal) that we refer to as the **ankle joint**. Movement in these two joints (two movements in each joint) result in the four movements attributable to the ankle joint. These

joints and movements are further discussed later in this chapter.

On the plantar surface of the head of the first metatarsal bone are two **sesamoid** bones (see figure 13.4), which are not typically counted with the other 26 bones of the foot. Like the patella and sesamoid bones in general, by definition, these sesamoid bones are embedded in the tendon of a muscle and are free floating. They not only protect the structures superior to them but also provide a biomechanical advantage for the function of the muscle in which they are embedded.

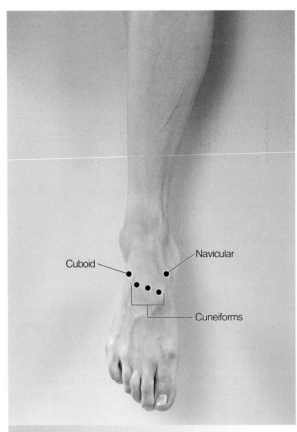

Figure 13.5 A dorsal view of the navicular, cuneiform, and cuboid bones.

Cuboid

Navicular

Cuneiforms

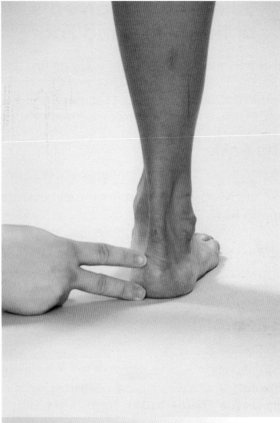

Figure 13.7 Identifying the calcaneus.

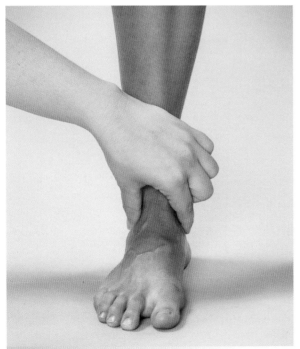

Figure 13.6 Locating the talus.

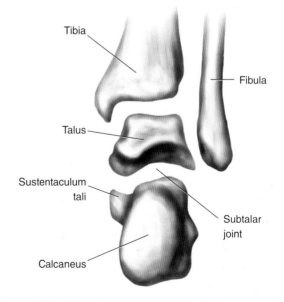

Tibia

Fibula

Talus

Sustentaculum tali

Subtalar joint

Calcaneus

Figure 13.8 Posterior view of the ankle, illustrating the sustentaculum tali of the calcaneus and the subtalar joint.

Joints and Ligaments of the Ankle and Foot

In the following sections, we examine first the joints and ligaments of the ankle and then those of the foot. Finally, we look at the arches of the feet, essential to any discussion of the foot.

The Ankle

The ankle is not really a single well-defined joint like many other articulations throughout the human body. Some authors call it the ankle joint complex because there is more than one joint where the movement we commonly refer to as ankle joint motion takes place.

The major ligaments of the lower leg are the **interosseous ligament** (interosseous membrane) (figure 13.9; similar to the interosseous ligament of the forearm) found between the medial border of the shaft of the fibula and the lateral border of the shaft of the tibia (from proximal to distal ends) and the **anterior** and **posterior tibiofibular ligaments** at the distal end of the leg (figure 13.10). The interosseous ligament serves as a source of attachment for numerous anterior and posterior muscles of the lower leg.

The talocrural aspect of the talocrural–talocalcaneal (ankle) joint is a hinge joint that permits movement in the sagittal plane (dorsiflexion and plantar flexion). In anatomy, the talocrural joint is often referred to as a loosely formed mortise-and-tenon joint. The mortise (recess or hole) of the joint is formed by the lateral malleolus of the fibula and the medial malleolus of the tibia. The tenon (peg) of the joint is the talus, which fits into the mortise (figure 13.11).

In addition to the capsular ligament present in all synovial joints, there are four major ligaments of the ankle joint (figure 13.10). For the most part, the names of ligaments indicate the bones that the ligaments bring together to form articulations.

Medially, the major ankle ligament is known as the **deltoid ligament**, made up of three superficial ligaments and one deep ligament. Superficially, the anterior portion of the deltoid ligament is the **tibionavicular ligament**, the middle portion is the **calcaneotibial ligament**, and the posterior portion is the **posterior talotibial ligament**. The deep ligament of the deltoid ligament is the **anterior talotibial**.

Laterally, the ankle has three major ligamentous structures. The shortest of these three ligaments is the **anterior talofibular** (ATF)

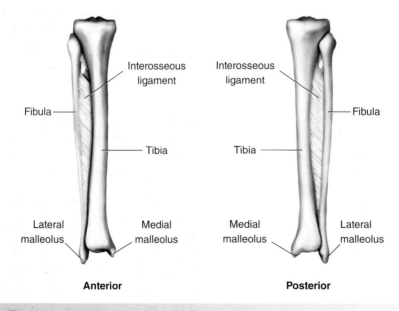

Anterior   **Posterior**

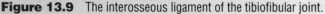

Figure 13.9 The interosseous ligament of the tibiofibular joint.

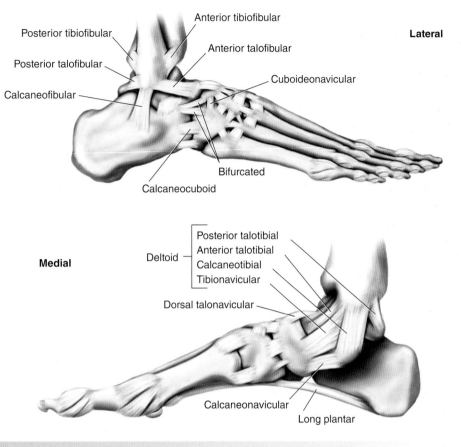

Figure 13.10 The ligaments of the foot and ankle.

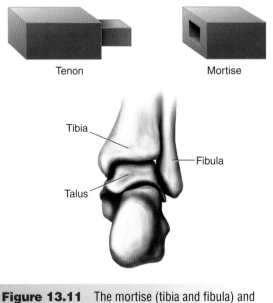

Figure 13.11 The mortise (tibia and fibula) and tenon (talus) of the talocrural joint.

ligament. It runs from the anterior aspect of the distal end of the lateral malleolus to the talus. The ATF ligament is the most commonly sprained ligament of the ankle joint.

✋ Hands On

Placing your finger just anterior and distal to the end of your fibula, you find a soft depression, the **sinus tarsi** (*sinus* = "cavity," *tarsi* = "tarsal bone"). The ATF ligament crosses through this area.

The strongest of the three lateral ankle ligaments is the **posterior talofibular ligament**, which runs from the posterior aspect of the lateral malleolus to the talus. The longest of the three lateral ankle joint ligaments is the **calcaneofibular ligament**. It runs from the lateral malleolus of the fibula to the lateral aspect of the calcaneus.

Ankle Joint Problems

Any of the ligaments of the ankle can be sprained, depending on the exact mechanics of the stress. However, the anterior talofibular (ATF) ligament is by far the most commonly sprained ligament of the ankle joint. "Rolling over" on the ankle (turning the foot inward excessively toward the other foot) places stress on all lateral ligaments of the ankle, but the ATF in particular is under the greatest amount of stress in this typical ankle sprain. The term *inversion sprain* is frequently used to describe this injury.

Although much more rare than the common inversion sprain, a high ankle sprain is caused by the same mechanism. A more anatomically correct term would be a *syndesmosis sprain*. The term *syndesmosis* is another way to define a joint: an articulation between bones tied together by ligaments. In the case of the ankle joint, when the ankle is excessively inverted, the talus may force against the fibula, causing the joint between the fibula and tibia (the tibiofibular syndesmosis) to spread apart. This spreading can possibly sprain the ATF ligament, the posterior tibiofibular ligament, the interosseous membrane, or any combination of these. The spraining of any of these three structures can be defined as a high ankle sprain.

Preventing ankle joint problems is important for many people involved in physical activity. Prescribing specific exercises for strengthening or rehabilitating the ankle, designing footwear, using preventive measures (such as taping, wrapping, or bracing), and understanding the effects of specific playing surfaces all rely on knowledge of the anatomy of the ankle joint. Prevention strategies rely on knowing the types of structures that are stressed during injury, the types of forces involved, how to counteract those potentially damaging forces, and the musculature that can be strengthened to better resist the damaging forces. The disciplines of biomechanics, athletic training, sports medicine, physical therapy, and exercise science can provide the necessary background for addressing this important aspect of sport.

✋ Hands On

After observing the anatomical arrangements of the bones and ligaments of the ankle, attempt to turn your ankle inward toward your other ankle (inversion), and then attempt to turn your ankle outward away from your other ankle (eversion). Which movement created the greatest motion? What structures limited the movement? Observing these actions and the structural arrangement of the bones and ligaments of the ankle should reveal why the ATF is the most likely ligament of the ankle joint to sprain. Inversion and eversion are discussed later in this chapter.

Ligaments of the Foot

The ligaments of the foot (see figures 13.10 and 13.12) can be divided into five groups: the **intertarsal ligaments**, the **tarsometatarsal ligaments**, the **intermetatarsal ligaments**, the **metatarsophalangeal ligaments**, and the **interphalangeal ligaments**.

The intertarsal ligaments tie together the articulations between the tarsal bones of the hindfoot and the bones of the midfoot. The hindfoot joint between the talus and the calcaneus (**talocalcaneal** or **subtalar joint**, figure 13.8) produces medial and lateral gliding movements in the hindfoot that are identified as inversion (movement of the foot toward the midline of the body) and eversion (movement of the foot away from the midline).

Intertarsal ligaments, like the intercarpals of the wrist, join the seven tarsal bones of the foot together. Most of these ligaments are identifiable by their names, which represent the bones they tie together. The talocalcaneonavicular joint has a capsular ligament and a **dorsal talonavicular ligament**. The calcaneocuboid joint has a capsular ligament and the **calcaneocuboid ligament**. The **bifurcated ligament** runs from the calcaneus and divides into fibers that run to the cuboid and to the navicular bones. The **long plantar ligament** runs from the calcaneus to the cuboid, with fibers extending to the bases of the third, fourth, and fifth metatarsal bones (figures 13.10 and 13.12). The long plantar ligament and the **calcaneonavicular**

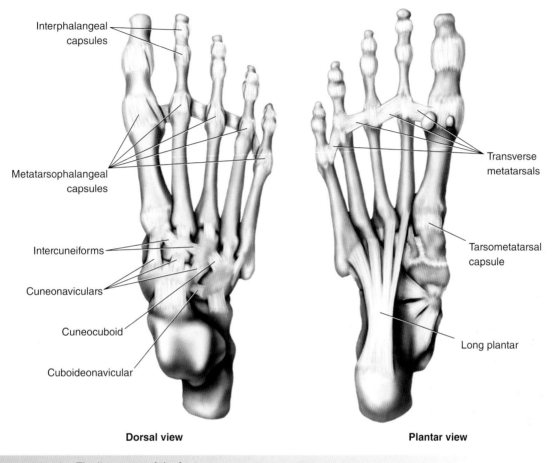

Dorsal view   Plantar view

Figure 13.12 The ligaments of the foot.

ligament are involved in one of the arches of the foot (figure 13.10). Additional ligaments of the intertarsal group include three dorsal and three plantar **cuneonaviculars**; the dorsal, plantar, and interosseous **cuboideonaviculars**; the dorsal, plantar, and interosseous **intercuneiforms**; and the dorsal, plantar, and interosseous **cuneocuboids**.

The tarsometatarsal ligaments are the dorsal and plantar capsular ligaments and the **interosseous (collateral) ligaments** (under the capsules on the medial and lateral aspects of the joints) joining the five metatarsal bones to the tarsal bones of the midfoot to form the tarsometatarsal joints (figures 13.10, 13.12, and 13.13). The intermetatarsal ligaments join the bases of the five metatarsal bones together at the tarsometatarsal joints.

The metatarsophalangeal (MP) joints that join the metatarsal bones with the proximal phalanges also have dorsal and plantar capsular

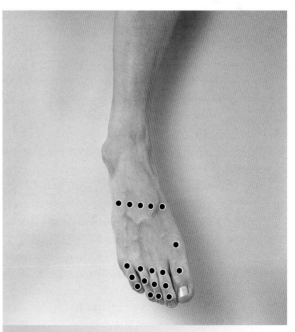

Figure 13.13 Locating the tarsometatarsal, metatarsophalangeal, and interphalangeal joints.

ligaments and interosseous (collateral) ligaments under the capsules on the medial and lateral aspects of the joints (figures 13.12 and 13.13). Additionally, there is a **transverse metatarsal ligament** connecting all five heads of the metatarsal bones (figure 13.12). Note that the first and fifth metatarsophalangeal joints are often referred to as the "large ball" and the "small ball" of the foot.

✋ Hands On

Observe the plantar surface of your foot, and determine for yourself which ball is which.

The interphalangeal joints include the four proximal interphalangeal (PIP) joints and four distal interphalangeal (DIP) joints of the four toes and the single interphalangeal (IP) joint of the great toe. The interphalangeal joints are joined by dorsal and plantar capsular ligaments and interosseous (collateral) ligaments.

Arches

Anatomically, an arch is defined as the structures forming a curved or bow-shaped object. The foot contains either two or three arches, depending on one's viewpoint.

The **longitudinal arch** runs from the calcaneus to the heads of the metatarsal bones on the plantar surface of the foot (figure 13.14). The arch is formed by a combination of the shapes of the bones and the ligamentous structures supporting the bones. The space created beneath the arch allows the muscles, tendons, blood vessels, and nerves of the plantar surface of the foot to pass without being crushed against the ground.

The metatarsal bones form an arch (or arches) as the result of their ligamentous attachments to both the phalanges and the tarsal bones. An anterior view of the foot reveals that the metatarsophalangeal joints form the **metatarsal arch**. This same arch appears at the other (proximal) end of the metatarsal bones as they articulate with the tarsal bones. Some authors refer to this arch also as the metatarsal arch, whereas others refer to it as the **transverse arch** of the foot (figure 13.14).

Five other structures (similar to the two found in the wrist) in the ankle and foot are the **extensor retinaculum** (**superior** and **inferior**), the

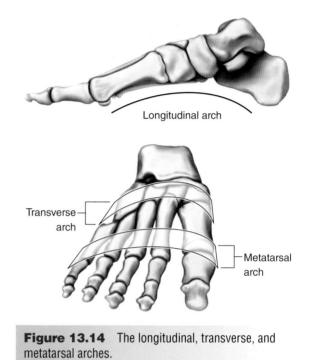

Figure 13.14   The longitudinal, transverse, and metatarsal arches.

peroneal retinaculum (**superior** and **inferior**), and the **flexor retinaculum** (see figures 13.19 and 13.24). All these structures function primarily to keep the tendons of muscles in their appropriate positions.

Fundamental Movements of the Lower Leg, Ankle, and Foot

Upward movement of the foot toward the anterior leg is known as **dorsiflexion** of the ankle joint. Downward movement of the foot is known as **plantar flexion** (figure 13.15). Lateral (outward) movement of the foot at the talocalcaneal (subtalar) joint produces a movement known as **eversion**. Medial (inward) movement of the foot at the talocalcaneal joint produces a movement known as **inversion**. The motion between the hindfoot (talus) and midfoot (navicular) bones also contributes to these movements (figure 13.16).

Although the talocrural joint is considered the true ankle joint, most authors cite plantar flexion, dorsiflexion, inversion, and eversion as the four fundamental movements of the ankle joint. Because plantar flexion and dorsiflexion occur in the sagittal plane and eversion and inversion in the frontal plane, this makes the ankle joint a biaxial joint. As a biaxial joint, the ankle is capable of circumduction (a combination of the fundamental movements of a biaxial or triaxial joint).

Movement between the tarsal bones of the foot is similar to that of the carpal bones of the wrist. The bones glide over each other to produce slight movement. These gliding movements of the tarsals combined with the movements of the ankle joint result in either **supination** of the foot (inversion of the ankle and adduction of the foot) or **pronation** of the foot (eversion of the ankle and abduction of the foot). Again, the motion between the hindfoot and midfoot bones also contributes to these movements.

Movements at the tarsometatarsal, metatarsophalangeal (MP), proximal interphalangeal (PIP), interphalangeal (IP), and distal interphalangeal (DIP) joints are limited to flexion and extension, except that the MP joints are also able to abduct and adduct (figure 13.17).

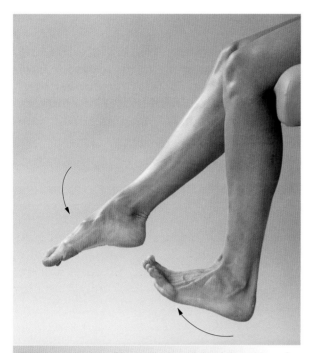

Figure 13.15   Plantar flexion of the right ankle and dorsiflexion of the left.

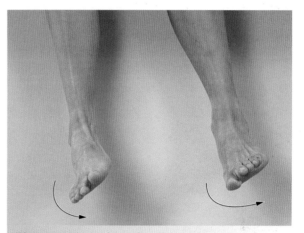

Figure 13.16   Inversion of the right ankle and eversion of the left.

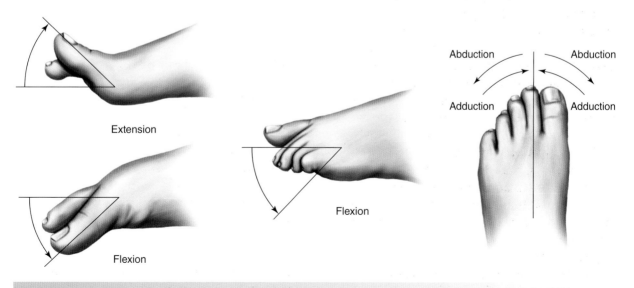

Figure 13.17   Extension and flexion of the toe joints. Abduction and adduction of the metatarsophalangeal joints.

Turf Toe

Spraining the ligaments of the first MP joint (great toe), often from hyperextension, is frequently referred to as "turf toe," depending on the mechanism of injury. Knowledge of anatomy, physics, and biomechanics, combined with the experience of engineers familiar with the effects of the interaction of footwear and the playing surface, can assist in preventing a sprain of the first MP joint.

Muscles of the Lower Leg, Ankle, and Foot

The muscles of the leg, ankle, and foot, like those of the hand, are typically divided into the **extrinsic muscles** (those originating outside the foot and inserting within the foot) and the **intrinsic muscles** (those originating and inserting within the foot).

Now that we've considered the movements possible in the ankle (talocrural and talocalcaneal) joint and in the MP, PIP, IP, and DIP joints, learning the actions and locations of these muscles should be easier.

Extrinsic Muscles

There are 12 extrinsic muscles of the foot and ankle that are contained in four well-defined compartments of the lower leg (figure 13.18). Four muscles are found in the **anterior compartment**, two in the **lateral compartment**, three in the **superficial posterior compartment**, and three in the **deep posterior compartment**.

Anterior Compartment

The four muscles of the lower-leg anterior compartment are the tibialis anterior, the extensor

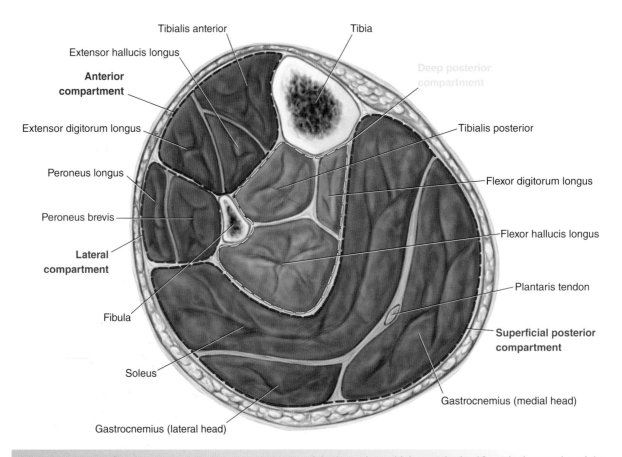

Figure 13.18　Cross section of the four compartments of the lower leg, which contain the 12 extrinsic muscles of the ankle and foot.

digitorum longus, the extensor hallucis longus, and the peroneus tertius (figure 13.19).

- **Tibialis anterior**: The tibialis anterior originates from the upper two-thirds of the lateral side of the tibia and inserts on the medial side of the medial cuneiform and the base of the first metatarsal bone (figure 13.19). If you consider the origin and insertion of the tibialis anterior, what actions of the ankle joint are produced by contraction of this muscle? (Answer: dorsiflexion and inversion)

🖐 Hands On

Find your anterior tibial shaft, and move your fingers just lateral to the tibia. The first soft tissue you palpate is your tibialis anterior muscle (figure 13.20). Additionally, as you dorsiflex your ankle, observe the tendon of the tibialis anterior on the anterior medial surface of the ankle joint.

- **Extensor digitorum longus**: The second extrinsic muscle of the anterior compartment is the extensor digitorum longus (figure 13.19). The extensor digitorum longus originates from the lateral condyle of the tibia, the proximal three-quarters of the fibula, and the interosseous membrane and inserts on the middle and distal phalanges of the lateral four toes. From this muscle's name and its origin and insertion, what are the actions of this muscle at the ankle joint and the MP, PIP, and DIP joints? (Answer: dorsiflexion of the ankle and extension of the MP, PIP, and DIP joints of the four lateral toes)

🖐 Hands On

As you dorsiflex your ankle, observe the tendons of the extensor digitorum longus on the anterior surface of the ankle joint (lateral to the tibialis anterior) (figure 13.21). Also observe how the tendon splits into four tendons of insertion that should be visible on the superior (dorsal) aspect of your foot.

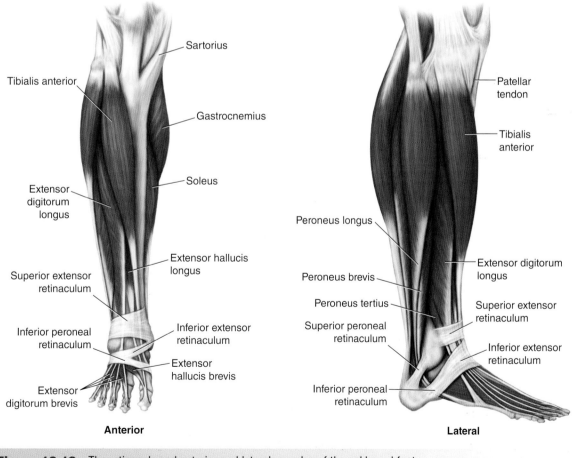

Anterior

Lateral

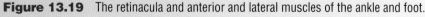

Figure 13.19 The retinacula and anterior and lateral muscles of the ankle and foot.

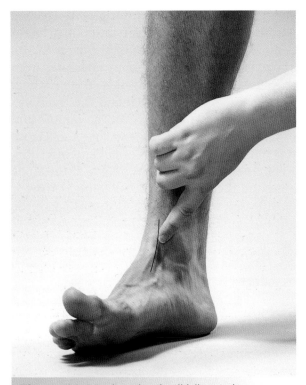

Figure 13.20   Locating the tibialis anterior.

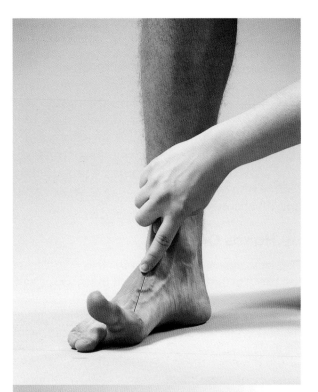

Figure 13.22   Locating the extensor hallucis longus.

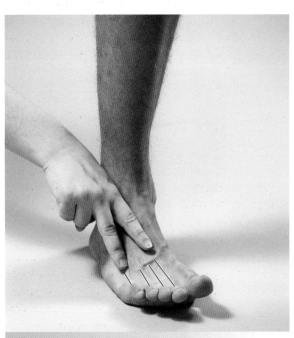

Figure 13.21   Locating the extensor digitorum longus.

• **Extensor hallucis longus**: The third muscle of the lower-leg anterior compartment is the extensor hallucis longus (figure 13.19). Just as

structures in the upper extremity use the term *pollicis* to mean thumb, in the lower extremity, the term *hallucis* refers to the great toe. The extensor hallucis longus originates on the middle half of the medial aspect of the fibula and the adjoining interosseous membrane and inserts on the distal phalanx of the great toe. As you observe the joints that this muscle crosses, what would you expect are its actions at the ankle and the MP and IP joints of the great toe? (Answer: dorsiflexion and inversion of the ankle and extension of the MP and IP joint of the great toe)

Hands On

Extend your great toe, and observe the tendon of insertion of the extensor hallucis longus (figure 13.22). Tracing it from the great toe to the ankle joint, note that it quickly disappears as it moves beneath the muscle fibers of the two previously discussed anterior compartment muscles, the tibialis anterior and the extensor digitorum longus (see figure 13.19).

• **Peroneus tertius**: The fourth and final muscle of the anterior compartment of the lower leg is the peroneus tertius (one of three peroneal muscles in the lower leg; the term *tertius* means "third"). The peroneus tertius originates from the lower third of the fibula and adjoining interosseous membrane and inserts on the superior (dorsal) surface of the base of the fifth metatarsal bone (figure 13.19). Because the peroneus tertius crosses the anterior aspect of the ankle and is the most lateral of the four anterior compartment muscles, what would you expect the actions of the peroneus tertius muscle to be? (Answer: dorsiflexion and eversion of the ankle joint)

Lateral Compartment

Two muscles are contained in the lateral compartment of the lower leg: the peroneus longus and the peroneus brevis (figures 13.19 and 13.23).

• **Peroneus longus**: The peroneus longus originates from the head and proximal two-thirds of the lateral aspect of the fibula and the lateral condyle of the tibia and inserts on the inferior lateral surfaces of the medial cuneiform and the first metatarsal bones. It was previously noted in this chapter that the cuboid bone has a concavity (groove) along its lateral border. This concavity is where the tendon of insertion of the peroneus longus passes from the lateral aspect of the foot to the inferior plantar surface.

• **Peroneus brevis**: The peroneus brevis (*brevis* means short) originates from the lower

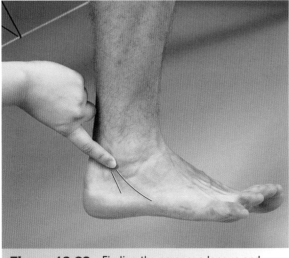

Figure 13.23 Finding the peroneus longus and peroneus brevis.

half of the lateral aspect of the fibula and inserts on the tuberosity on the base of the fifth metatarsal bone.

Note that the tendons of insertion of both the peroneus longus and peroneus brevis muscles run posterior to the lateral malleolus of the fibula. To ensure that these tendons remain in their appropriate positions while the ankle joint dorsiflexes and plantar flexes, the tendons are held in place by two structures previously discussed in this chapter: the superior and inferior peroneal retinacula (see figure 13.19). The action of both of these muscles is the same. If we consider the path that these muscles and their tendons of insertion follow, what are these actions? (Answer: plantar flexion and eversion of the ankle joint)

Superficial Posterior Compartment

The three muscles of the superficial posterior compartment of the lower leg are the gastrocnemius, the soleus, and the plantaris. Collectively, these are often referred to as the muscles of the calf (figure 13.24) and also are referred to as the **triceps surae**. Two of these muscles originate superior to the lower leg on the femur of the thigh (gastrocnemius and plantaris), whereas the third (soleus) originates on the lower leg (tibia only).

• **Gastrocnemius**: The gastrocnemius is a two-headed muscle (figures 13.24 and 13.25). Its lateral head originates on the popliteal surface of the femur just medial to the lateral condyle, whereas the medial head originates on the popliteal surface of the medial condyle of the femur. Both heads combine to form the belly of the muscle, and a common tendon of insertion, known as the Achilles tendon, attaches the muscle to the posterior surface of the calcaneus. Because the gastrocnemius crosses both the knee joint and the ankle joint, extension of the knee combined with dorsiflexion of the ankle can contribute to an Achilles tendon strain.

• **Soleus**: The soleus (figure 13.24) originates from the head and proximal third of the posterior aspect of the fibula and the middle third of the posterior aspect of the tibia. It inserts on the calcaneus via the Achilles tendon, the same tendon of insertion as the gastrocnemius.

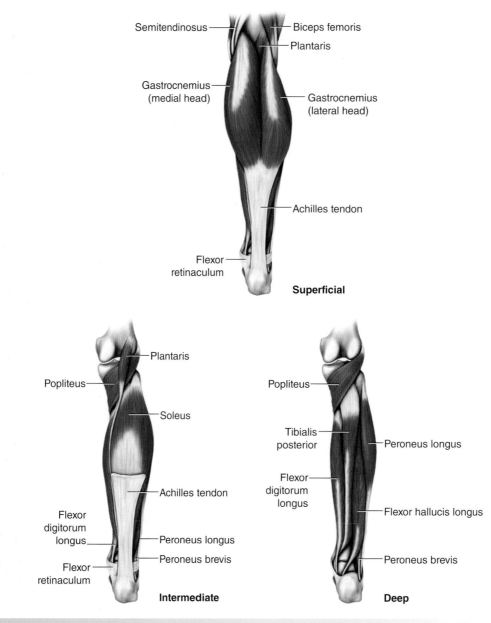

Figure 13.24 The posterior muscles of the ankle and foot.

Achilles Tendon

The tendon of insertion of both the gastrocnemius and soleus muscles is the **Achilles tendon**, which is often a source of concern in athletics. The gastrocnemius and soleus muscles are subject to strain as a result of overuse, and these strains (often referred to as "tennis leg") are common in middle-aged tennis players, sports officials, and runners. Overuse injuries often result from increased activity (running, jumping), defined as an increase in frequency, intensity, duration, or any combination of these factors. Any of this can lead to inflammation of the tendon, which is poorly vascularized to begin with, making recovery more complicated. Achilles tendinitis (inflammation) can lead to a rupture of the tendon as a result of a progressive degeneration of the tissue. Additionally, a rupture of the tendon can result from excessive force in the form of (1) direct force to, (2) forceful contraction of, or (3) forceful loading of the gastrocnemius and soleus muscles.

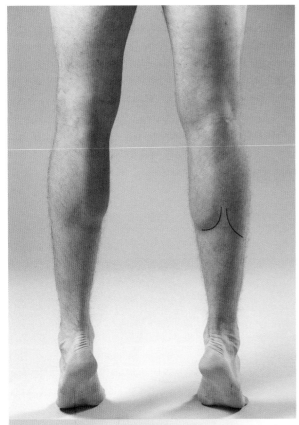

Figure 13.25 Locating the gastrocnemius muscle.

Because all three muscles of the superficial posterior compartment insert on the posterior aspect of the calcaneus, their action at the ankle joint is limited to one movement. What action do all three of these muscles perform at the ankle joint? (Answer: plantar flexion of the ankle)

Deep Posterior Compartment

Three muscles, which bear names very similar to those of the anterior compartment, are found in the deep posterior compartment of the lower leg. These three muscles are the tibialis posterior, the flexor digitorum longus, and the flexor hallucis longus (see figure 13.24).

- **Tibialis posterior**: The tibialis posterior originates from the middle third of the posterior aspect of the tibia, the proximal two-thirds of the medial aspect of the fibula, and the interosseous membrane. It inserts on the inferior (plantar) surfaces of the navicular and medial cuneiform bones and the bases of the second, third, fourth, and fifth metatarsal bones (figures 13.24 and 13.26). Observe that the tendon of insertion passes posterior to the medial malleolus of the tibia: What are the likely actions of this muscle?

✋ Hands On

You can easily palpate the Achilles tendon just proximal to your calcaneus.

- **Plantaris**: The third muscle that occurs in the superficial posterior compartment is the plantaris muscle (figure 13.24). This muscle originates from the lateral linea aspera and the oblique popliteal ligament and inserts on the posterior aspect of the calcaneus (separately from the Achilles tendon). Although this short-bellied muscle lays claim to having the longest tendon of any muscle in the body, it is of minor significance in assisting other muscles of the knee and ankle joints. The plantaris crosses the knee and ankle in the same fashion as the gastrocnemius and is strained by the same mechanisms. Its function is that of an assistant to the knee flexors and ankle plantar flexors. It is not a major joint mover and is analogous in function to the palmaris longus in the forearm and wrist.

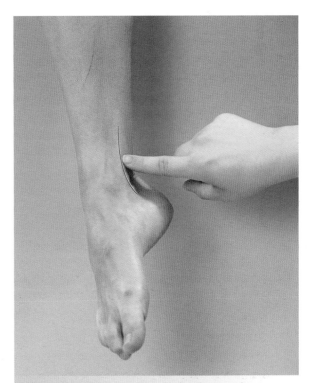

Figure 13.26 Locating the tibialis posterior.

(Answer: plantar flexion and inversion of the ankle joint)

• **Flexor digitorum longus**: The flexor digitorum longus originates from the lower two-thirds of the posterior aspect of the tibia and inserts on the bases of the distal phalanges of the four lateral toes (figures 13.24 and 13.27). Because it crosses the medial aspect of the ankle and splits into four tendons running through the plantar surface of the foot to the distal phalanges,

what are the actions performed by this muscle? Don't forget to look carefully at the name of the muscle. (Answer: plantar flexion and inversion of the ankle joint and flexion of the MP, PIP, and DIP joints of the four lateral toes)

• **Flexor hallucis longus**: The flexor hallucis longus originates from the lower two-thirds of the posterior aspect of the fibula and interosseous membrane and inserts on the base of the distal phalanx of the great toe (figures 13.24 and

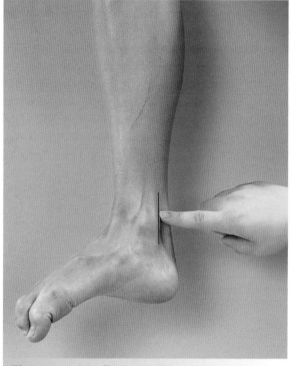

Figure 13.27   Finding the flexor digitorum longus.

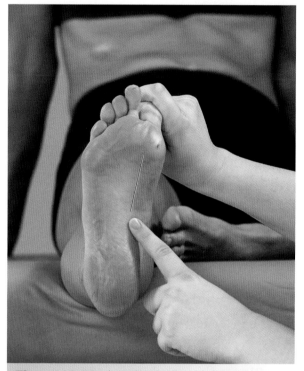

Figure 13.28   Locating the flexor hallucis longus.

FOCUS ON

Shin Splints

The term *shin splints* has been called a "wastebasket term" by many because it is used to describe a variety of conditions involving the lower leg, including inflammation of the tendons of both the anterior and posterior tibialis muscles, stress fractures of the tibia and the fibula, inflammation of the periosteum of the tibia and the fibula (periostitis), and inflammation of the interosseous membrane between the fibula and the tibia. Another accepted description is *osteoperiostitis* (inflammation of the periosteum of the tibia) resulting from overuse (repetitive loading) of the posterior tibialis and soleus muscles. Causes include improper footwear, problems of the longitudinal arch, playing surfaces, and repetitive training activities (resulting in multiple symptoms referred to as an overuse syndrome). Whatever the definition, whatever the cause, attention must be given to lower-leg pain. "Running it out" is not the answer. Ignoring a leg injury can result in continuing inflammation that can develop into ischemia (an anemia, lack of blood, caused by obstruction of circulation), which can lead to serious conditions such as necrosis (death of tissue) and stress fractures.

13.28). Because it crosses the medial aspect of the ankle and runs through the plantar surface of the foot to the distal phalanx, what are the actions performed by this muscle? Don't forget to look carefully at the name of the muscle. (Answer: plantar flexion and inversion of the ankle joint and flexion of the MP and IP joints of the great toe)

Intrinsic Muscles of the Foot

The majority of intrinsic muscles of the foot are found on the plantar surface, where the bony and ligamentous arrangements create space to accommodate these structures. These muscles are found in four distinct layers and are very similar in arrangement to the intrinsic muscles of the hand. Note that the names of many of these muscles indicate their actions.

Superficial to all muscles on the plantar surface of the foot is a structure known as the **plantar** **fascia** (figure 13.29). This fibrous band originates from the calcaneus; inserts on the five MP joints; and helps to protect the muscles, blood vessels, and nerves running through the plantar surface of the foot.

FOCUS ON

Plantar Fasciitis

A common problem for runners is inflammation of the area of origin of the plantar fascia from repeated stress. This condition is known as plantar fasciitis. The point of attachment of the plantar fascia is just anterior to the tubercle of the inferior aspect of the calcaneus. Stress placed on the foot through the repetitive action of running can cause inflammation of this area. Prevention and treatment of this condition are often concerns for orthopedists and podiatrists.

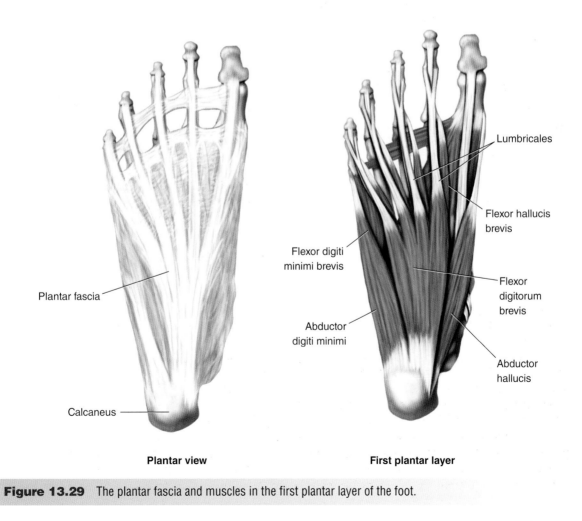

Plantar fascia

Calcaneus

Plantar view

Lumbricales

Flexor hallucis brevis

Flexor digiti minimi brevis

Flexor digitorum brevis

Abductor digiti minimi

Abductor hallucis

First plantar layer

Figure 13.29 The plantar fascia and muscles in the first plantar layer of the foot.

First Layer

Beneath the plantar fascia, the first plantar layer contains three muscles (figure 13.29): the flexor digitorum brevis, the abductor hallucis, and the abductor digiti minimi.

- **Flexor digitorum brevis**: The flexor digitorum brevis originates from the tuberosity on the plantar surface of the calcaneus and inserts on the middle phalanges of the four lateral toes. This muscle flexes the MP and PIP joints of the four lateral toes.

- **Abductor hallucis**: The abductor hallucis originates from the medial aspect of the plantar surface of the calcaneus and inserts on the medial surface of the proximal phalanx of the great toe. The muscle abducts (moves away from the second toe) the MP joint of the great toe and also assists with flexion of the MP joint of the great toe.

- **Abductor digiti minimi**: The abductor digiti minimi originates from the tuberosity on the plantar surface of the calcaneus and inserts on the lateral aspect of the base of the proximal phalanx of the fifth toe. The muscle abducts the MP joint of the fifth (little) toe and also assists with flexion of the fifth MP joint.

Second Layer

The second plantar layer contains two muscles, quadratus plantae and the lumbricales (figure 13.30), with some unique characteristics.

- **Quadratus plantae**: The quadratus plantae muscle has two heads—one from the medial surface of the calcaneus and the other from the lateral surface of the calcaneus—that combine to attach to the tendons of the flexor digitorum longus. This muscle assists with flexion of the MP, PIP, and DIP joints of the lateral four toes.

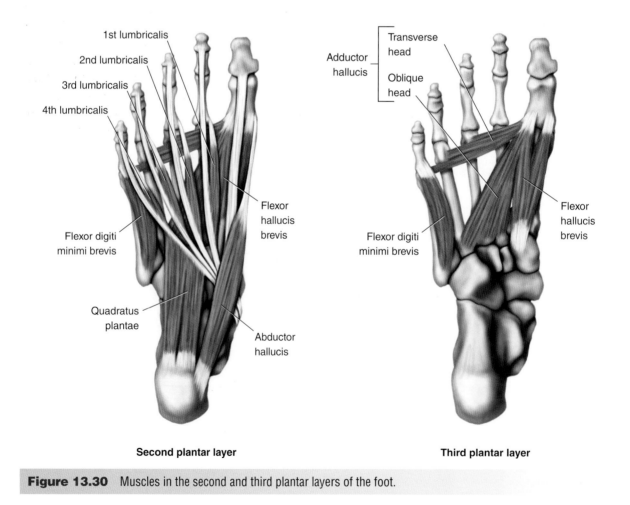

Figure 13.30 Muscles in the second and third plantar layers of the foot.

- **Lumbricales**: The other muscle of the second plantar layer is really a group of four muscles collectively known as the lumbricales. These four muscles have no bony attachments: They originate from the tendon of the flexor digitorum longus and insert on the tendon of the extensor digitorum longus. The lumbricales flex the MP joints and extend the PIP and DIP joints of the lateral four toes.

Third Layer

The third plantar layer contains three muscles whose names indicate their functions (figure 13.30): the flexor hallucis brevis, the adductor hallucis, and the flexor digiti minimi brevis.

- **Flexor hallucis brevis**: The flexor hallucis brevis originates from the three cuneiform bones and inserts on the base of the proximal phalanx of the great toe. It flexes the MP joint of the great toe, as its name indicates.
- **Adductor hallucis**: The adductor hallucis has two heads: One is an oblique head, which originates from the bases of the second, third, and fourth metatarsals and inserts on the proximal phalanx of the great toe. The second head is a transverse head, which originates from the plantar surface of the MP capsular ligaments of the third, fourth, and fifth toes and inserts on the proximal phalanx of the great toe. As its name indicates, this muscle adducts the great toe.

- **Flexor digiti minimi brevis**: The flexor digiti minimi brevis originates from the base of the fifth metatarsal bone and inserts on the lateral aspect of the base of the proximal phalanx of the fifth (little) toe. Its function, as its name indicates, is to flex the MP joint of the little toe.

Fourth Layer

The fourth layer on the plantar surface of the foot contains two groups of muscles: the dorsal interossei and plantar interossei found between the bones (interosseous) of the foot (figure 13.31).

- **Dorsal interossei**: There are four dorsal interossei muscles. The dorsal interossei originate from the adjacent sides of all metatarsal bones and insert on the bases of the second, third, and fourth proximal phalanges. Note that the midline of the foot is considered the second toe (as opposed to the middle finger in the hand). The action of the dorsal interossei results in abduction of the third and fourth MP joints. Because there are dorsal interossei muscles on either side of the second toe, contraction of the dorsal interossei results in no movement of the second toe.

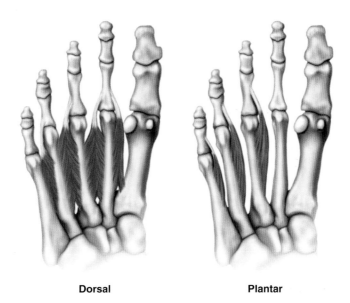

Dorsal   Plantar

Figure 13.31   The dorsal and plantar interossei.

• **Plantar interossei**: There are three plantar interossei muscles. The plantar interossei originate from the medial surfaces of the third, fourth, and fifth metatarsal bones and insert on the medial aspect of the bases of the proximal phalanges of the third, fourth, and fifth toes. The action of the plantar interossei results in the adduction of the MP joints of the third, fourth, and fifth toes. Note again that the second toe does not move on contraction of the plantar interossei muscles. Is this for the same reason that the second toe did not move when the dorsal interossei contracted? If not, why does the second toe not move when the plantar interosseous muscles contract?

Dorsal Intrinsic Muscles of the Foot

Although there is not as much space on the dorsal aspect of the foot as there is on the plantar aspect, there are two intrinsic muscles (extensor digitorum brevis and extensor hallucis brevis) on the dorsal aspect (figure 13.19).

• **Extensor digitorum brevis**: The extensor digitorum brevis (figure 13.32) originates from the calcaneus and inserts on the tendons of the extensor digitorum longus, which in turn inserts on the second, third, and fourth toes.

• **Extensor hallucis brevis**: A fourth tendon of the extensor digitorum brevis branches to the distal phalanx of the great toe and is often called the extensor hallucis brevis muscle.

As their names indicate, these two muscles assist with extension of the MP and IP joints of the great toe and the MP, PIP, and DIP joints of the second, third, and fourth toes.

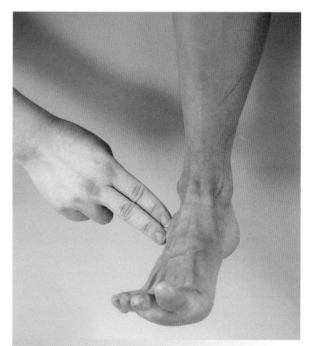

Figure 13.32 Locating the extensor digitorum brevis.

LEARNING AIDS

REVIEW OF TERMINOLOGY

The following terms are discussed in this chapter. Define or describe each term, and where appropriate, identify the location of the named structure either on your body or in an appropriate illustration.

abductor digiti minimi
abductor hallucis
Achilles tendon
adductor hallucis
ankle joint
anterior compartment
anterior talofibular ligament
anterior talotibial ligament
anterior tibiofibular ligament
base of a metatarsal
bifurcated ligament

calcaneocuboid ligament
calcaneofibular ligament
calcaneonavicular ligament
calcaneotibial ligament
calcaneus
cuboid
cuboideonavicular ligament
cuneiform
cuneocuboid ligament
cuneonavicular ligament
deep posterior compartment

deltoid ligament
dorsal interossei
dorsal talonavicular ligament
dorsiflexion
eversion
extensor digitorum brevis
extensor digitorum longus
extensor hallucis brevis
extensor hallucis longus
extensor retinaculum (superior and inferior)

extrinsic muscle
fibula
fibular notch of the tibia
flexor digiti minimi brevis
flexor digitorum brevis
flexor digitorum longus
flexor hallucis brevis
flexor hallucis longus
flexor retinaculum
forefoot
gastrocnemius
head of a metatarsal
hindfoot
intercuneiform ligament
intermetatarsal ligament
interosseous ligament
interosseous (collateral) ligament
interphalangeal ligament
intertarsal ligament
intrinsic muscle
inversion
lateral compartment
lateral malleolus

long plantar ligament
longitudinal arch
lumbricales
malleoli
medial malleolus
metatarsal arch
metatarsal bone
metatarsophalangeal ligament
midfoot
navicular
peroneal retinaculum (superior and inferior)
peroneus brevis
peroneus longus
peroneus tertius
phalange
plantar fascia
plantar flexion
plantar interossei
plantaris
posterior talofibular ligament
posterior talotibial ligament
posterior tibiofibular ligament

pronation
quadratus plantae
scaphoid
sesamoid
shaft of a metatarsal
sinus tarsi
soleus
subtalar joint
superficial posterior compartment
supination
sustentaculum tali
talocalcaneal joint
talus
tarsal bone
tarsometatarsal ligament
tibia
tibialis anterior
tibialis posterior
tibiofibular joint
tibionavicular ligament
transverse arch
transverse metatarsal ligament
triceps surae

SUGGESTED LEARNING ACTIVITIES

1. While standing in the anatomical position, perform the following:

   a. Stand on your toes. In what position does this place your ankle joints, and what muscles produced this movement?

   b. Stand on your heels. In what position does this place your ankle joints, and what muscles produced this movement?

2. Someone you know has a sprained ankle. The physician says that this is the most common of all ankle sprains and that part of the rehabilitation routine should concentrate on strengthening the everter muscles.

   a. What ligament did this person likely sprain?

   b. What specific muscles does the physician recommend be strengthened?

3. From a standing position, go down into a baseball or softball catcher's position.

   a. Did your heels rise up off the floor?

   b. If so, why? What muscles caused the heels to rise?

4. Look carefully at the peroneal tendons on the lateral side of the ankle as the ankle dorsiflexes and plantar flexes. If, for some reason, the peroneal retinacula failed to perform their function and the peroneal tendons were allowed to move forward in front of the lateral malleolus of the fibula, how would the function of the peroneal muscles be altered?

5. While you are walking, what is the position of the ankle joint when your heel strikes the walking surface, and what is the position of the ankle joint when your foot "toes off" and no longer bears weight? How would a weakness in either the posterior compartment muscles or the anterior compartment muscles affect walking?

MULTIPLE-CHOICE QUESTIONS

1. Which of the following muscles attaches to the base of the fifth metatarsal bone?

    a. peroneus longus
    b. peroneus brevis
    c. tibialis anterior
    d. tibialis posterior

2. Which malleolus at the ankle joint is the most distal in relationship to the other?

    a. lateral
    b. medial
    c. anterior
    d. posterior

3. How many tarsal bones are found in the foot?

    a. 2
    b. 5
    c. 7
    d. 26

4. Which of the tarsal bones sits on top of (superior to) the calcaneus?

    a. scaphoid
    b. talus
    c. cuboid
    d. lateral cuneiform

5. The quadratus plantae muscle assists which of the following muscles in flexing the toes?

    a. flexor digitorum longus
    b. flexor digitorum brevis
    c. flexor hallucis longus
    d. flexor hallucis brevis

6. Pronation of the foot combines abduction of the foot and which of the following ankle joint movements?

    a. rotation
    b. circumduction
    c. inversion
    d. eversion

7. How many phalanges are found in the forefoot?

    a. 5
    b. 14
    c. 15
    d. 19

8. Supination of the foot combines adduction of the foot and which of the following ankle joint movements?

    a. rotation
    b. circumduction
    c. inversion
    d. eversion

9. Which of the following muscles is considered a muscle of the lower leg's anterior compartment?

    a. peroneus magnus
    b. peroneus longus
    c. peroneus brevis
    d. peroneus tertius

10. Which of the following muscles is not found in the superficial posterior compartment of the leg?

    a. plantaris
    b. tibialis posterior
    c. gastrocnemius
    d. soleus

11. Of the following muscles, which is not found in the deep posterior compartment of the leg?

    a. tibialis posterior
    b. flexor digitorum longus
    c. plantaris
    d. flexor hallucis longus

FILL-IN-THE-BLANK QUESTIONS

1. The subtalar joint is the joint between the talus and the _____.

2. The largest bone of the foot is the _____.

3. The only function of the soleus muscle is ankle _____.

4. The gastrocnemius muscle originates on the _____.

5. The medial ligaments of the ankle joint are often referred to collectively as the _____ ligament.

6. Fundamental ankle joint movements in the sagittal plane include plantar flexion and _____.

7. The medial malleolus is found at the distal end of the _____.

8. The abductor hallucis muscle's function is to _____ the _____ joint.

9. Eversion of the ankle joint is an attempt to move the plantar surface of the foot _____.

10. The bases of the long metatarsal bones of the foot form the _____ arch.

11. The most lateral bone in the midfoot is the _____.

12. Muscles originating on the leg, crossing the ankle, and inserting on the foot are considered _____ muscles.

13. The adductor hallucis muscle has two heads: the oblique and the _____.

14. The Achilles tendon consists of the soleus and _____ tendons.

15. The dorsal interosseous muscles of the foot _____ the toes.

16. The anatomical name of the large ball of the foot is the _____ joint.

17. In addition to dorsiflexing the ankle joint, the tibialis anterior muscle _____ the _____.

18. The plantar interossei of the foot _____ the toes.

19. The longitudinal arch of the foot runs from the calcaneus to the _____.

20. In addition to being a major inverter of the ankle, the tibialis posterior muscle _____ the _____.

FUNCTIONAL MOVEMENT EXERCISE

Standing on your tiptoes requires muscle actions at joints throughout the lower leg, ankle, and foot. List one muscle acting as a prime mover, one as an antagonist, one as a fixator, and one as a synergist at the ankle joint complex and the joints of the foot (MP, PIP, and DIP).

	Ankle	MP, PIP, DIP
Prime mover		
Antagonist		
Fixator		
Synergist		

Nerves and Blood Vessels of the Lower Extremity

As in both the upper extremity and the spinal column and thorax, the musculature of the lower extremity is innervated by spinal nerves that are formed into plexuses. The spinal nerves that innervate the muscles of the lower extremity arise from the **lumbosacral plexus**, which is typically divided into the **lumbar plexus** (T12, L1, L2, L3, L4; figure 14.1), the **sacral plexus** (L4, L5, S1, S2, S3; figure 14.1), and the **pudendal (coccygeal) plexus** (S2, S3, S4, S5, C1, C2). The pudendal plexus, which often is considered part of the sacral plexus, innervates many of the structures of the abdominal cavity and the reproductive systems but none of the muscles of the lower extremity. Therefore, the pudendal plexus is not considered in this discussion of the lower extremity.

Nerves of the Lumbosacral Plexus

The nerves of the lumbosacral plexus are divided into three parts according to their location relative to the spinal column (superior, middle, and inferior).

Superior Portion

The most superior portion of the lumbosacral plexus, known as the lumbar plexus, consists of the **femoral**, **anterior femoral cutaneous**, **saphenous**, **genitofemoral**, **iliohypogastric**, **ilioinguinal**, and **obturator nerves** and the lateral cutaneous nerve of the thigh, the **lateral femoral cutaneous** (figure 14.2).

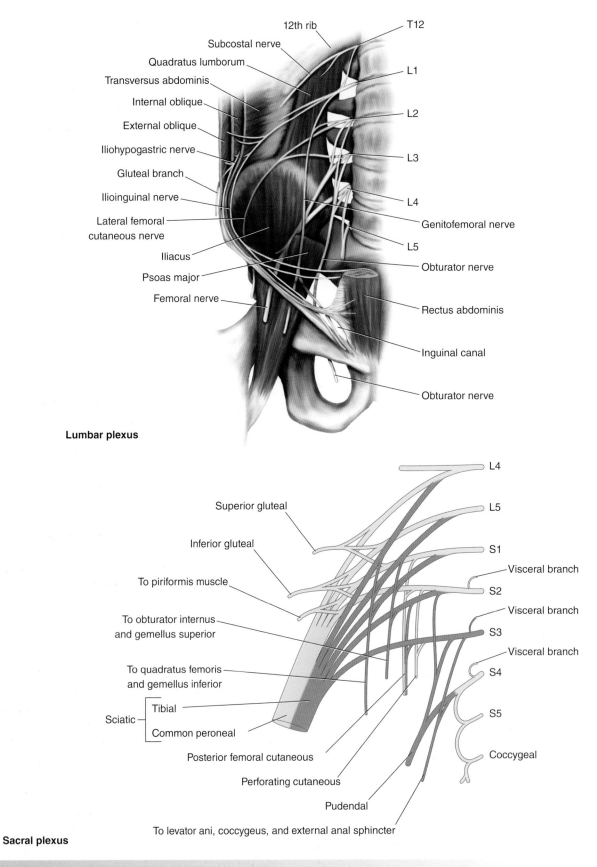

Lumbar plexus

- 12th rib
- Subcostal nerve
- Quadratus lumborum
- Transversus abdominis
- Internal oblique
- External oblique
- Iliohypogastric nerve
- Gluteal branch
- Ilioinguinal nerve
- Lateral femoral cutaneous nerve
- Iliacus
- Psoas major
- Femoral nerve
- T12
- L1
- L2
- L3
- L4
- Genitofemoral nerve
- L5
- Obturator nerve
- Rectus abdominis
- Inguinal canal
- Obturator nerve

Sacral plexus

- Superior gluteal
- Inferior gluteal
- To piriformis muscle
- To obturator internus and gemellus superior
- To quadratus femoris and gemellus inferior
- Sciatic — Tibial / Common peroneal
- Posterior femoral cutaneous
- Perforating cutaneous
- Pudendal
- To levator ani, coccygeus, and external anal sphincter
- L4
- L5
- S1
- Visceral branch
- S2
- Visceral branch
- S3
- Visceral branch
- S4
- S5
- Coccygeal

Figure 14.1 The lumbar and sacral plexuses.

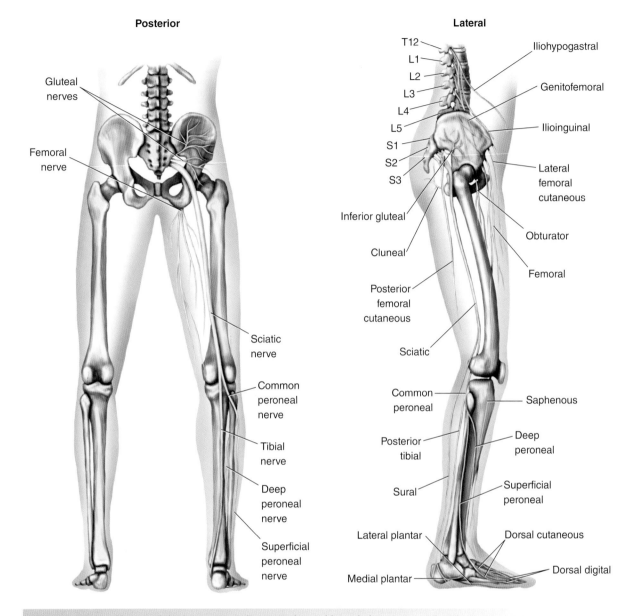

Posterior

Gluteal
nerves

Femoral
nerve

Sciatic
nerve

Common
peroneal
nerve

Tibial
nerve

Deep
peroneal
nerve

Superficial
peroneal
nerve

Lateral

T12
L1
L2
L3
L4
L5
S1
S2
S3

Inferior gluteal

Cluneal

Posterior
femoral
cutaneous

Sciatic

Common
peroneal

Posterior
tibial

Sural

Lateral plantar

Medial plantar

Iliohypogastral

Genitofemoral

Ilioinguinal

Lateral
femoral
cutaneous

Obturator

Femoral

Saphenous

Deep
peroneal

Superficial
peroneal

Dorsal cutaneous

Dorsal digital

Figure 14.2 Nerves of the lower extremity, posterior and lateral views.

The femoral nerve (L2, L3, L4) innervates the pectineus, the four muscles of the quadriceps group, and the sartorius. The cutaneous branches of the femoral nerve, the anterior femoral cutaneous nerve, and the saphenous nerve do not innervate muscles. The genitofemoral nerve (L1, L2) supplies the male and female genitalia and has a muscular branch innervating the cremaster muscle in the scrotum and therefore is not considered a nerve of the lower extremity. Likewise, both the iliohypogastric (L1 and, in some people, T12) and ilioinguinal (T12 and, in some people, L1) nerves, although part of the lumbar plexus, have muscular branches that innervate only the abdominal muscles and are considered nerves of the trunk and not of the lower extremity. The lateral femoral cutaneous nerve (L1, L2, L3, L4) has muscular branches to the psoas major and minor muscles as well as the quadratus lumborum. Again, these muscles are typically considered muscles of the trunk and not the lower extremity. The obturator nerve (L2, L3, L4) has anterior and posterior muscular branches that innervate the adductor magnus, adductor longus, adductor brevis, gracilis, pectineus, and external obturator muscles.

Sciatic Neuritis

The sciatic nerve, because of its position and length, is often subjected to trauma in several places as a result of a direct force, stretching, or impingement. Impingement at the lumbar or sacral disc space and muscular strains in the low back and lower extremities often cause an inflammation of the sciatic nerve called sciatic neuritis. Rest, anti-inflammatory agents, and eventually exercise are often prescribed for this condition.

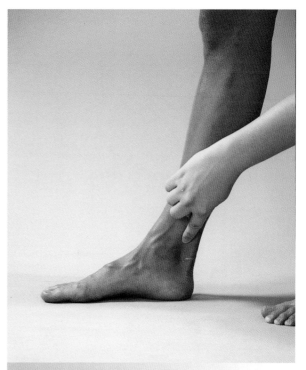

Figure 14.3   Locating the tibial nerve.

Middle Portion

The middle portion of the lumbosacral plexus, known as the sacral plexus, consists of the **superior gluteal**, **inferior gluteal**, **sciatic**, **common peroneal**, **tibial**, and **sural nerves** and the **nerve to the piriformis**.

The superior and inferior gluteal nerves (L4, L5, S1, S2) innervate the three gluteal muscles (the gluteus maximus, the gluteus medius, and the gluteus minimus), as well as the tensor fasciae latae muscle. The sciatic nerve (L4, L5, S1, S2, S3) is the longest and largest nerve in the body and is a combination of two nerves: the tibial (L4, L5, S1, S2, S3; figure 14.3) and the common peroneal (L4, L5, S1, S2; figure 14.4). A cutaneous branch of the common peroneal nerve, the **lateral sural nerve**, combines with the **medial sural branch of the tibial nerve** to form the sural nerve. The sural nerve and these branches do not innervate any lower-extremity muscles.

Just distal to the head of the fibula, the common peroneal nerve divides to form three terminal branches: the **deep peroneal nerve**, the **superficial peroneal nerve**, and the **recurrent tibial nerve** (figure 14.5). The deep peroneal nerve has muscular branches innervating the four muscles of the anterior compartment in the lower leg, the extensor digitorum brevis, and the interosseous muscles of the foot. The superficial peroneal nerve innervates the peroneus longus and peroneus brevis muscles of the lateral compartment of the lower leg. The

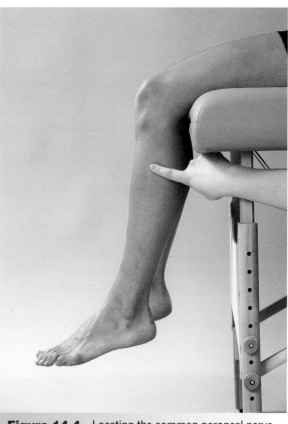

Figure 14.4   Locating the common peroneal nerve.

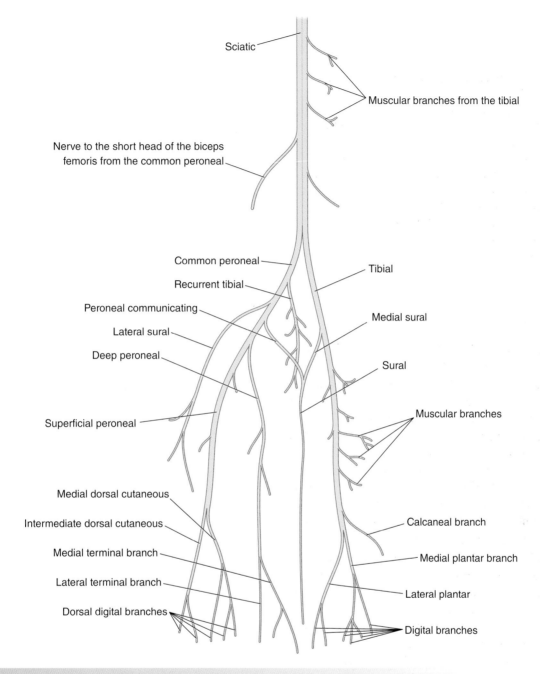

Sciatic

Muscular branches from the tibial

Nerve to the short head of the biceps femoris from the common peroneal

Common peroneal

Recurrent tibial

Peroneal communicating

Lateral sural

Deep peroneal

Superficial peroneal

Medial dorsal cutaneous

Intermediate dorsal cutaneous

Medial terminal branch

Lateral terminal branch

Dorsal digital branches

Tibial

Medial sural

Sural

Muscular branches

Calcaneal branch

Medial plantar branch

Lateral plantar

Digital branches

Figure 14.5 The sciatic nerve and its branches (on the right side), anterior view.

recurrent tibial nerve innervates the proximal portion of the tibialis anterior muscle. The tibial nerve has muscular branches innervating the three hamstring muscles (but not the short head of the biceps femoris). It also innervates four of the six deep external rotators of the hip joint (the superior and inferior gemelli muscles, the internal obturator muscle, and the quadratus femoris)

and the muscles in both the superficial and deep posterior compartments of the lower leg.

The tibial nerve has two main terminal branches: the lateral plantar nerve and the medial plantar nerve. The **lateral plantar nerve** has branches innervating the abductor digiti minimi, the quadratus plantae, the adductor hallucis, the lumbricales, the flexor digiti minimi brevis, and

Lower-Leg Trauma

Trauma (e.g., contusion, overuse) to the anterior compartment of the lower leg may cause so much swelling within the compartment that pressure on the deep peroneal nerve reduces its ability to transmit an impulse to the muscles of the compartment. Loss of strength and atrophy (wasting away) of the muscle tissue can lead to complications such as a "dropped foot," resulting from an inability to properly dorsiflex the ankle. In addition to deep peroneal nerve trauma, the function of blood vessels in the compartment may be compromised, which can result in possible necrosis (death) of the tissues within the compartment.

the interosseous muscles of the foot. The **medial plantar nerve** has branches innervating the abductor hallucis, the flexor digitorum brevis, and the lumbricales. The last nerve of the sacral plexus, the nerve to the piriformis (S2), innervates the piriformis muscle, one of the six deep external rotators of the hip joint.

Inferior Portion

The most inferior portion of the lumbosacral plexus, known as the pudendal plexus (see figure 14.1), consists of the **posterior femoral cutaneous nerve** (S2, S3), the **anococcygeal nerve** (S4, S5, C1), the **perforating cutaneous nerve** (S2, S3), and the **pudendal nerve** (S2, S3, S4). None of these nerves innervate muscles of the lower extremity, and only the pudendal nerve and anococcygeal nerve innervate muscles at all (those of the abdominal cavity and genitalia).

Major Arteries of the Lower Extremity

As discussed in chapter 6, the blood vessels of the body are divided into arteries, arterioles, capillaries, venules, and veins (figure 14.6).

The major arteries of the lower extremity include the **femoral**, the **popliteal**, the **ante-**

rior tibial, and the **posterior tibial**, all with numerous branches. These arteries supply blood to the lower extremity. The femoral artery has many branches, including the **genicular arteries** that supply the adductor magnus, the gracilis, the three vastus muscles, the adductor group, the sartorius, and the vastus medialis. Another branch of the femoral artery, the **superficial circumflex iliac artery**, supplies the iliacus, tensor fasciae latae, and sartorius muscles. The largest branch of the femoral artery is the **profunda artery**. It has three branches: the **lateral femoral circumflex**, **the medial circumflex**, and four **perforating arteries**, which supply the three gluteal muscles, the tensor fasciae latae, the sartorius, the vastus lateralis and intermedius muscles, the rectus femoris, the adductors, the external obturator, the pectineus, and the hamstrings (figure 14.7).

Hands On

Apply pressure with your fingertips on your femoral triangle, and attempt to feel your pulse from the femoral artery (figure 14.8).

At the level of the knee joint, the femoral artery becomes known as the popliteal artery. The branches of the popliteal artery (the **lateral superior genicular** and **lateral inferior genicular arteries**, the **medial superior genicular** and **medial inferior genicular arteries**, the **middle genicular artery**) and

Femoral Triangle

One of the pressure points stressed in first-aid classes for control of bleeding in the lower extremity is the femoral artery. The femoral artery is easily located within the anatomical structure known as the **femoral triangle** (see figure 14.7). The triangle is formed by the inguinal ligament (superior border), the sartorius muscle (lateral border), and the adductor longus muscle (medial border). Also passing through this area are the femoral vein and the femoral nerve.

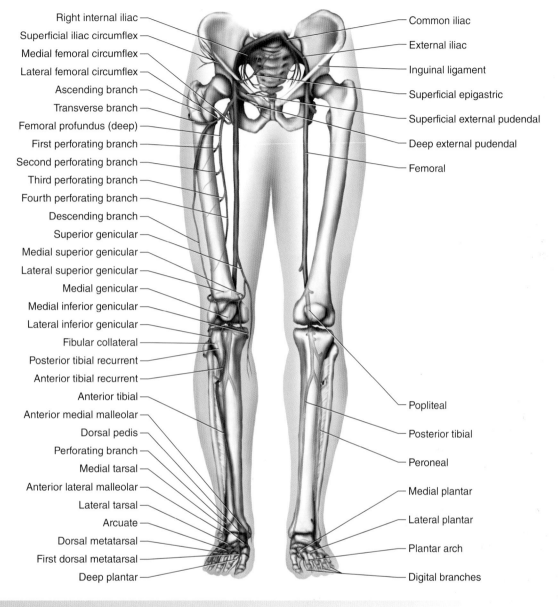

Right internal iliac
Superficial iliac circumflex
Medial femoral circumflex
Lateral femoral circumflex
Ascending branch
Transverse branch
Femoral profundus (deep)
First perforating branch
Second perforating branch
Third perforating branch
Fourth perforating branch
Descending branch
Superior genicular
Medial superior genicular
Lateral superior genicular
Medial genicular
Medial inferior genicular
Lateral inferior genicular
Fibular collateral
Posterior tibial recurrent
Anterior tibial recurrent
Anterior tibial
Anterior medial malleolar
Dorsal pedis
Perforating branch
Medial tarsal
Anterior lateral malleolar
Lateral tarsal
Arcuate
Dorsal metatarsal
First dorsal metatarsal
Deep plantar

Common iliac
External iliac
Inguinal ligament
Superficial epigastric
Superficial external pudendal
Deep external pudendal
Femoral
Popliteal
Posterior tibial
Peroneal
Medial plantar
Lateral plantar
Plantar arch
Digital branches

Figure 14.6 Major arteries of the lower extremity.

its **muscular branches** supply the vastus lateralis, adductor magnus, biceps femoris, semimembranosus, semitendinosus, popliteus, gastrocnemius, plantaris, and soleus muscles. As the popliteal artery passes the popliteal muscle, the artery divides into the anterior tibial and posterior tibial arteries. The anterior tibial artery has branches—**anterior** and **posterior tibial recurrent** (figures 14.6 and 14.9) and **fibular**—that supply the popliteus, extensor digitorum longus and brevis, tibialis anterior, peroneus longus, and soleus muscles. Below the ankle and into the foot, the anterior tibial artery becomes the **dorsal pedis artery** and ends on the dorsal (top) side of the foot (figures 14.6, 14.9, and 14.10). The branches of the dorsal pedis artery (**arcuate**, **deep plantar**, **first dorsal metatarsal**, **lateral tarsal**, and **medial tarsal**) supply the dorsal interosseous and extensor digitorum brevis muscles (figure 14.10).

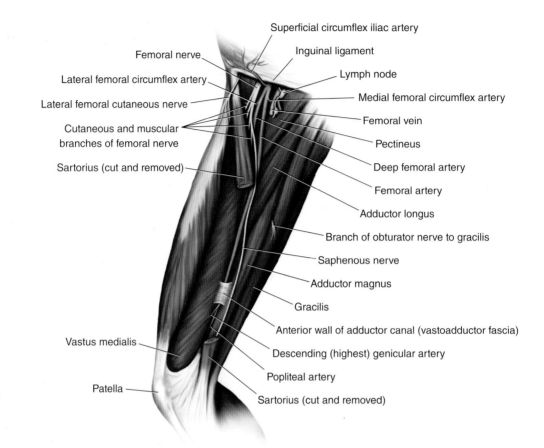

- Femoral nerve
- Lateral femoral circumflex artery
- Lateral femoral cutaneous nerve
- Cutaneous and muscular branches of femoral nerve
- Sartorius (cut and removed)
- Superficial circumflex iliac artery
- Inguinal ligament
- Lymph node
- Medial femoral circumflex artery
- Femoral vein
- Pectineus
- Deep femoral artery
- Femoral artery
- Adductor longus
- Branch of obturator nerve to gracilis
- Saphenous nerve
- Adductor magnus
- Gracilis
- Anterior wall of adductor canal (vastoadductor fascia)
- Vastus medialis
- Descending (highest) genicular artery
- Popliteal artery
- Patella
- Sartorius (cut and removed)

Figure 14.7 The femoral triangle.

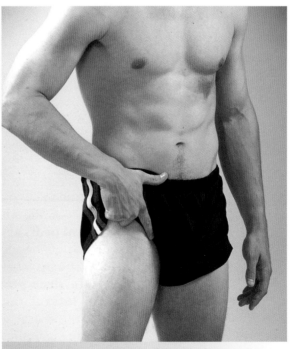

Figure 14.8 Locating the femoral artery.

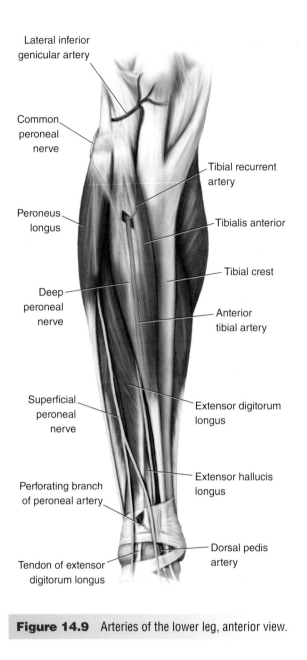

Figure 14.9 Arteries of the lower leg, anterior view.

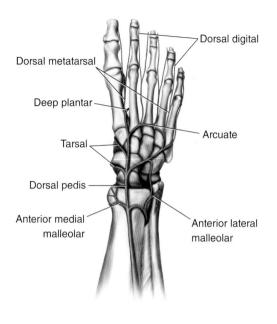

Figure 14.10 Branches of the dorsal pedis artery.

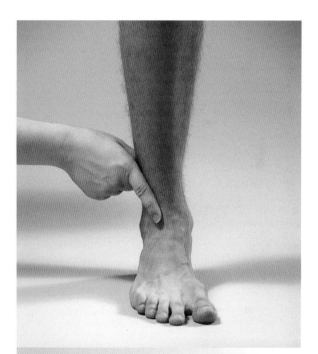

Figure 14.11 Locating the dorsal pedis artery.

🖐 Hands On

In trying to establish whether the blood supply of the foot is impeded for some reason, health care professionals attempt to palpate a dorsal pedis pulse. This pulse is normally found on the dorsal side of the foot between the proximal ends of the first and second metatarsal bones (between the extensor hallucis longus and extensor digitorum longus tendons). Use figure 14.11 to assist you in checking your dorsal pedis pulse. Not everyone can easily feel this pulse, because it has been found to be absent in 10% to 15% of all people.

The posterior tibial artery and its branches (**medial calcaneal**, **peroneal**, and **perfo-** rating branches) (figure 14.12) supply the posterior and lateral lower-leg muscles, most of the muscles of the foot, and the peroneus tertius in the lower portion of the anterior lower-leg compartment. The posterior tibial artery has two terminating branches (the **lateral** and **medial plantar arteries**) that supply most of the muscles of the foot.

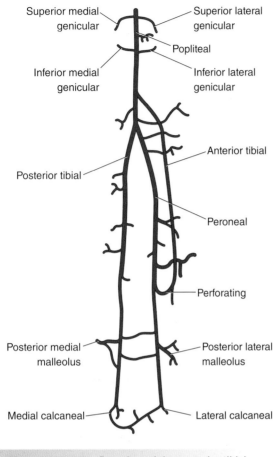

Figure 14.12 Branches of the posterior tibial artery.

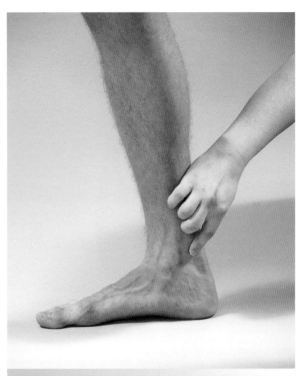

Figure 14.13 Locating the posterior tibial artery.

✋ Hands On

When you are trying to determine blood flow to the foot, a more reliable pulse to take than the dorsal pedis pulse is the posterior tibial pulse. The artery is found between the flexor digitorum longus and flexor hallucis longus tendons just posterior to the medial malleolus of the tibia. Refer to figure 14.13 to palpate your posterior tibial pulse.

Major Veins of the Lower Extremity

The veins that return blood to the heart are commonly divided into **deep veins** and **superficial veins** (figure 14.14). With a few exceptions, the deep veins have the same names as the arteries they parallel, such as the femoral, popliteal, and tibial. The superficial veins have specific names and are located near the skin. Unlike the arter-

ies, veins do not appear exactly where one might expect, and often they are absent altogether.

The major deep veins of the lower extremity are the femoral and popliteal. The **femoral vein** drains the entire thigh and, in the area of the inguinal ligament, becomes the **external iliac vein**. The **popliteal vein**, which unites with the anterior and posterior tibial arteries, drains the structures from the foot to the lower edge of the popliteus muscle. It has tributaries known as the **anterior** and **posterior tibial veins** and the **lesser saphenous vein**, and it has branches that correspond to the anterior and posterior tibial arteries and their branches.

The major superficial veins of the lower extremity include the **great saphenous** and the lesser saphenous veins (figures 14.15 and 14.16). The great saphenous vein runs between the medial aspect of the **dorsal venous arch** in the foot to the femoral triangle at the groin. This makes the great saphenous vein the longest vein in the body. Major tributaries of the great saphenous vein in the lower extremity include the **lateral** and **medial superficial femoral veins** and the **superficial circumflex iliac vein**. The lesser saphenous vein is found at

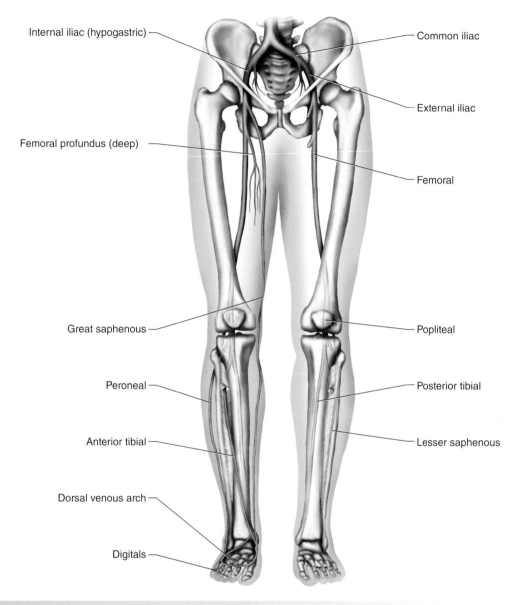

Internal iliac (hypogastric)

Common iliac

External iliac

Femoral profundus (deep)

Femoral

Great saphenous

Popliteal

Peroneal

Posterior tibial

Anterior tibial

Lesser saphenous

Dorsal venous arch

Digitals

Figure 14.14 The major veins of the lower extremity.

Varicose Veins

Varicose veins are simply defined as enlarged veins. The condition can appear in any vein in the body, but it is mentioned in this chapter because it most often is found in the lower extremity (thigh and lower leg). There are many causes for the condition to develop such as heredity, pregnancy, obesity, and long periods of standing. Any of these causes can lead to pressure on the veins of the lower extremity, particularly in the groin area, which can restrict blood flow return from the legs. This can cause a pooling of blood in the lower extremities that can weaken valves in the veins and actually stretch the walls of the veins. Varicose veins usually involve incompetent valves. Blood tends to pool in lower extremities, weakening the valves and eventually stretching the venous walls.

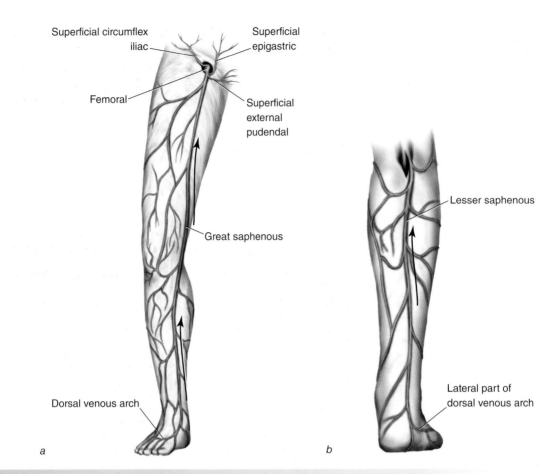

Superficial circumflex iliac

Superficial epigastric

Femoral

Superficial external pudendal

Great saphenous

Lesser saphenous

Dorsal venous arch

Lateral part of dorsal venous arch

a

b

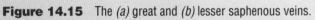

Figure 14.15 The *(a)* great and *(b)* lesser saphenous veins.

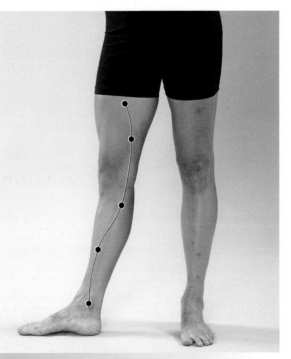

Figure 14.16 The path of the great saphenous vein.

the lateral aspect of the dorsal venous arch and empties into the popliteal vein in the area of the popliteal space posterior to the knee joint. The lesser saphenous vein drains the lateral and posterior aspects of the foot and lower leg. Interest-

ingly, sections of the saphenous veins are often surgically removed and used to replace damaged or diseased sections of other blood vessels. The many peripheral branches of the saphenous vein assume the function of the removed portion.

LEARNING AIDS

REVIEW OF TERMINOLOGY

The following terms are discussed in this chapter. Define or describe each term, and where appropriate, identify the location of the named structure either on your body or in an appropriate illustration.

anococcygeal nerve
anterior femoral cutaneous nerve
anterior tibial artery
anterior tibial recurrent artery
anterior tibial vein
arcuate artery
common peroneal nerve
deep peroneal nerve
deep plantar artery
deep vein
dorsal pedis artery
dorsal venous arch
external iliac vein
femoral artery
femoral nerve
femoral triangle
femoral vein
fibular artery
first dorsal metatarsal artery
genicular artery
genitofemoral nerve
great saphenous vein
iliohypogastric nerve
ilioinguinal nerve
inferior gluteal nerve
lateral femoral circumflex artery

lateral femoral cutaneous nerve
lateral inferior genicular artery
lateral plantar artery
lateral plantar nerve
lateral superficial femoral vein
lateral superior genicular artery
lateral sural nerve
lateral tarsal artery
lesser saphenous vein
lumbar plexus
lumbosacral plexus
medial calcaneal artery
medial circumflex artery
medial inferior genicular artery
medial plantar artery
medial plantar nerve
medial superficial femoral vein
medial superior genicular artery
medial sural branch of the tibial nerve
medial tarsal artery
middle genicular artery
muscular branch of popliteal artery
nerve to the piriformis
obturator nerve
perforating artery

perforating branch of posterior tibial artery
perforating cutaneous nerve
peroneal artery
popliteal artery
popliteal vein
posterior femoral cutaneous nerve
posterior tibial artery
posterior tibial recurrent artery
posterior tibial vein
profunda artery
pudendal (coccygeal) plexus
pudendal nerve
recurrent tibial nerve
sacral plexus
saphenous nerve
sciatic nerve
superficial circumflex iliac artery
superficial circumflex iliac vein
superficial peroneal nerve
superficial vein
superior gluteal nerve
sural nerve
tibial nerve

SUGGESTED LEARNING ACTIVITIES

1. Locate the anatomical area identified as the femoral triangle.

   a. Identify the three structures defining the triangle.

   b. Apply pressure to this area until you feel a pulse. What structure is providing this pulse? In typical first-aid courses, what is this area designated in terms of controlling bleeding?

2. Apply pressure with your fingers on the dorsal (top) side of the foot. Depending on the soft-tissue structures present, a pulse may be felt. If so, what structure is creating that pulse?

3. Determine whether a pulse can be established posterior to the medial malleolus of the tibia (between the medial malleolus and the Achilles tendon). If so, what structure is creating that pulse?

MULTIPLE-CHOICE QUESTIONS

1. The longest vein in the body is the
   a. femoral
   b. popliteal
   c. great saphenous
   d. tibial

2. The longest nerve in the body is the
   a. femoral
   b. popliteal
   c. sciatic
   d. common peroneal

3. The dorsal venous arch drains blood from the
   a. toes
   b. anterior tibial compartment
   c. ankle
   d. posterior tibial compartment

4. The anterior and posterior tibial veins drain into the
   a. great saphenous vein
   b. lesser saphenous vein
   c. femoral vein
   d. popliteal vein

FILL-IN-THE-BLANK QUESTIONS

1. The major nerve innervating the muscles of the anterior tibial compartment is the _____ nerve.

2. The _____ branch of the posterior tibial artery provides the blood supply to the peroneal muscles of the lateral compartment of the lower leg.

3. The medial adductors of the hip joint are innervated by the _____ nerve.

Articulations of the Lower Extremity

Joint	Type	Bones	Ligaments	Movements
Hip	Ball and socket	Ilium, ischium pubic, and femur	• Capsule • Acetabular labrium (glenoid lip) • Iliofemoral (Y) • Pubofemoral • Transverse acetabular • Ligamentum capitis femoris (teres)	Flexion, extension, abduction, adduction, internal (or medial or inward) rotation, external (or lateral or outward) rotation
Knee				
Tibiofemoral	Modified hinge	Femur and tibia	• Capsule • Medial (tibial) collateral • Lateral (fibular) collateral • Anterior cruciate • Posterior cruciate • Oblique popliteal • Arcuate (popliteal) • Wrisberg • Coronary • Transverse • Patellar (considered a continuation of the quadriceps tendon)	Flexion, extension (lower leg internally rotates on knee flexion and externally rotates on knee extension in non-weight-bearing movements)
Patellofemoral	Plane	Patella and femur	• Patellar	Gliding
Lower leg				
Proximal tibiofibular	Diarthrodial (plane)	Tibia and fibula	• Capsule • Anterior of head of fibula • Posterior of head of fibula	Slight gliding
Middle tibiofibular	Synarthrodial	Tibia and fibula	• Interosseous	None
Distal tibiofibular	Syndesmodial	Tibia and fibula	• Interosseous • Anterior tibiofibular • Posterior tibiofibular • Inferior transverse • (Anterior to the posterior tibiofibular)	Slight give on ankle dorsiflexion

(continued)

Joint	Type	Bones	Ligaments	Movements
Ankle				
Talotibial (talocrural)	Hinge	Tibia and talus	• Capsule • Deltoid • Tibionavicular • Calcaneotibial • Anterior talotibial • Posterior talotibial	Dorsiflexion, plantar flexion
Talofibular	Hinge	Talus and fibula	• Capsule • Calcaneofibular • Anterior talofibular • Posterior talofibular	Dorsiflexion, plantar flexion
Intertarsal joints				
Subtalar[a] (talocalcaneal)	Arthrodial	Talus and calcaneus	• Capsule • Anterior talocalcaneal • Posterior talocalcaneal • Lateral talocalcaneal • Medial talocalcaneal • Interosseous talocalcaneal	Gliding
Transverse tarsal[a]	Arthrodial	Talus, navicular, calcaneus, and cuboid	• Bifurcated (the calcaneonavicular part) • Plantar calcaneonavicular (or spring ligament)	Gliding
Talocalcaneonavicular	Arthrodial	Talus, navicular, and calcaneus	• Capsule • Dorsal talonavicular • Plantar calcaneonavicular	Gliding
Calcaneocuboid	Arthrodial	Calcaneus and cuboid	• Capsule • Plantar calcaneocuboid (short plantar) • Dorsal calcaneocuboid • Bifurcated • Long plantar	Gliding
Cuneonavicular	Arthrodial	Cuneiforms (3) and navicular	• Dorsal cuneonaviculars • Plantar cuneonaviculars	Gliding
Cuboideonavicular	Arthrodial	Cuboid and navicular	• Dorsal cuboideonavicular • Plantar cuboideonavicular • Interosseous cuboideonavicular	Gliding
Intercuneiform and cuneocuboid	Arthrodial	Cuneiforms (3) and cuboid	• Dorsal intercuneiforms • Plantar intercuneiforms • Interosseous intercuneiforms	Gliding

Joint	Type	Bones	Ligaments	Movements
Tarsometatarsophalangeal joints				
Tarsometatarsals	Arthrodial (plane)	Cuneiforms (3), cuboid, and bases of metatarsals	• Capsules • Plantar tarsometatarsal and dorsal tarsometatarsal • Interosseous collaterals	Gliding
Intermetatarsals	Arthrodial (plane)	Bases of metatarsals	• Dorsal capsules • Plantar capsules • Interosseous collaterals	Gliding
Metatarsophalangeal (5)	Condyloid	Heads of metatarsals and bases of proximal phalanges	• Dorsal capsules • Plantar capsules • Interosseous collaterals • Transverse metatarsal	Flexion, extension, abduction, adduction
Interphalangeals (4 proximal interphalangeal, 4 distal interphalangeal, and 1 interphalangeal)	Hinge	Proximal, middle, and distal phalanges	• Dorsal capsules • Plantar capsules • Interosseous collaterals	Flexion, extension

[a]The combined movements of the subtalar and transverse tarsal joints produce inversion (supination, adduction, and plantar flexion) and eversion (pronation, abduction, and dorsiflexion) of the ankle joint.

Muscles, Nerves, and Blood Supply of the Lower Extremity

Muscle	Origin	Insertion	Action	Nerve	Blood supply
Hip, anterior					
Psoas major (Iliopsoas)	L1–L5 transverse processes, T12 and L1–L5 intervertebral discs and bodies	Lesser trochanter of femur	Flexion and rotation of spinal column; flexion, adduction, and external rotation of hip	Femoral	External iliac, internal iliac, and lumbar
Psoas minor	T12 and L1 intervertebral discs and bodies	Iliopectineal line	Flexion of pelvis	Lumbar	Lumbar
Iliacus (Iliopsoas)	Iliac fossa, iliolumbar, and sacroiliac ligaments	Just distal to lesser trochanter of femur	Flexion, adduction, and external rotation of hip	Femoral	External iliac and hypogastric
Sartorius	Anterior superior iliac spine	Inferior to medial condyle of tibia	Flexion, abduction, and external rotation of hip; flexion of knee; internal rotation of lower leg	Femoral	Profunda and femoral
Rectus femoris	Anterior inferior iliac spine and area superior to the acetabulum	Patella and tibial tuberosity	Flexion of hip, extension of knee	Femoral	Profunda
Tensor fasciae latae	Outer lip of iliac crest and between anterior superior and anterior inferior iliac spines	Greater trochanter of femur and (as iliotibial band) inferior and anterior to lateral condyle of tibia	Flexion, abduction, and rotation of hip	Superior gluteal	Lateral circumflex
Pectineus	Iliopectineal line and superior aspect of pubic bone	Pectineal line on femur	Flexion, adduction, and external rotation of hip	Femoral	Femoral and femoral circumflex

Hip, lateral					
Gluteus medius	Between iliac crest and anterior and posterior gluteal lines	Posterolateral edge of greater trochanter	Abduction of hip; anterior portion: flexion and internal rotation of hip; posterior portion: extension and external rotation of hip	Superior gluteal	Superior gluteal
Gluteus minimus	Anterior and inferior gluteal lines of ilium	Anterior edge of greater trochanter	Abduction and internal rotation of hip; anterior portion: flexion of hip; posterior portion: extension of hip	Superior gluteal	Superior gluteal
Hip, posterior					
Biceps Femoris (Long head)	Ischial tuberosity	Head of fibula and lateral tibial condyle	Extension, adduction, and external rotation of hip; flexion of knee	Tibial	Profunda branch of femoral
Biceps Femoris (Short head)	Linea aspera	Head of fibula and lateral tibial condyle	Flexion of knee, external rotation of lower leg	Peroneal	Profunda branch of femoral
Semitendinosus	Ischial tuberosity	Inferior to medial tibial condyle	Extension, adduction, and internal rotation of hip; flexion of knee; internal rotation of lower leg	Tibial	Profunda branch of femoral and popliteal
Semimembranosus	Ischial tuberosity	Medial tibial condyle	Extension, adduction, and internal rotation of hip; flexion of knee; internal rotation of lower leg	Tibial	Profunda branch of femoral and popliteal
Gluteus maximus	Sacrum and coccyx, posterior gluteal line, and iliac crest	Gluteal line of femur and (as iliotibial band) inferior and anterior to lateral condyle of tibia	Extension, adduction, and external rotation of hip; abduction of flexed hip	Inferior gluteal	Superior and inferior gluteals and medial circumflex

(continued)

Muscle	Origin	Insertion	Action	Nerve	Blood supply
Hip, posterior *(continued)*					
Piriformis	Superior 2/3 of sacrum and greater sciatic notch	Greater trochanter	External rotation, abduction, and extension of hip	1st and 2nd sacral	Superior and inferior gluteals
Superior gemellus	Ischial spine	Greater trochanter	External rotation of hip	1st and 2nd sacral	Inferior gluteal and obturator
Internal obturator	Inner rim of obturator foramen	Greater trochanter	External rotation of hip	1st and 2nd sacral	Inferior gluteal and obturator
Inferior gemellus	Ischial tuberosity	Greater trochanter	External rotation of hip	Lumbosacral to quadratus femoris	Inferior gluteal and obturator
External obturator	Outer aspect of pubic and ischial bones at the obturator foramen	Trochanteric fossa of femur	External rotation of hip	Obturator	Inferior gluteal and obturator
Quadratus femoris	Ischial tuberosity	Intertrochanteric crest of femur	External rotation, adduction, and extension of hip	Branch from lumbosacral plexus	Inferior gluteal and medial femoral circumflex
Hip, medial					
Adductor longus	Pubic bone	Middle 1/3 of linea aspera	Adduction, flexion, and external rotation of hip	Obturator	Femoral
Adductor brevis	Pubic bone	Proximal 1/3 of linea aspera	Adduction, flexion, and external rotation of hip	Obturator	Femoral
Adductor magnus	Pubic bone and ischial tuberosity	Linea aspera and adductor tubercle	Anterior portion: adduction, flexion, and external rotation of hip; posterior portion: adduction, extension, and internal rotation of hip	Obturator and sciatic	Femoral
Gracilis	Anterior inferior aspect of pubic symphysis and inferior pubic bone	Distal to medial tibial condyle	Adduction, flexion, and external rotation of hip; flexion of knee; internal rotation of lower leg	Obturator	Femoral

Knee, anterior					
Sartorius	See: Hip, anterior				
Quadriceps femoris					
Rectus femoris	See: Hip, anterior				
Vastus lateralis	Greater trochanter; intertrochanteric line, proximal to lateral lip of linea aspera	Lateral patella, anterior aspect of lateral tibial condyle, and rectus femoris tendon	Extension of knee	Femoral	Lateral circumflex
Vastus intermedius	Proximal 2/3 of anterior of femur	Inferior aspect of patella and tendons of vastus lateralis and medialis	Extension of knee	Femoral	Lateral circumflex and profunda
Vastus medialis	Intertrochanteric line and linea aspera	Medial tibial condyle, medial patella, and medial aspect of rectus femoris tendon	Extension of knee	Femoral	Femoral
Knee, posterior					
Biceps femoris (hamstring)	See: Hip, posterior				
Semimembranosus (hamstring)	See: Hip, posterior				
Semitendinosus (hamstring)	See: Hip, posterior				
Gracilis	See: Hip, medial				
Popliteus	Lateral femoral condyle	Proximal tibia at popliteal line	Flexion of knee; internal rotation of lower leg	Tibial	Popliteal
Gastrocnemius (Lateral head)	Popliteal area medial to lateral femoral condyle	Calcaneus	Flexion of knee; plantar flexion of ankle	Tibial	Posterior tibial and peroneal
Gastrocnemius (Medial head)	Medial condyle of femur	Calcaneus	Flexion of knee; plantar flexion of ankle	Tibial	Posterior tibial and peroneal
Plantaris	Lateral linea aspera and oblique popliteal ligament	Calcaneus	Flexion of knee; plantar flexion of ankle	Tibial	Popliteal

(continued)

PART IV Summary Tables *(continued)*

Muscle	Origin	Insertion	Action	Nerve	Blood supply
Ankle and foot (extrinsic), anterior					
Tibialis anterior	Lateral condyle of tibia and proximal 1/2 of lateral aspect of tibia	1st cuneiform and base of 1st metatarsal	Dorsiflexion of ankle; inversion of foot	Deep peroneal	Anterior tibial
Extensor digitorum longus	Lateral condyle of tibia, proximal 3/4 fibula, and interosseous membrane	Phalanges of toes 2–5	Dorsiflexion and eversion of ankle; extension of toes 2–5	Deep peroneal	Anterior tibial
Extensor hallucis longus	Middle 1/2 of fibula and interosseous membrane	Base of distal phalanx of great toe	Dorsiflexion and inversion of ankle; extension of great toe	Deep peroneal	Anterior tibial
Peroneus tertius	Distal 1/3 of fibula and interosseous membrane	Dorsal aspect of base of 5th metatarsal	Dorsiflexion and eversion of ankle	Deep peroneal	Anterior tibial and peroneal
Ankle and foot (extrinsic), lateral					
Peroneus longus	Lateral condyle of tibia, head and proximal 2/3 of lateral fibula	Lateral aspect of base of 1st metatarsal and medial cuneiform	Eversion and plantar flexion of ankle	Superficial peroneal	Peroneal
Peroneus brevis	Distal 1/2 of fibula	5th metatarsal tuberosity	Eversion and plantar flexion of ankle	Superficial peroneal	Peroneal
Ankle and foot (extrinsic), posterior					
Gastrocnemius	See: Knee				
Soleus	Head and proximal 1/3 of fibula, middle 1/3 of tibia	Calcaneus	Plantar flexion of ankle	Tibial	Posterior tibial and peroneal
Plantaris	See: Knee, posterior				
Tibialis posterior	Middle 1/3 of posterior tibia, head and proximal 2/3 of medial aspect of fibula, and interosseous membrane	Navicular, medial cuneiform, and 2nd–5th metatarsals	Plantar flexion and inversion of ankle; adduction of foot	Tibial	Posterior tibial
Flexor digitorum longus	Posterior tibia: popliteal line to within a few inches of distal end of tibia	Base of distal phalanges of toes 2–5	Plantar flexion and inversion of ankle; adduction of foot; flexion of toes 2–5	Tibial	Posterior tibial

Muscle	Origin	Insertion	Action	Nerve	
Flexor hallucis longus	Lower 2/3 of posterior fibula and interosseous membrane	Distal phalanx of great toe	Plantar flexion and inversion of ankle; flexion of great toe	Tibial	Posterior tibial
Ankle and foot (intrinsic), 1st plantar layer					
Flexor digitorum brevis	Calcaneal tuberosity	Sides of middle phalanges of toes 2–5	Flexion of toes 2–5	Medial plantar	Posterior tibial
Abductor hallucis	Medial aspect of calcaneus	Base of proximal phalanx of great toe	Abduction and flexion of great toe	Medial plantar	Posterior tibial
Abductor digiti minimi	Calcaneal tuberosity	Lateral aspect of base of proximal phalanx of 5th toe	Abduction and flexion of 5th toe	Lateral plantar	Posterior tibial
Ankle and foot (intrinsic), 2nd plantar layer					
Quadratus plantae (Medial head)	Medial aspect of calcaneus	Flexor digitorum longus tendon	Assists flexion of toes 2–5	Lateral plantar	Posterior tibial
Quadratus plantae (Lateral head)	Lateral aspect of calcaneus	Flexor digitorum longus tendon	Assists flexion of toes 2–5	Lateral plantar	Posterior tibial
Lumbricales	Flexor digitorum longus tendon	Extensor digitorum longus tendon	Flexion of PIP joints (toes 2–5); extension of DIP joints (toes 2–5)	Lateral and medial plantar	Plantar
Ankle and foot (intrinsic), 3rd plantar layer					
Flexor hallucis brevis	1st, 2nd, and 3rd cuneiforms	Base of proximal phalanx of great toe	Flexion of MP joint of great toe	Medial plantar	Posterior tibial
Adductor hallucis (Transverse head)	Plantar surface of MP ligament of toes 3–5	Proximal phalanx of great toe	Adduction of great toe	Lateral plantar	Posterior tibial
Adductor hallucis (Oblique head)	Bases of metatarsals 2–4	Proximal phalanx of great toe	Adduction of great toe	Lateral plantar	Posterior tibial
Flexor digiti minimi brevis	Base of 5th metatarsal	Lateral aspect of base of proximal phalanx of 5th toe	Flexion of 5th toe	Lateral plantar	Posterior tibial

(continued)

PART IV Summary Tables (continued)

Muscle	Origin	Insertion	Action	Nerve	Blood supply
Ankle and foot (intrinsic), 4th plantar layer					
Dorsal interossei (4)	Adjacent aspects of all metatarsals	Base of proximal phalanges of toes 2–4	Abduction of toes 3 and 4	Lateral plantar	Anterior tibial
Plantar interossei (3)	Inferior medial aspect of metatarsals 3–5	Medial aspect of base of proximal phalanges of toes 3–5 and extensor digitorum longus tendon	Adduction of toes 3–5	Lateral plantar	Deep and lateral plantars
Ankle and foot (intrinsic), dorsal					
Extensor digitorum brevis	Calcaneus	Long extensor tendons of toes 2–5	Extension of toes 2–5	Deep peroneal	Anterior tibial and peroneal
Extensor hallucis brevis	Calcaneus	Distal phalanx of great toe	Extension and adduction of great toe	Deep peroneal	Anterior tibial

PIP = proximal interphalangeal; DIP = distal interphalangeal; MP = metatarsophalangeal.

ANSWERS TO END-OF-CHAPTER QUESTIONS

CHAPTER 1

Multiple-Choice Questions

1. c	**5.** c
2. a	**6.** c
3. b	**7.** c
4. a	**8.** c

Fill-in-the-Blank Questions

1. suture
2. bursa
3. muscle fibers
4. toward
5. away from
6. systole
7. viscosity
8. third-class

CHAPTER 2

Multiple-Choice Questions

1. b
2. d
3. b
4. b

Fill-in-the-Blank Questions

1. flexion
2. plane
3. sagittal horizontal
4. proximal

CHAPTER 3

Multiple-Choice Questions

1. d	**15.** d
2. d	**16.** d
3. a	**17.** b
4. c	**18.** a
5. a	**19.** a
6. a	**20.** d
7. d	**21.** b
8. d	**22.** b
9. d	**23.** d
10. b	**24.** d
11. b	**25.** d
12. c	**26.** a
13. a	**27.** a
14. c	

Fill-in-the-Blank Questions

1. coracoclavicular
2. trapezius
3. subscapularis
4. supraspinatus
5. acromion process
6. coracobrachialis
7. scapula
8. abduction
9. clavicular
10. scapula
11. latissimus dorsi

Functional Movement Exercise

The following answers are only some of the correct responses. Multiple answers can be considered correct. Please either consult *Musculoskeletal Anatomy Review* online or, if currently enrolled in a course on human anatomy, consult with your instructor and classmates to list additional correct answers.

Shoulder joint

Prime mover: Middle deltoid
Antagonist: Pectoralis major
Fixator: Supraspinatus
Synergist: Anterior and posterior deltoid

Shoulder girdle

Prime mover: Levator scapulae
Antagonist: Upper and lower trapezius
Fixator: Middle trapezius
Synergist: Rhomboids

CHAPTER 4

Multiple-Choice Questions

1. c	**7.** b
2. c	**8.** b
3. b	**9.** d
4. d	**10.** a
5. b	**11.** c
6. a	**12.** a

Fill-in-the-Blank Questions

1. humerus
2. annular
3. ulna
4. elbow flexion
5. trochlea
6. straight line
7. pronation
8. triceps brachii
9. pronator quadratus

Functional Movement Exercise

The following answers are only some of the correct responses. Multiple answers can be considered correct. Please either consult *Musculoskeletal Anatomy Review* online or, if currently enrolled in a course on human anatomy, consult with your instructor and classmates to list additional correct answers.

Elbow joint flexion

Prime mover: Biceps brachii
Antagonist: Triceps brachii
Fixator: Brachioradialis
Synergist: Brachialis

Forearm pronation

Prime mover: Pronator teres
Antagonist: Supinator
Fixator: Biceps brachii
Synergist: Pronator quadratus

CHAPTER 5

Multiple-Choice Questions

1. a	**9.** b
2. b	**10.** c
3. a	**11.** a
4. b	**12.** c
5. a	**13.** c
6. b	**14.** a
7. a	**15.** d
8. c	

Fill-in-the-Blank Questions

1. palmaris longus
2. midcarpal
3. flexors
4. thumb
5. radiocarpal
6. adduction
7. index
8. scaphoid
9. opposition
10. scaphoid
11. flexor retinaculum
12. collateral
13. carpometacarpal
14. scaphoid

Functional Movement Exercise

The following answers are only some of the correct responses. Multiple answers can be considered correct. Please either consult *Musculoskeletal Anatomy Review* online or, if currently enrolled in a course on human anatomy, consult with your instructor and classmates to list additional correct answers.

Wrist joints

Prime mover: Extensor digitorum communis
Antagonist: Palmaris longus
Fixator: Extensor digiti minimi
Synergist: Extensor carpi ulnaris

Finger joints (MP, PIP, DIP)

Prime mover: Flexor digitorum superficialis
Antagonist: Extensor digitorum
Fixator: Palmaris longus
Synergist: Flexor digitorum profundus

CHAPTER 6

Multiple-Choice Questions

1. b
2. c
3. c
4. a

Fill-in-the-Blank Questions

1. median cubital
2. dorsal scapular
3. arteries

CHAPTER 7

Multiple-Choice Questions

1. a	**5.** b
2. c	**6.** c
3. c	**7.** d
4. b	**8.** d

Fill-in-the-Blank Questions

1. middle
2. synarthrodial
3. maxillary
4. extrinsic
5. hyoid
6. sinus
7. foramen magnum
8. vitreous humor

CHAPTER 8

Multiple-Choice Questions

1. c 9. d
2. d 10. b
3. b 11. b
4. a 12. d
5. b 13. d
6. c 14. b
7. b 15. a
8. d 16. a

Fill-in-the-Blank Questions

1. lumbar
2. thoracic
3. posterior longitudinal
4. atlas
5. intervertebral disc
6. pubic symphysis
7. ischium
8. lordosis
9. kyphosis
10. scoliosis

Functional Movement Exercise

The following answers are only some of the correct responses. Multiple answers can be considered correct. Please either consult *Musculoskeletal Anatomy Review* online or, if currently enrolled in a course on human anatomy, consult with your instructor and classmates to list additional correct answers.

Spinal extension

Prime mover: Erector spinae
Antagonist: Rectus abdominis
Fixator: Internal and external obliques
Synergist: Semispinalis group

Spinal rotation

Prime mover: Internal and external obliques (same side)
Antagonist: Internal and external obliques (opposite side)
Fixator: Rectus abdominis
Synergist: Semispinalis group (same side)

CHAPTER 9

Multiple-Choice Questions

1. b 6. a
2. b 7. b
3. a 8. d
4. a 9. c
5. d 10. b

Fill-in-the-Blank Questions

1. sternum
2. manubrium
3. diaphragm
4. three
5. liver
6. bronchi
7. mediastinum
8. interatrial septum
9. two; cusps

CHAPTER 10

Multiple-Choice Questions

1. a 5. c
2. b 6. a
3. b 7. a
4. b

Fill-in-the-Blank Questions

1. phrenic
2. subclavian
3. brain stem
4. frontal
5. venae cordis minimae
6. epidural
7. precentral

CHAPTER 11

Multiple-Choice Questions

1. c 6. b
2. a 7. b
3. b 8. d
4. d 9. d
5. b 10. b

Fill-in-the-Blank Questions

1. hip joint
2. iliofemoral
3. gluteus maximus
4. ischium
5. thoracic and lumbar spine
6. posterior

Functional Movement Exercise

The following answers are only some of the correct responses. Multiple answers can be considered correct. Please either consult *Musculoskeletal Anatomy Review* online or, if currently enrolled in a course on human anatomy, consult with your instructor and classmates to list additional correct answers.

Hip flexion

Prime mover: Iliacus
Antagonist: Gluteus maximus
Fixator: Adductor longus
Synergist: Psoas major

Forward pelvic tilt

Prime mover: Rectus abdominis
Antagonist: Erector spinae
Fixator: Transversus abdominis
Synergist: Internal and external obliques

CHAPTER 12

Multiple-Choice Questions

1. a	**8.** b
2. d	**9.** c
3. b	**10.** d
4. b	**11.** a
5. c	**12.** b
6. a	**13.** c
7. d	**14.** b

Fill-in-the-Blank Questions

1. medial condyle
2. tibia
3. transverse
4. femur
5. anterior cruciate
6. knee
7. gluteus maximus

Functional Movement Exercise

The following answers are only some of the correct responses. Multiple answers can be considered correct. Please either consult *Musculoskeletal Anatomy Review* online or, if currently

enrolled in a course on human anatomy, consult with your instructor and classmates to list additional correct answers.

Knee flexion

Prime mover: Biceps femoris
Antagonist: Rectus femoris
Fixator: Vastus medialis
Synergist: Semimembranosus

Knee extension

Prime mover: Rectus femoris
Antagonist: Biceps femoris
Fixator: Gracilis
Synergist: Vastus intermedius

CHAPTER 13

Multiple-Choice Questions

1. b	**7.** b
2. a	**8.** c
3. c	**9.** d
4. b	**10.** b
5. a	**11.** c
6. d	

Fill-in-the-Blank Questions

1. calcaneus
2. calcaneus
3. plantar flexion
4. femur
5. deltoid
6. dorsiflexion
7. tibia
8. abduct; first MP
9. laterally
10. transverse
11. cuboid
12. extrinsic
13. transverse
14. gastrocnemius
15. abduct
16. first MP
17. inverts; ankle
18. adduct
19. metatarsal heads
20. plantar flexes; ankle

Functional Movement Exercise

The following answers are only some of the correct responses. Multiple answers can be considered correct. Please either consult *Musculoskeletal Anatomy Review* online or, if currently enrolled in a course on human anatomy, consult with your instructor and classmates to list additional correct answers.

Ankle

Prime mover: Soleus
Antagonist: Tibialis anterior
Fixator: Tibialis posterior
Synergist: Gastrocnemius

MP, PIP, DIP

Prime mover: Flexor digitorum longus
Antagonist: Extensor digitorum longus
Fixator: Plantar interossei
Synergist: Flexor digitorum brevis

CHAPTER 14

Multiple-Choice Questions

1. c
2. c
3. a
4. d

Fill-in-the-Blank Questions

1. deep peroneal
2. peroneal
3. obturator

Agur, A.M.R., and A.F. Daily. 2008. *Grant's atlas of anatomy*. 12th ed. Baltimore: Lippincott, Williams & Wilkins.

Drake, R.A., W. Vogl, and A.W.M. Mitchell. 2009. *Gray's anatomy for students*. 2nd ed. New York: Churchill Livingstone.

Gilroy, A., B. MacPherson, L. Ross, M. Schuenke, E. Schulte, and U. Schmacher. 2008. *Atlas of anatomy*. New York: Thieme.

Gray, H. (1918) 2000. *Anatomy of the human body*. Philadelphia: Lea & Febiger; Bartleby.com.

Logan, B.M., P. Reynolds, and R.T. Hutchings. 2009. *McMinn's color atlas of head and neck anatomy*. 4th ed. Philadelphia: Elsevier (Mosby).

Logan, B.M., and R.T. Hutchings. 2011. *McMinn's color atlas of foot and ankle anatomy*. 4th ed. Philadelphia: Elsevier (Saunders).

Marieb, E.N., and K.N. Hoehn. 2010. *Human anatomy and physiology*. 8th ed. San Francisco: Benjamin/Cummings.

Morton, D., K.B. Foreman, and K. Albertine. 2011. *Gross anatomy: The big picture*. New York: McGraw-Hill Medical.

Netter, F.H. 2010. *Atlas of human anatomy*. 5th ed. Philadelphia: Elsevier (Saunders).

Whiting, W., and S. Rugg. 2006. *Dynatomy: Dynamic human anatomy*. Champaign, IL: Human Kinetics.

INDEX

Note: The italicized *f* and *t* following page numbers refer to figures and tables, respectively.

Robert S. Behnke, HSD, is retired after 39 years of teaching anatomy, kinesiology, physical education, and athletic training courses. Behnke has been honored on several occasions for excellence in teaching—including receiving the Educator of the Year Award from the National Athletic Trainers' Association. During his 11-year tenure as chair of the NATA Professional Education Committee, he initiated the petition to the American Medical Association that led to the national accreditation process for entry-level athletic training education programs.

Dr. Behnke spent most of his career at Indiana State University, where he was a full professor of physical education and athletic training and director of undergraduate and graduate athletic training programs. Earlier in his career he spent five years teaching at the secondary level. He has extensive experience as an athletic trainer, both in high schools and at the university level,

Courtesy of Robert Behnke

serving as head athletic trainer at Illinois State University from 1966 to 1969 and the University of Illinois from 1969 to 1973. He was an athletic trainer at the 1983 World University Games and for the U.S. men's Olympic basketball trials in 1984. During sabbaticals in 1982 and 1989, he served as an athletic trainer for boxing, men's field hockey, team handball, ice skating, roller hockey, gymnastics, judo, and cycling at the United States Olympic Training Center in Colorado Springs. He has been in demand as a speaker and an athletic trainer throughout the United States and internationally.

These broad experiences enabled Dr. Behnke to understand the needs of undergraduate students—and to develop an unparalleled grasp of which pedagogical approaches work and which do not. *Kinetic Anatomy* is the culmination of his unique understanding; it should be a staple in undergraduate courses for years to come.